Design, Optimization and Analysis of Agricultural Machinery

Design, Optimization and Analysis of Agricultural Machinery

Editor

Massimiliano Varani

Basel • Beijing • Wuhan • Barcelona • Belgrade • Novi Sad • Cluj • Manchester

Editor
Massimiliano Varani
Università di Bologna
Bologna
Italy

Editorial Office
MDPI AG
Grosspeteranlage 5
4052 Basel, Switzerland

This is a reprint of articles from the Special Issue published online in the open access journal *Agriculture* (ISSN 2077-0472) (available at: https://www.mdpi.com/journal/agriculture/special_issues/8U9Q8HGM0M).

For citation purposes, cite each article independently as indicated on the article page online and as indicated below:

Lastname, A.A.; Lastname, B.B. Article Title. *Journal Name* **Year**, *Volume Number*, Page Range.

ISBN 978-3-7258-1457-2 (Hbk)
ISBN 978-3-7258-1458-9 (PDF)
doi.org/10.3390/books978-3-7258-1458-9

Contents

About the Editor

Massimiliano Varani

Massimiliano Varani graduated with a M.Sc degree in Vehicle Engineering from the University of Modena and Reggio Emilia in 2013, and a Ph.D. degree in Agricultural Engineering at Alma Mater Studiorum - University of Bologna in 2018. Currently, Massimiliano is a senior assistant professor in Agricultural Mechanics at the Department of Agricultural and Food Sciences of the Alma Mater Studiorum - University of Bologna. He won the "Guarnieri-Montel" prize for the best doctoral thesis of the Agricultural Mechanics sector within the "1st Workshop on Innovation in Mechanics and Plant Engineering Applied to Agrifood and Forest Biosystems". Moreover, he won the prize "ISHS Young mind awards - Best poster" at the "VI International Symposium on Applications of Modelling as an Innovative Technology in the Horticultural Supply Chain". Massimiliano has participated in 17 research projects, with most of them being conducted in partnership with major agricultural machinery manufacturers, and he was the coordinator in 3 of them. His main research interests include analyzing the efficiency of tractors and agricultural machinery, implementation of electrified solutions in agriculture to lower their environmental impact, and methods for analyzing agricultural machinery fleet data via the CAN-Bus network.

Article

Extrapolation of Tractor Traction Resistance Load Spectrum and Compilation of Loading Spectrum Based on Optimal Threshold Selection Using a Genetic Algorithm

Meng Yang [1], Xiaoxu Sun [1], Xiaoting Deng [1], Zhixiong Lu [1,*] and Tao Wang [2,*]

[1] College of Engineering, Nanjing Agricultural University, Nanjing 210031, China
[2] College of Emergency Management, Nanjing Tech University, Nanjing 210009, China
* Correspondence: luzx@njau.edu.cn (Z.L.); wang-tao@njtech.edu.cn (T.W.)

Abstract: To obtain the load spectrum of the traction resistance of the three-point suspension device under tractor-plowing conditions, a load spectrum extrapolation method based on a genetic algorithm optimal threshold selection is proposed. This article first uses a pin force sensor to measure the plowing resistance of the tractor's three-point suspension device under plowing conditions and preprocesses the collected load signal. Next, a genetic algorithm is introduced to select the threshold based on the Peak Over Threshold (POT) extremum extrapolation model. The Generalized Pareto Distribution (GPD) fits the extreme load distribution that exceeds the threshold range, generating new extreme points that follow the GPD distribution to replace the extreme points in the original data, achieving the extrapolation of the load spectrum. Finally, the loading spectrum that can be achieved on the test bench is obtained based on the miner fatigue theory and accelerated life theory. The results show that the upper threshold of the time-domain load data obtained by the genetic algorithm is 10.975 kN, and the grey correlation degree is 0.7249. The optimal lower threshold is 8.5455 kN, the grey correlation degree is 0.7722, and the fitting effect of the GPD distribution is good. The plowing operation was divided into five stages: plowing tool insertion, acceleration operation, constant speed operation, deceleration operation, and plowing tool extraction. A traction resistance loading spectrum that can be achieved on the test bench was developed. The load spectrum extrapolation method based on the genetic algorithm optimal threshold selection can improve the accuracy of threshold selection and achieve the extrapolation and reconstruction of the load spectrum. After processing the extrapolated load spectrum, it can be transformed into a load spectrum that can be recognized by the test bench.

Keywords: tractor; peak over threshold (POT) model; generalized pareto distribution (GPD); genetic algorithm; miner fatigue theory

Citation: Yang, M.; Sun, X.; Deng, X.; Lu, Z.; Wang, T. Extrapolation of Tractor Traction Resistance Load Spectrum and Compilation of Loading Spectrum Based on Optimal Threshold Selection Using a Genetic Algorithm. *Agriculture* **2023**, *13*, 1133. https://doi.org/10.3390/agriculture13061133

Academic Editor: Massimiliano Varani

Received: 10 May 2023
Revised: 26 May 2023
Accepted: 26 May 2023
Published: 28 May 2023

1. Introduction

During farming operations in the field, the load spectrum of a tractor is subjected to random loads due to the complex and variable environment. The load spectrum is a load time history that reflects the loading situation of the entire structure or key components [1], containing load information and the distribution law of the tractor under operating conditions [2]. By extrapolating and reconstructing the load spectrum, a full-life-cycle load spectrum can be obtained within a finite detection time of the load spectrum, thereby reducing time and testing costs. This is of great significance for predicting fatigue life and conducting reliability testing of various components of tractors [3]. The extrapolated load spectrum cannot directly guide relevant performance tests. To facilitate the loading of the test bench, it is necessary to convert the load spectrum into a constant stress spectrum, i.e., to compile the load spectrum.

In recent years, the application of load spectrum extrapolation in agricultural machinery has become increasingly widespread. Shao et al. established a load transfer model for tractor plowing operations and analyzed the load characteristics of the tractor transmission system for field plowing operations through rainfall basin extrapolation [4]. Roberto Tovo proposed a new method to evaluate the single Weibull distribution of the period generated by the rainflow count of random processes, which provides a basis for reliability analysis of fatigue behavior of actual components under service loads [5]. Wang et al. proposed a tractor power take-off (PTO) torque load spectrum extrapolation method based on FDR (False Discovery Rate) threshold automatic selection and optimized the time-domain extrapolation threshold selection method [6]. Yang et al. proposed a time-domain load extrapolation method based on the EMD-POT (Empirical mode decomposition-peaks over threshold) model to address the two issues of insufficient adaptability of traditional POT extrapolation methods to non-stationary loads and the lack of discussion on extrapolation reconstruction. The stability of the mean and standard deviation of this extrapolation reconstruction method has been improved by 28.5% and 31.2%, respectively. Compared with the random reconstruction method, the damage consistency has been improved by 9.4% [7]. He et al. address the problems that the conventional time-domain extrapolation ignores: the interval time of extreme adjacent values and the high sensitivity of the POT model to extreme thresholds. A computer numerical control machine tool load extrapolation method based on the GRA-POT model (Gray relational analysis-peak over threshold mode) is proposed, which can obtain a POT extrapolation model with high fitting accuracy. In addition, the accuracy of the load spectrum of CNC (computerized numerical control) machine tools is improved [8]. Yang et al. studied the time-domain extrapolation method of tractor drive shaft load under static working conditions and proposed a time-domain extrapolation method of tractor drive shaft load based on the MCMC-POT (Markov chain Monte Carlo-peak over threshold) model [9]. Yang et al. obtained the load spectrum of a high-power tractor drive shaft under field working conditions, and a time-domain extrapolation method of high-power tractor-drive shaft load was proposed based on the POT model, aiming at the limitations of rain flow counting and rain basin extrapolation methods in the compilation of traditional driveline load spectrum. This method can not only obtain the load time-domain sequence of any mileage, but also preserve the order of measured load cycles to a great extent, providing real and reliable data support for future indoor load spectrum loading tests of high-power tractor transmission systems [10]. Dai et al. proposed the CEEMDAN-POT (Complete Ensemble Empirical Mode Decomposition with Adaptive Noise -Peak Over Threshold) model to comprehensively construct the ground load spectrum of tractor vibration in its full life cycle under six ground conditions and different field operating conditions. After extrapolation, the overall distribution of rain–flow matrix is more consistent, and the mean value and amplitude of spectral data increase. This study unifies the load spectrum of tractors operating and transporting under various farm surface conditions and provides the real load data of the laboratory four-column drill test [11]. Wang et al. proposed a PTO loading method based on the dynamic load spectrum obtained in field work, taking PTO torque load as the object. The load extremity was extended from (63.24, 469.50) to (60.88, 475.18) by the time-domain extrapolation method, and the coverage was extended by 1.98%. This study provides a reference for the practical application of PTO load spectrum of tractor [12]. Wang et al., in view of the problems that the traditional parameter extrapolation compilation method fails to verify, or the poor fitting effect on the operating loads with multiple peaks and unclear probability distribution, and taking the measured tractive loads of three-point suspension of tractors as the object, proposed a load spectrum compilation method based on optimal distribution fitting, and conducted an indoor bench test to verify the reliability of the load spectrum [13].

Comparing and analyzing existing research results shows that load spectrum extrapolation is mainly divided into time-domain extrapolation methods [14–16] and rain-flow extrapolation methods [17–19]. Rain-flow extrapolation methods are widely used and have high computational efficiency, but there is a loss of time sequence information for the load and the need to reconstruct the load-time history to obtain the load is a problem. Conversely, time-domain extrapolation methods can retain the time sequence of the load, making it more suitable for extrapolating steady loads [20]. The time-domain extrapolation method is more appropriate for the traction resistance load spectrum of the tractor plowing operation studied in this paper. However, for time-domain extrapolation, selecting an appropriate threshold is crucial for determining the effectiveness of the load spectrum extrapolation. Currently, the method combines image and grey correlation analysis to select the threshold. However, this method requires calculating the grey correlation degree for the threshold within the initial selection range, which is computationally intensive and has low accuracy. Therefore, optimizing the selection method for threshold values and selecting the optimal threshold is of great significance for improving the fitting effect and rationality of the load spectrum extrapolation.

In this study, the traction resistance signal of the three-point suspension device of the tractor was collected under the plowing operation condition, and the load data was preprocessed. The time-domain extrapolation method was used to extrapolate and reconstruct the load spectrum of the tractor's traction resistance under the plowing operation condition. To address the problem of high computational intensity and low accuracy in selecting the threshold range of the POT extreme value extrapolation model, this paper proposes a load spectrum extrapolation method based on genetic algorithm optimization of the threshold value. It verifies the rationality of the extrapolated load spectrum. Based on the Miner fatigue theory and accelerated life theory, the plowing operation condition was divided into five working stages, and the load spectrum was converted into a constant stress spectrum to obtain a loading spectrum that can be realized on the test bench, laying the foundation for predicting the fatigue life and conducting reliability testing of tractors.

2. Materials and Methods

2.1. Load Spectrum Extrapolation Principle and Process

Using the time-domain load extrapolation method, the preprocessed time-domain load data is directly extrapolated to obtain the long-term load time history. The extreme load is the center of gravity for extrapolation, and the tail data of load distribution is mainly described through extreme value theory [21]. The process of time-domain load extrapolation mainly includes the following: removing small cycles from time-domain load data and extracting inflection points of extreme values; selecting an appropriate extreme load model and establishing an extreme load distribution model; and randomly generating a new time-domain extreme load sequence using the extreme load distribution to obtain extrapolated long-term time-domain loads. According to different extreme value models, common time-domain load extrapolation methods can be divided into time-domain extrapolation based on the BMM (Block Maximum Method) model, time-domain extrapolation based on the POT model, and time-domain extrapolation based on the MIS (Management Information System) model.

The process of load spectrum extrapolation is shown in Figure 1, which mainly includes the following steps: determining typical test conditions, load measurement, signal preprocessing, statistical counting [22], load extrapolation, program loading spectrum, and reliability testing.

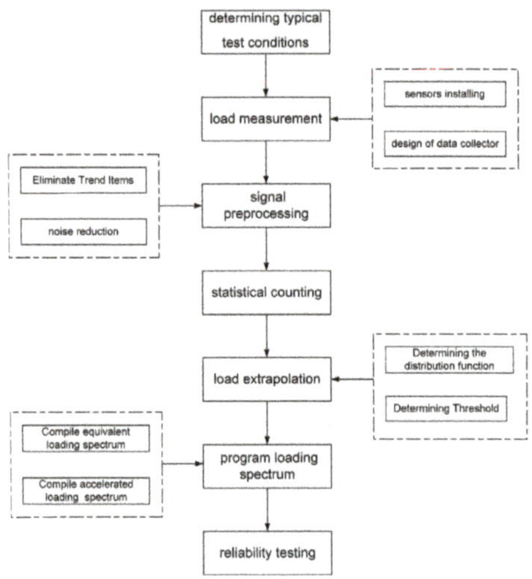

Figure 1. Load spectrum extrapolation process.

2.2. Collection and Data Preprocessing of Traction Resistance Load Signal

2.2.1. Collection of Traction Resistance Load Signal

The experimental site for collecting traction resistance load signals is located in Liuhe District, Nanjing City, Jiangsu Province (Figure 2a). The soil specific resistance was about 3.5–4 N/cm^2 and the experiment area was 5000 m^2. The Dongfeng DF1004 tractor is taken as the research object, and the tractor's three-point suspension is connected to the L1-435 moldboard plow to collect traction resistance signals under plowing conditions. The details of plow connection and sensor layout are shown in Figure 2b. The sensor adopts the XZNJNY-T3d30 KN electric quantity weighing sensor produced by Ningbo Keli Sensing Technology Co., Ltd. Ningbo China. This sensor is a pin-type force sensor, arranged at the pull-down rod of the three-point suspension device and collects the tractor traction resistance load signal through a wired collection system. The field test and sensor layout are shown in Figure 2. The technical specifications of the tractor, the sensor, and the moldboard plow are shown in Table 1 [23].

(a)

(b)

Figure 2. Field experiment and sensor arrangement of the three-point linkage: (**a**) the tractor plowing resistance load test setup; (**b**) three-point linkage and sensor arrangement.

Table 1. The technical specifications of the sensor.

Name of Part	Parameter	Parameter Value
The tractor	Model name	DF1004X
	Outer dimension (mm × mm × mm)	4555 × 2270 × 2775
	Wheel pitch (Front wheel mm/rear wheel mm)	1550–2010, 1650
	Engine-calibrated power (kW)	73.5
	Minimum ground spacing (mm)	425
	Minimum service quality (kg)	4340
	Power output shaft power (kW)	63
The sensor	Model name	XZNJNY-T3d30 KN
	Supply voltage (V)	12
	Output signal (V)	2.5–4.5
The moldboard plow	Model name	1L-435
	Matching power (kW)	66.1–88.2
	Outer dimension (mm × mm × mm)	3400 × 1650 × 1350 mm
	Total weight (kg)	1050
	Depth range (mm)	200–350
	Plow number	4
	Adjustable range of total tillage (mm)	1400
	Plow spacing (mm)	880
	Operating speed (km/h)	8–12
	Matching tire spacing (mm)	1700–1900
	Connection type	three-point suspension

2.2.2. Preprocessing of Traction Resistance Load Signal

According to the traction resistance data of the three-point suspension system collected from field experiments, the original load signal can be obtained by processing the data, as shown in Figure 3.

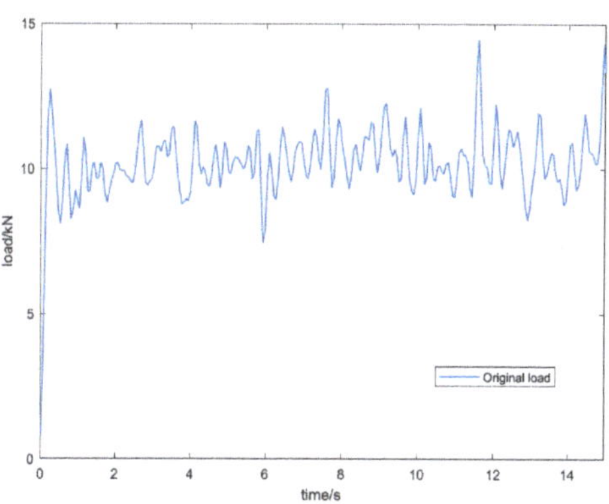

Figure 3. Original load signal.

Due to the complex field environment, there may be interference signals in the measured load signals during field experiments. If the original data is directly used to compile the load spectrum, the reliability of the compiled load spectrum will not be very high. Therefore, for the original load signal of tractor traction resistance collected by the testing system, preprocessing is necessary, and the processed signal needs to be verified and prepared for the compilation of load spectra for subsequent tractor field operations [24].

The collected raw signal data Itself has a certain range of oscillations, and in addition, there may be some low-frequency components that affect our calculations or observations.

The polynomial least squares method is used to eliminate the trend term from the original load signal, and the results are shown in Figure 4.

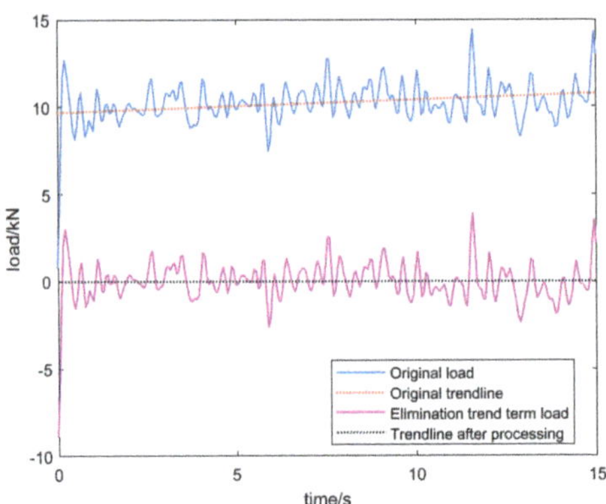

Figure 4. Load signal elimination trend term.

Due to the harsh working conditions in the field, there are some abnormal peaks in the lines drawn after dispersion due to external interference or human error during the tractor field test. In order to obtain true and reliable load data, outlier of the original signal should be detected and smoothed before spectrum compilation, and the abnormal peaks should be eliminated [25,26], and the results are shown in Figure 5.

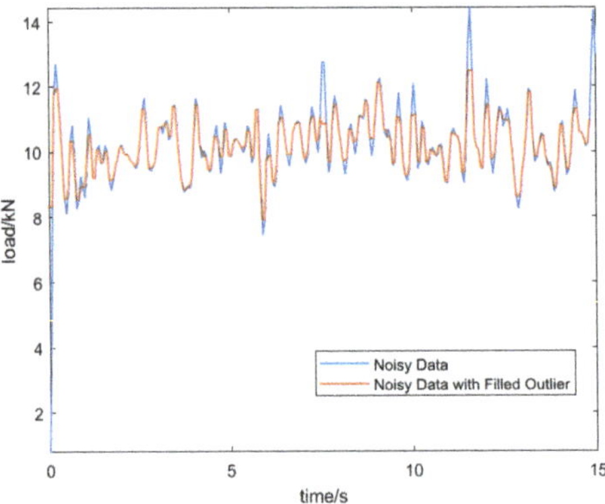

Figure 5. Load signal after noise reduction.

2.3. Extrapolation of Traction Resistance Load Spectrum

2.3.1. Determine the Distribution Function

Assuming $\{x_i\}(i = 1, 2, \ldots, n)$ is the load spectrum sample data, and its distribution is $F(x)$. For a specific threshold, samples that are greater than the threshold are referred to as over-threshold samples, and $z_i = x_i - \mu(i = 1, 2, \ldots, n)$ is the excess amount. The distribution functions of the exceeding amount and exceeding threshold are shown in Equations (1) and (2), respectively:

$$F_\mu(z) = P\{X - \mu \le z | X \ge \mu\} = \frac{P\{\mu \le X - \mu \le z\}}{X \ge \mu} = \frac{F(z + \mu) - F(\mu)}{1 - F(\mu)}, z \ge 0 \quad (1)$$

$$F_\mu(x) = P\{X \le x | X \ge \mu\} == \frac{F(x) - F(\mu)}{1 - F(\mu)}, x \ge \mu \quad (2)$$

According to the load characteristic analysis data, the threshold's candidate interval is [7.898, 12.47] kN. Research has shown that the excess distribution tends to follow the generalized Pareto distribution (GPD) when the threshold is sufficiently large. Therefore, this article fits the excess distribution based on the GPD distribution.

The expression for the GPD cumulative distribution function is:

$$G(z, \mu, \sigma, \xi) = \begin{cases} 1 - (1 + \xi \frac{z}{\sigma})^{-\frac{1}{\xi}}, \xi \neq 0, x > \mu \\ 1 - \exp(-\frac{z}{\sigma}), \xi = 0, x > \mu \end{cases} \tag{3}$$

The expression of the GPD probability density function is:

$$g(z, \mu, \sigma, \xi) = \begin{cases} \frac{1}{\sigma}(1 + \xi \frac{z}{\sigma})^{-\frac{1+\xi}{\xi}}, \xi \neq 0, x > \mu \\ \frac{1}{\sigma}\exp(-\frac{z}{\sigma}), \xi = 0, x > \mu \end{cases} \tag{4}$$

In the equation, $z_i = x_i - \mu (i = 1, 2, \ldots, n)$ is the excess, x_i (kN) is the observed load value, μ (kN) is the threshold value, σ is the scale parameter, and ξ is the shape parameter.

2.3.2. Determining the Threshold

The mean function of the excess of the random variable X is defined as $e(\mu)$, and its expression is:

$$e(\mu) = E(x - \mu | X > \mu) = \frac{\sigma + \xi \mu}{1 - \xi} \tag{5}$$

When the scale parameter σ and shape parameter ξ are determined, there is a linear relationship between $e(\mu)$ and the threshold. Each threshold has a mean of the excess corresponding to it, as shown in Equation (6).

$$e_n(\mu) = \frac{1}{N}\sum_{i=1}^{n} X_i - \mu \tag{6}$$

In the formula, μ is the threshold, and N is the number of excess samples.

The graph of the mean function of the upper threshold exceedance is shown in Figure 6:

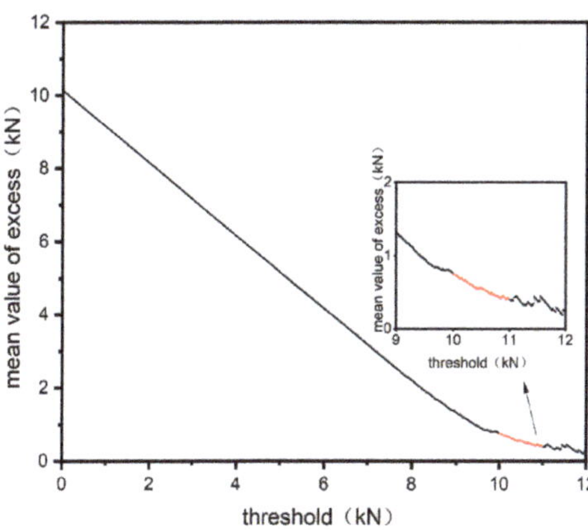

Figure 6. Function chart of the mean value of upper threshold exceeding.

From the graph, it can be seen that as the threshold increases, the tail means oscillates violently. The linear change interval [10.10, 11.00] that is closest to the oscillation before the oscillation is selected as the initial upper threshold interval.

Similarly, the graph of the mean function of threshold exceedance can be obtained, as shown in Figure 7, with [8.50, 9.50] as the initial threshold interval.

- Threshold selection method based on grey correlation analysis;

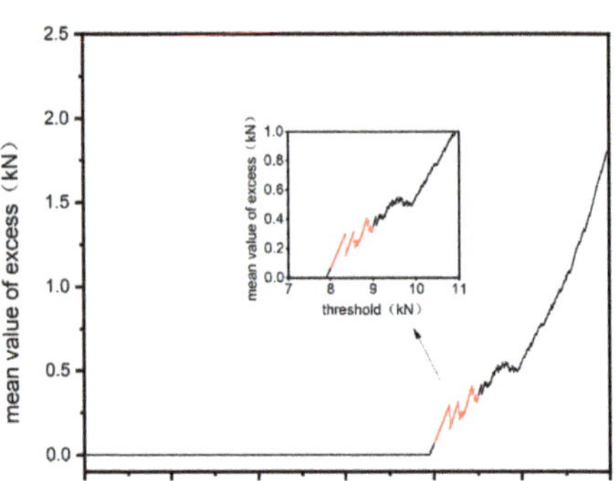

Figure 7. Function diagram of the mean value of threshold exceedance.

Grey correlation analysis is a common threshold selection method. Grey correlation analysis is used to quantify the goodness of fit of the GPD distribution of excess quantities with different thresholds. The greater the grey correlation, the better the fitting effect [27]. The correlation analysis process is as follows [28]:

(1) Given an alternative threshold, the corresponding excess sample data, and GPD fitting data are shown in Equation (7).

$$\begin{cases} f(x_i) = f_1, f_2, \cdots, f_n \\ \hat{f}(x_i) = \hat{f}_1, \hat{f}_2, \cdots, \hat{f}_n \end{cases}, (i = 1, 2, 3, \cdots, n) \tag{7}$$

In the formula, $f(x_i)$ is the quantile of the sample point distribution of the original load data, $\hat{f}(x_i)$ is the quantile of the fitting GPD distribution, and n is the number of excess samples.

(2) Using the averaging method to $\hat{f}(x_i)$ perform dimensionless processing on the load sample data $f(x_i)$ and fitting data, as shown in (8).

$$\begin{cases} f(x_i) = \dfrac{f_i}{\sum\limits_{i=1}^{n} f_i/n} \\ \hat{f}(x_i) = \dfrac{\hat{f}_i}{\sum\limits_{i=1}^{n} \hat{f}_i/n} \end{cases}, (i = 1, 2, 3, \cdots, n) \tag{8}$$

(3) Calculate the absolute difference between the sample and the fitted data after normalization and calculate the correlation coefficient based on the extreme values.

$$\Delta(x_i) = \left| \hat{f}(x_i) - f(x_i) \right| \tag{9}$$

$$\begin{cases} M = \max(\Delta(x_i)) \\ m = \min(\Delta(x_i)) \end{cases} \tag{10}$$

$$\omega(x_i) = \frac{m + \eta M}{\Delta(x_i) + \eta M} \tag{11}$$

In the formula, $\Delta(x_i)$ is the absolute difference, η is the resolution coefficient, and is taken as $\eta = 0.5$.

(1) Calculate alternative thresholds μ Corresponding grey correlation degree λ.

$$\lambda(\mu) = \frac{1}{n}\sum_{i=1}^{n}\omega(x_i) \tag{12}$$

- Optimal threshold selection based on genetic algorithm;

The threshold obtained by using the grey correlation analysis method was calculated using only 10 data points to reduce computational complexity. Therefore, this method's accuracy of the threshold obtained is relatively low. This section uses genetic algorithm to optimize threshold selection and find the optimal threshold. The process is as follows [29]:

(2) Determine the threshold range and initialize the population. The sample accuracy of the input data is 10^{-2}. If the input data is converted into binary encoding, the encoding length of the individual is:

$$l_G = 1 + \log_2 \frac{(\mu_2 - \mu_1)}{eps} \tag{13}$$

In the formula, l_G is the data encoding length, eps is the sample accuracy, μ_1 and μ_2 is the upper and lower intervals of the initial threshold range.

(3) Calculate individual fitness. Individual fitness is the survival probability of an individual under given environmental conditions. Based on the optimization objective, select the fitness function as the "environmental condition" in the genetic algorithm. This article aims to find the threshold that best fits the GPD function. Therefore, the goodness of fit of each candidate threshold is calculated as the fitness of the individual in the "environment". This paper takes the grey relational degree as the fitness function. The greater the grey relational degree, the higher the probability that individuals can survive in the environment and pass on genes to the next generation and vice versa.

(4) Individual survival rate. The survival rate of individuals in the environment is essentially the principle of "survival of the fittest" proposed by evolutionary theory. According to the choice function, the individuals with high fitness will have a higher survival rate, and the good genes will be passed on to the next generation, whereas the inferior genes in the population will be eliminated. The common roulette wheel method is selected as the choice function, in which the survival rate of each individual in the population is proportional to its fitness.

$$p(i) = \frac{f_g(i)}{\sum\limits_{j=1}^{N} f_i(x_j)} \tag{14}$$

In the formula, $f_g(i)$ $(i = 1,2,3 \ldots , n)$ is the fitness of each individual in the population, and $p(i)$ is the survival probability of the i-th individual in this inheritance.

(5) Intersection and variation. The crossover and mutation process in genetic algorithms simulates the pairing and mutation of two pairs of chromosomes in nature. During the crossover process, two sets of data exchange "chromosomes" through certain crossover methods to form new individuals. This article adopts a single-point crossover. Mutation refers to the phenomenon where a certain "gene" within the data has a certain probability of being transformed into an opposite gene.

2.4. Equivalence of Traction Resistance Load Signal

The load spectrum obtained from the GPD distribution cannot directly guide the reliability tests related to plowing operations and the fatigue life prediction of key functional components. In order to facilitate the loading of the test bench, the principle of equal damage can be adopted to convert the load spectrum of each working stage into a constant stress spectrum. Then, according to the test requirements, the loading time of each working stage can be reasonably divided to obtain the loading curve of each cycle, which is called the loading spectrum [30,31].

Miner fatigue theory believes that each pressure applied to a part will cause certain damage to the part, and the magnitude of the damage is determined by the combined magnitude of the applied stress and the characteristics of the material itself [32]. When the damage accumulates to a certain value, the part fails, and vice versa, no failure occurs. The expression is:

$$D = \sum_i \frac{n_i}{N_i} \tag{15}$$

Among them, D is the amount of damage, and $D \in [0,1)$ when the part does not fail. When $D = 1$, the part experiences fatigue failure. n_i is the S_i number of loadings under stress S_i, and N_i is the number of loadings where fatigue damage occurs under the equivalent force. The expression for N_i is:

$$N_i = CS_i^{-\beta} \tag{16}$$

where C is a constant and β is the inverse slope coefficient of the material S-N curve, which is related to the material's own properties, $\beta = 7.1$ [33,34].

From Equations (15) and (16), the equivalent loading stress S_k and equivalent loading frequency n expressions can be obtained as:

$$S_k = \sqrt[\beta]{\frac{\sum_{i=1}^{n} n_i S_i^{\beta}}{n}} \tag{17}$$

$$n = \frac{\sum_{i=1}^{n} n_i S_i^{\beta}}{S_k^{\beta}} \tag{18}$$

where S_k is the equivalent loading stress, n is the equivalent loading frequency, S_i is the load spectrum extrapolation data, and n_i is the frequency of loading stress.

3. Results and Discussions

3.1. Load Characterization and Smoothness Testing after Pre-Processing

The load characteristics before and after pretreatment are shown in Table 2. By comparing the data in Table 2, the data of maximum and minimum values before and after preprocessing changed significantly, which is due to the elimination of abnormal spikes during our preprocessing, and the variance and standard deviation were reduced after preprocessing, which indicates the good noise reduction effect of preprocessing. The mean and median values remain unchanged, which indicates that the data characteristics of the original load signal are also well preserved after preprocessing.

Table 2. Comparison of characteristics before and after load signal preprocessing.

	Minimum /kN	Maximum /kN	Mean /kN	Median /kN	Std /kN	Range /kN
Before	0.7946	14.44	10.22	10.2	1.232	13.65
After	7.898	12.47	10.2	10.16	0.8382	4.57

Using the adtest function in MATLAB, the ADF test (Augmented Dickey–Fuller test) can be performed on the smoothness of the load signal. The function's output is "1", and the test proves that the pre-processed load signal is smooth.

3.2. Threshold Selection Results and Analysis

3.2.1. Threshold Selection Method Based on Grey Correlation Analysis

Based on the initial range of upper and lower thresholds obtained by the image method, 10 thresholds are selected at equal intervals, and the gray correlation degree is calculated. The gray correlation degrees of different thresholds are shown in Table 3 below.

Table 3. Grey correlation degree corresponds to each threshold within the threshold interval.

Upper Threshold		Lower Threshold	
Threshold/kN	**Gray Correlation**	**Threshold/kN**	**Gray Correlation**
10.10	0.6763	8.60	0.6942
10.20	0.6762	8.70	0.6630
10.30	0.6855	**8.80**	**0.7153**
10.40	0.6788	8.90	0.6443
10.50	0.6887	9.00	0.6962
10.60	0.6832	9.10	0.6738
10.70	0.7005	9.20	0.6598
10.80	0.7092	9.30	0.7001
10.90	**0.7182**	9.40	0.6620
11.00	0.7173	9.50	0.6540

Based on the calculation results, the bolded thresholds in the red rectangular boxes in Table 3 were the best-fitting thresholds. The threshold interval was determined to be [8.80, 10.90] kN.

Using the extracted exceedance samples for parameter estimation, the exceedance samples were fitted to the GPD distribution using the great likelihood estimation method to find the corresponding scale parameter σ and shape parameter ξ, as shown in Table 4 [35].

Table 4. GPD fitting results corresponding to threshold based on grey correlation analysis.

Threshold/kN	Scale Parameter σ	Shape Parameter ξ
8.80	117.7353	−1.3046
10.90	61.2523	−0.2479

3.2.2. Optimal Threshold Selection Based on Genetic Algorithm

By writing the program in MATLAB, 50 iterations yielded an optimal upper threshold of 10.975 kN, at which time the gray correlation was 0.7249; the optimal lower threshold was 8.5455 kN, at which time the gray correlation was 0.7722, and the evolutionary process of the upper and lower threshold genetic algorithm is shown in Figure 8.

The comparison of the data with the gray correlation analysis method is shown in Table 5. The results show that the gray correlation of the upper and lower thresholds selected by the genetic algorithm increased by 0.933% and 7.950%, respectively, compared with the traditional gray correlation analysis method, and the gray correlation of the thresholds obtained by this method was greater, indicating that the GPD fit of the threshold exceeded the amount was better.

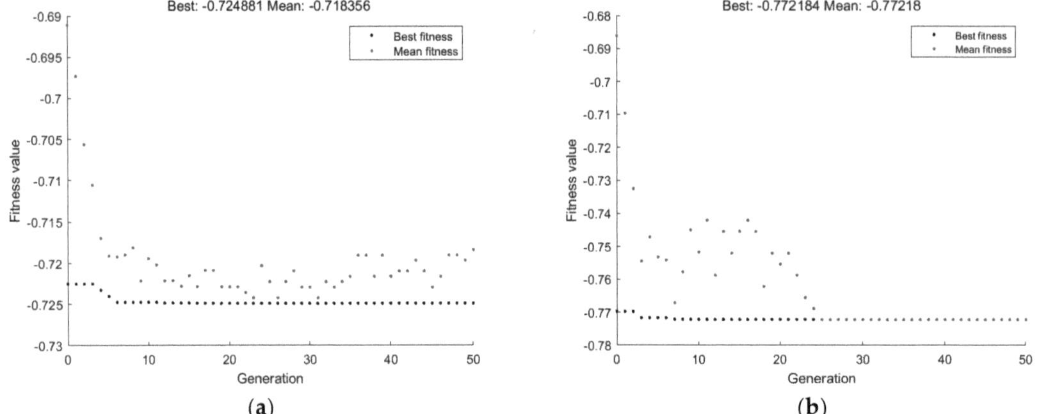

Figure 8. The evolution process of genetic algorithms: (**a**) the upper threshold evolution process of genetic algorithm; (**b**) the lower threshold evolution process of genetic algorithm.

Table 5. Thresholds and gray correlations obtained by the two methods.

	Upper Threshold/kN	Gray Correlation	Lower Threshold/kN	Gray Correlation
Threshold selection method based on grey correlation analysis	10.90	0.7182	8.80	0.7153
Threshold selection method based on genetic algorithm	10.9750	0.7249	8.5455	0.7722

Using the extracted exceedance samples for parameter estimation, the exceedance samples were fitted to the GPD distribution using the great likelihood estimation method to find the corresponding scale parameter σ and shape parameter ξ, as shown in Table 6.

Table 6. GPD fitting results corresponding to the optimal threshold.

Threshold/kN	Scale Parameter σ	Shape Parameter ξ
8.54558	146.8319	-0.4468
10.975	58.5881	-0.2369

After obtaining the scale parameter σ and shape parameter ξ for the upper and lower thresholds, the cumulative distribution functions and probability density functions corresponding to the upper and lower thresholds can be written:

$$\begin{cases} G(z,\mu,\sigma,\xi) = 1 - (1 - 0.2369\frac{x-1097.5}{58.5881})^{4.2212} \\ g(z,\mu,\sigma,\xi) = 0.0171(1 - 0.2369\frac{x-1097.5}{58.5881})^{3.2212} \end{cases} \tag{19}$$

$$\begin{cases} G(z,\mu,\sigma,\xi) = 1 - (1 - 0.4468\frac{x-854.55}{146.8319})^{2.2381} \\ g(z,\mu,\sigma,\xi) = 0.0068(1 - 0.4468\frac{x-854.55}{146.8319})^{1.2381} \end{cases} \tag{20}$$

The goodness of fit can be observed more visually by plotting the Q–Q plot (Quantile–Quantile) of the original sample points against the fitted sample points. As shown in Figure 9, the fitted data points for the upper and lower thresholds (as the blue crosses in the figure) have a slight deviation around the reference line at both ends, whereas in the middle region, they fit the reference line very well, and the Q–Q plot of the final fitted sample approximates a straight line, which indicates a good fit of GDP at this threshold.

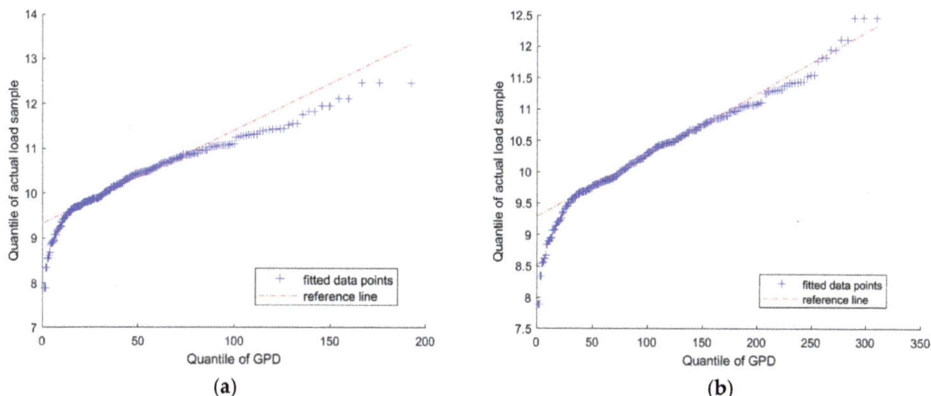

(a) **(b)**

Figure 9. Upper and lower threshold goodness of fit detection: (**a**) upper threshold Q–Q graph; (**b**) lower threshold Q–Q graph.

3.3. Load Spectrum Extrapolation Reconstruction Results and Validation

Combined with the probability density function of the GPD fitted distribution of the excess amount, a random load sequence consistent with the number of samples is generated, and the extrapolated time-domain signal is obtained by replacing the original excess amount at the original time point with the generated load sequence [36,37]. The original load is compared with extrapolated time courses of $1\times$, $2\times$, and $10\times$, as shown in Figure 10.

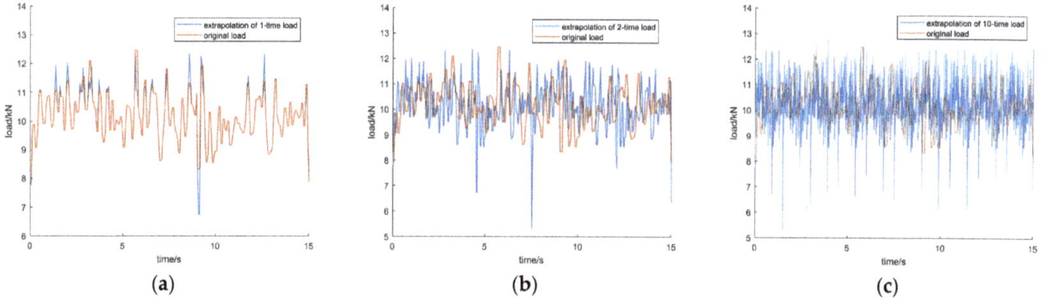

(a) **(b)** **(c)**

Figure 10. Time history of extrapolated load and original load: (**a**) extrapolation of 1-time load history; (**b**) extrapolation of 2-time load history; (**c**) extrapolation of 10-time load history.

The 10-time original load and 10-time extrapolated load are counted, and the comparison of the accumulated frequency data of 10-time original load and 10-time extrapolated load is shown in Figure 11. It can be seen that the changing trend of the accumulated frequency of extrapolated load and original load is basically the same, and this extrapolation method is reasonable.

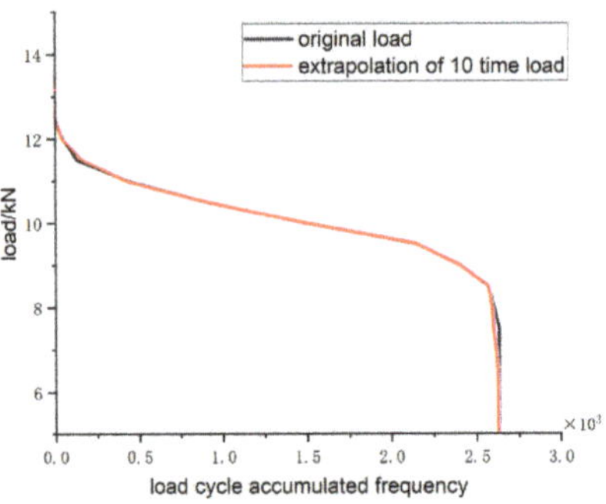

Figure 11. Comparison of cumulative frequency between original load and 10-time extrapolated load.

The load spectra obtained by extrapolation and the original load spectra are counted separately for rainfall, and the mean frequency histogram can be obtained, as shown in Figure 12. The load cycle distribution obtained by extrapolation in the time domain has similarity with the original data load cycle distribution, and the correlation coefficients of its magnitude and mean value are 0.95913 and 0.99187, respectively, indicating that the load cycle distribution can better simulate the real distribution law of the load under the plowing-operation conditions of the tractor.

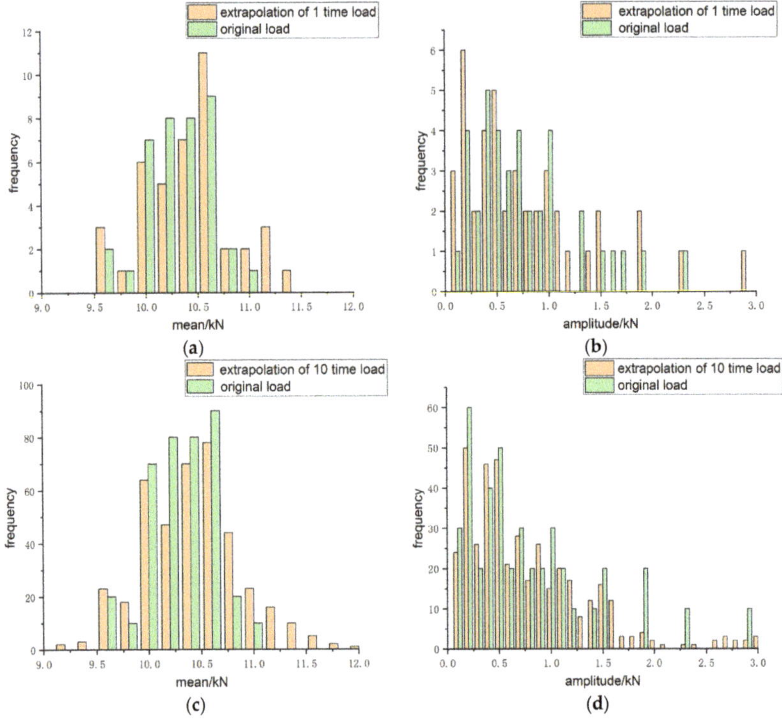

Figure 12. Histogram of mean and amplitude frequency: (**a**) extrapolated of 1-time load mean; (**b**) extrapolation of 1-time load amplitude; (**c**) extrapolated of 10-time load mean; (**d**) 1 extrapolation of 10-time load amplitude.

3.4. Plotting of Loading Spectra and Accelerated Loading Spectra

3.4.1. Analysis of the Plowing Process

The operating conditions of plowing operations consist of five stages: plow tool entry, accelerated operation, uniform speed operation, deceleration operation, and plow tool exit. The boundary of each working phase was distinguished by observing the change in load signal of each moving structure. The time consumption of each working stage is 4.1 s, 1.9 s, 5.6 s, 1.6 s, and 1.8 s. The working stages are shown in Figure 13.

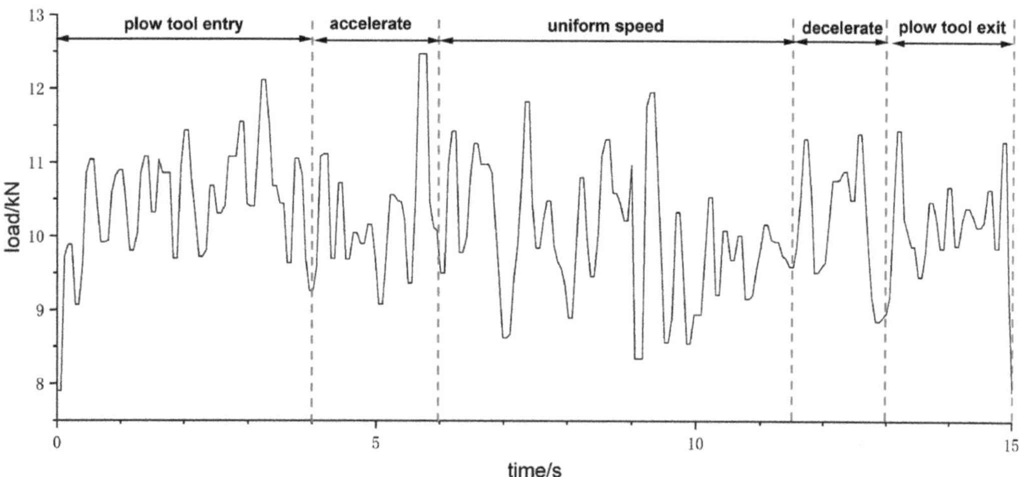

Figure 13. Work phase division diagram.

3.4.2. Division of Loading Loops

In order to facilitate the preparation of the subsequent loading spectra and the easy implementation of the test bench loading, the time periods were rounded, and the elapsed time of each working phase in one loading cycle is shown in Table 7.

Table 7. Time consumption of each work stage.

	Plow Tool Entry	Accelerated Operation	Uniform Speed Operation	Deceleration Operation	Plow Tool Exit
Time/s	4	2	5.5	1.5	2

Preparing the loading spectrum requires the calculation of equivalent loading stresses for each working stage and the extrapolation of the load spectrum for rainflow counting statistics. According to the literature, the load level divided into 8 levels can accurately reflect the fatigue characteristics of the material, so the amplitude load interval of each working stage of the plowing load is divided into 8 levels with the ratio coefficients of 0.125, 0.275, 0.425, 0.575, 0.725, 0.850, 0.950, and 1.000 [38]. The amplitude equivalent of different working stage load spectrum is shown in Table 8 below.

Table 8. Amplitude Equivalent Load Spectrum.

		1	2	3	4	5	6	7	8
Plow tool entry	Amplitude/kN	0.4963	0.9925	1.4888	1.985	2.4813	2.9775	3.4738	3.9658
	Frequency	3	1	3	3	2	0	0	1
Accelerated operation	Amplitude/kN	0.4129	0.8258	1.2387	1.6512	2.0646	2.4774	2.8903	3.3032
	Frequency	1	2	2	0	0	1	1	0
Uniform speed operation	Amplitude/kN	0.4563	0.9125	1.3688	1.8250	2.2813	2.7375	3.1938	3.6412
	Frequency	12	1	5	0	2	1	2	1
Deceleration operation	Amplitude/kN	0.4380	0.8761	1.3141	1.7522	2.1902	2.6282	3.0663	3.5043
	Frequency	0	1	0	0	1	0	0	1
Plow tool exit	Amplitude/kN	0.4488	0.8975	1.3463	1.795	2.2438	2.6925	3.1413	3.5891
	Frequency	0	3	2	1	1	0	0	1

According to the measured limited load data, the load cycle accumulation frequency is extended to 10^6 times according to Equation (21) so as to obtain the extrapolation factor. Keeping the loading times constant, the loading times of each working stage after extrapolation are 236,364, 127,273, 436,364, 54,545, and 145,455, and the amplitude equivalent load spectrum after extrapolation is shown in Table 9.

$$N_i'/N_i = 10^6/N \tag{21}$$

where N is the total frequency of load, N_i is the frequency of load in a certain operation phase, and N_i' is the frequency of load after expansion.

Table 9. Equivalent load spectrum of amplitude after frequency extrapolation.

		1	2	3	4	5	6	7	8
Plow tool entry	Amplitude/kN	0.4963	0.9925	1.4888	1.985	2.4813	2.9775	3.4738	3.9658
	Frequency	54,546	18,182	54,546	54,546	36,364	0	0	18,182
Accelerated operation	Amplitude/kN	0.4129	0.8258	1.2387	1.6512	2.0646	2.4774	2.8903	3.3032
	Frequency	18,182	36,364	36,364	0	0	18,182	18,182	0
Uniform speed operation	Amplitude/kN	0.4563	0.9125	1.3688	1.8250	2.2813	2.7375	3.1938	3.6412
	Frequency	218,182	18,182	90,909	0	36,364	18,182	36,364	18,182
Deceleration operation	Amplitude/kN	0.4380	0.8761	1.3141	1.7522	2.1902	2.6282	3.0663	3.5043
	Frequency	0	18,182	0	0	18,182	0	0	18,182
Plow tool exit	Amplitude/kN	0.4488	0.8975	1.3463	1.795	2.2438	2.6925	3.1413	3.5891
	Frequency	0	54,546	36,364	18,182	18,182	0	0	18,182

3.4.3. Preparation of Loading Spectra

The equivalent force of each working stage is calculated according to Equation (17), as shown in Table 10.

Table 10. Equivalent stress at each working stage.

	Plow Tool Entry	Accelerated Operation	Uniform Speed Operation	Deceleration Operation	Plow Tool Exit
Time/s	4	2	5.5	1.5	2
Equivalent Amplitude/kN	2.799476	2.289872	2.565511	3.016759	2.694487
Equivalent Stress /kN	9.698	10.379	9.857	9.621	9.855

The loading spectrum is plotted according to the time consumed in each working phase and the equivalent force obtained from Table 9, as shown in Figure 14.

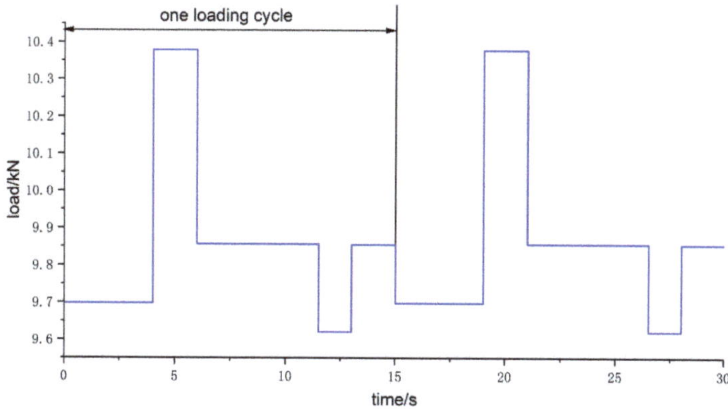

Figure 14. Loading spectrum for plowing operations.

3.4.4. Development of Accelerated Loading Spectrum

The loading test is time-consuming, and the human, financial, and time costs are too great if the test is loaded directly. Therefore, choosing a suitable acceleration factor to accelerate the equivalent force and make the part fail quickly on the test bench is necessary.

The acceleration factor is related to the material of the part. The fatigue characteristics of the material can be described by the S-N curve, through the S-N curve can get the fatigue strength ratio K_n of the part under different stresses. However, the material used in the process, as well as variations in its composition, can also influence the selection of the acceleration factor. To account for the inherent dispersion of the material, it is necessary to introduce the material dispersion correction factor, denoted as K_v. Therefore, it can be seen that the acceleration factor K can be expressed by the formula (22).

$$K = K_n K_v \tag{22}$$

where K is the acceleration factor, K_n is the fatigue strength ratio of the material, 1.15, and K_v is the discrete correction factor of the material, 1.3. Thus, it can be calculated that $K = 1.5$ [39].

After the acceleration, the equivalent force of each working stage is shown in Table 10. The accelerated loading spectrum is plotted according to the time consumed in each working stage and the equivalent force obtained from Table 11, as shown in Figure 15.

Table 11. Accelerated stress at each working stage.

	Plow Tool Entry	Accelerated Operation	Uniform Speed Operation	Deceleration Operation	Plow Tool Exit
Accelerated Stress/kN	14.547	15.5685	14.7855	14.4315	14.7825

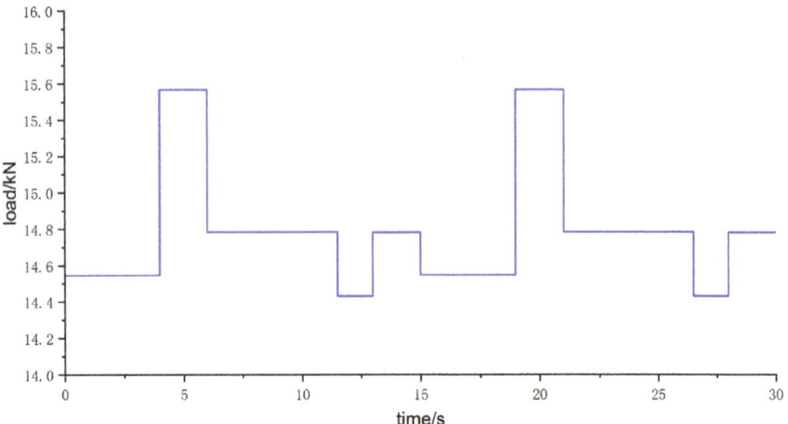

Figure 15. Accelerated loading spectrum for plowing operations.

Comparing Figures 14 and 15, it can be seen that the waveforms of the accelerated loading spectrum are consistent with those before the acceleration, and only the accelerated stress of each section is raised to the loading stress. This shows that when using accelerated loading spectra for reliability testing on a test stand, the test time can be shortened while achieving the desired results, which greatly reduces human, financial and time costs.

4. Conclusions

From the above research and analysis, this paper concludes the following five points:

(1) Taking the Dongfeng DF1004 tractor as the research object, the plowing operation was carried out in a test field with an area of about 5000 m^2, and the axle-pin force sensor was used to collect the traction resistance load signal of the three-point suspension device under the plowing condition. We stored sensor signals with the EPEC controller and transmitted them to the host computer. The foundation was laid for plotting the load time course curve later on;

(2) Based on the least squares method, the original load signal was processed by de-trending, the data was noise-reduced by smoothing, and the smoothness of the pre-processed traction resistance load signal was verified by ADF test to obtain the original load time history of traction resistance under plowing conditions, which laid the foundation for the extrapolation and reconstruction of the load spectrum later on;

(3) Based on the POT model, the time-domain extrapolation of the original load data is carried out, and the excess threshold values are selected using gray correlation analysis and genetic algorithm. The threshold values obtained are [8.80 kN, 10.90 kN] and [8.5455 kN, 10.975 kN], respectively, and the GPD distribution goodness-of-fit test is performed on the selected upper and lower thresholds. The fit superiority of the upper and lower thresholds obtained by the genetic algorithm is improved by 0.933% and 7.95%, respectively, which proves that the fitted curves obtained based on the optimal threshold selection of the genetic algorithm can better reflect the actual loading situation;

(4) Based on the threshold thresholds obtained from the optimal threshold selection of the genetic algorithm, the original load time histories were extrapolated and reconstructed. The extrapolated load spectra obtained from the original load signal and 10-fold extrapolation were compared and analyzed by rain flow counting. The results show that the changing trend of the extrapolated load and the accumulated frequency of the original load are basically the same, and the extrapolation method is reasonable; the load cycle distribution obtained by time-domain extrapolation is similar to the load cycle distribution of the original data, and the correlation coefficients of its

amplitude and mean value are 0.95913 and 0.99187, respectively. The load cycle distribution can better simulate the real load under the working condition of tractor plowing-distribution law;

(5) The plowing condition is divided into five working stages: plow tool entry, accelerated operation, uniform speed operation, deceleration operation, and plow tool exit. The accumulated frequency of load cycles was extended to 106 times to obtain the amplitude equivalent load spectrum. Based on Miner fatigue theory, the equivalent force of each working stage was calculated, and the loading spectrum under plowing working conditions was drawn. Based on the accelerated life theory, the acceleration factor of 1.5 is obtained according to the S-N curve, and the accelerated loading spectrum under plowing conditions is finally drawn. The accelerated loading spectrum is consistent with the loading spectrum waveform, and only the stress is accelerated to the loading stress, which is convenient for loading on the test bench.

Author Contributions: Methodology, M.Y.; software, M.Y. and X.S.; validation, M.Y.; investigation, M.Y. and X.S.; resources, Z.L. and X.S.; writing—original draft preparation, M.Y.; writing—review and editing, X.D., T.W., and Z.L.; supervision, Z.L.; project administration, Z.L. and T.W. All authors have read and agreed to the published version of the manuscript.

Funding: This study was funded by the National Key Research and Development Plan (2022YFD2001202), and the Open Project of the State Key Laboratory of Tractor Power System in China (SKT2022006).

Institutional Review Board Statement: Not applicable.

Informed Consent Statement: Not applicable.

Data Availability Statement: The data presented in this study are available on demand from the corresponding author or first author at luzx@njau.edu.cn or yangmeng@njau.edu.cn.

Acknowledgments: The authors thank the National Key Research and Development Plan (2016YFD0701103), and the Open Project of the State Key Laboratory of Tractor Power System in China (SKT2022006) for funding. We also thank anonymous reviewers for providing critical comments and suggestions that improved the manuscript.

Conflicts of Interest: The authors declare no conflict of interest.

References

1. Zheng, G.; Wang, Q.; Cai, C. Criterion to determine the minimum sample size for load spectrum measurement and statistical extrapolation. *Measurement* **2021**, *178*, 109387. [CrossRef]
2. He, C. Research on Extrapolation and Compilation Method of Pressure Load Spectrum of Excavator Main Pump. Master's Thesis, Jilin University, Jilin, China, 2022.
3. Wang, Q.; Zhou, J.; Gong, D.; Wang, T.; Sun, Y. Fatigue life assessment method of bogie frame with time-domain extrapolation for dynamic stress based on extreme value theory. *Mech. Syst. Signal Process.* **2021**, *159*, 107829. [CrossRef]
4. Shao, X.; Yang, Z.; Mowafy, S.; Zheng, B.; Song, Z.; Luo, Z.; Guo, W. Load characteristics analysis of tractor drivetrain under field plowing operation considering tire-soil interaction. *Soil Tillage Res.* **2023**, *227*, 105620. [CrossRef]
5. Tovo, R. A damage-based evaluation of probability density distribution for rain-flow ranges from random processes. *Int. J. Fatigue* **2000**, *22*, 425–429. [CrossRef]
6. Wang, Y.; Wang, L.; Lv, D.; Wang, S. Extrapolation of Tractor PTO Torque Load Spectrum Based on Automated Threshold Selection with FDR. Trans. *J. Agric. Mach.* **2021**, *52*, 364–372. [CrossRef]
7. Yang, Z.; Song, Z.; Luo, Z.; Zhao, X.; Yin, Y. Time-domain Load Extrapolation Method for Tractor Key Parts Based on EMD-POT Model. *J. Mech. Eng.* **2022**, *58*, 252–262. [CrossRef]
8. He, J.L.; Zhao, X.Y.; Li, G.F.; Chen, C.; Yang, Z.; Hu, L.; Xinge, Z. Time domain load extrapolation method for CNC machine tools based on GRA-POT model. *Int. J. Adv. Manuf. Technol.* **2019**, *103*, 3799–3812. [CrossRef]
9. Yang, Z.H.; Song, Z.H.; Zhao, X.Y.; Zhou, X. Time-domain extrapolation method for tractor drive shaft loads in stationary operating conditions. *Biosyst. Eng.* **2021**, *210*, 143–155. [CrossRef]
10. Yang, Z.; Song, Z.; Yin, Y.; Zhao, X.; Liu, J.; Han, J. Time domain extrapolation method for load of drive shaft of high-power tractor based on POT model. *Trans. Chin. Soc. Agric. Eng.* **2019**, *35*, 40–47. [CrossRef]
11. Dai, D.; Chen, D.; Wang, S.; Li, S.; Mao, X.; Zhang, B.; Wang, Z.; Ma, Z. Compilation and Extrapolation of Load Spectrum of Tractor Ground Vibration Load Based on CEEMDAN-POT Model. *Agriculture* **2023**, *13*, 125. [CrossRef]

12. Wang, Y.; Wang, L.; Zong, J.; Lv, D.; Wang, S. Research on Loading Method of Tractor PTO Based on Dynamic Load Spectrum. *Agriculture* **2021**, *11*, 982. [CrossRef]
13. Wang, L.; Zong, J.; Wang, Y.; Fu, L.; Mao, X.; Wang, S. Compilation and bench test of traction force load spectrum of tractor three-point hitch based on optimal distribution fitting. *Trans. Chin. Soc. Agric. Eng.* **2022**, *38*, 41–49. [CrossRef]
14. Johannesson, P. Extrapolation of load histories and spectra. *Fatigue Fract. Eng. Mater. Struct.* **2006**, *29*, 201–207. [CrossRef]
15. Yang, X.; Liu, X.; Tong, J.; Wang, Y.; Wang, X. Research on load spectrum construction of bench test based on automotive proving ground. *J. Test. Eval.* **2018**, *46*, 244–251. [CrossRef]
16. Yang, X.; Zhang, J.; Ren, W.X. Threshold selection for extreme strain extrapolation due to vehicles on bridges. *Procedia Struct. Integr.* **2017**, *5*, 1176–1183. [CrossRef]
17. Johannesson, P.; Thomas, J.J. Extrapolation of rainflow matrices. *Extremes* **2001**, *4*, 241–262. [CrossRef]
18. Wang, M.; Liu, X.; Wang, X.; Wang, Y.S. Research on load spectrum construction of automobile key parts based on monte carlo sampling. *J. Test. Eval.* **2018**, *46*, 1099–1110. [CrossRef]
19. Nagode, M.; Fajdiga, M. A general multi-modal probability density function suitable for the rainflow ranges of stationary random processes. *Int. J. Fatigue* **1998**, *20*, 211–223. [CrossRef]
20. Yu, L.; An, Y.; He, J.; Li, G.; Wang, S. Research Progress and Development Trend of Load Spectrum Extrapolation Technology for Mechanical and Electrical Equipment. *J. Jilin Univ. Eng. Technol.* **2023**, *53*, 941–953. [CrossRef]
21. Marty, C.; Blanchet, J. Long-term Changes in Annual Maximum Snow Depth and Snowfall in Switzerland Based on Extreme Value Statistics. *Clim. Change* **2012**, *111*, 705–721. [CrossRef]
22. Shao, X.; Song, Z.; Yin, Y.; Xie, B.; Liao, P. Statistical Distribution Modelling and Parameter Identification of the Dynamic Stress Spectrum of a Tractor Front Driven Axle. *Biosyst. Eng.* **2021**, *205*, 152–163. [CrossRef]
23. Wen, C.; Xie, B.; Song, Z.; Yang, Z.; Dong, N.; Han, J.; Yang, Q.; Liu, J. Methodology for designing tractor accelerated structure tests for an indoor drum-type test bench. *Biosyst. Eng.* **2021**, *205*, 59–67. [CrossRef]
24. Deng, X.; Sun, H.; Lu, Z.; Cheng, Z.; An, Y.; Chen, H. Research on Dynamic Analysis and Experimental Study of the Distributed Drive Electric Tractor. *Agriculture* **2023**, *13*, 40. [CrossRef]
25. Cheng, Z.; Lu, Z. Research on Dynamic Load Characteristics of Advanced Variable Speed Drive System for Agricultural Machinery during Engagement. *Agriculture* **2022**, *161*, 20161. [CrossRef]
26. Cheng, Z.; Lu, Z. Research on Load Disturbance Based Variable Speed PID Control and a Novel Denoising Method Based Effect Evaluation of HST for Agricultural Machinery. *Agriculture* **2021**, *960*, 960. [CrossRef]
27. Kayacan, E.; Ulutas, B.; Kaynak, O. Grey system theory-based models in time series prediction. *Expert Syst.* **2010**, *37*, 1784–1789. [CrossRef]
28. Abhang, L.B.; Hameedullah, M. Determination of optimum parameters for multi-performance characteristics in turning by using grey relational analysis. *Int. J. Adv. Manuf. Technol.* **2012**, *63*, 13–24. [CrossRef]
29. Cheng, Z.; Chen, Y.; Li, W.; Liu, J.; Li, L.; Zhou, P.; Chang, W.; Lu, Z. Full Factorial Simulation Test Analysis and I-GA Based Piecewise Model Comparison for Efficiency Characteristics of Hydro Mechanical CVT. *Machines* **2022**, *358*, 358. [CrossRef]
30. Liu, J.; Wen, C.; Xie, B.; Han, J.; Yuan, W. Study on Load Spectrum of Axial Parts Durability Test Bench for Tractor Transmission System. *Tractor Farm Transp.* **2021**, *48*, 29–35. [CrossRef]
31. Dong, G.; Zhang, M.; Wei, L.; Zhang, L. Study on High-cycle Fatigue Criteria of Chassis Components under Multi-axis Random Loads. *Chin. Mech. Eng.* **2021**, *32*, 2294–2304. [CrossRef]
32. Han, Y.; Lin, Y.; Zhang, C.; Wang, D. Customer-related durability test of semi-trailer engine based on failure mode. *Eng. Fail. Anal.* **2021**, *120*, 105393. [CrossRef]
33. Yang, Z.; Song, Z. MCMC Simulation of Agricultural Equipment Load Using Optimal State Number. *Trans. Chin. Soc. Agric. Eng.* **2021**, *37*, 15–22. [CrossRef]
34. An, Z.; Gao, J.; Liu, B. Strength degradation stochastic model based on P-S-N curve. *Chin. J. Comput. Mech.* **2015**, *32*, 118–122. [CrossRef]
35. Heidenreich, N.B.; Schindler, A.; Sperlich, S. Bandwidth Selection for Kernel Density Estimation: A Review of Fully Automatic Selectors. *AStA Adv. Stat. Anal.* **2013**, *97*, 403–433. [CrossRef]
36. Zheng, G.; Zhu, H.; Wu, C.; Xiao, P. Research on Load Spectrum Extrapolation Method Based on Generalized Pareto Distribution of Extreme Value Exceedance. *Chin. Mech. Eng.* **2020**, *31*, 2262–2267. [CrossRef]
37. Carboni, M.; Cerrini, A.; Johannesson, P.; Guidetti, M.; Beretta, S. Load spectra analysis and reconstruction for hydraulic pump components. *Fatigue Fract. Eng. Mater. Struct.* **2008**, *31*, 251–261. [CrossRef]
38. Qin, D.; Xie, L. *Fatigue Strength and Reliability Design*; Chemical Industry Press: Beijing, China, 2013; pp. 58–262.
39. Zhao, Y.; Song, Y. Study on Multi-axial Fatigue Experiment Spectrum Compilation Based on Damage Equivalence. *J. Aerosp. Power* **2009**, *24*, 2026–2032. [CrossRef]

Article

Reliability Study of an Intelligent Profiling Progressive Automatic Glue Cutter Based on the Improved FMECA Method

Heng Zhang [1], Yaya Chen [1], Jingyu Cong [1,2], Junxiao Liu [3,4], Zhifu Zhang [3] and Xirui Zhang [1,3,4,*]

[1] School of Information and Communication Engineering, Hainan University, Haikou 570228, China
[2] State Key Laboratory of Marine Resource Utilization in South China Sea, Hainan University, Haikou 570228, China
[3] School of Mechanical and Electrical Engineering, Hainan University, Haikou 570228, China
[4] Sanya Nanfan Research Institute, Hainan University, Sanya 572025, China
* Correspondence: zhangxr@hainanu.edu.cn

Abstract: This study introduces the fuzzy theory approach as an enhancement to the traditional failure mode, effect, and criticality analysis (FMECA) method in order to address its limitations, which primarily stem from subjectivity and a lack of quantitative analysis. The proposed method, referred to as FMECA improvement based on fuzzy comprehensive evaluation, aims to quantify the qualitative aspect of the analysis and provides a detailed outline of the analysis procedure. By applying the enhanced FMECA method to assess the reliability of an intelligent profiling progressive automatic rubber cutter, the hazard ranking for each failure mode of the cutter can be determined, thereby identifying areas that require reliability improvement. The analysis outcomes demonstrate that this method establishes a theoretical foundation for subsequent cutter improvement designs, enables early identification of potential failures, and consequently leads to a reduced failure rate and an enhanced reliability level for the intelligent profiling progressive automatic cutter. Furthermore, this innovative agricultural equipment reliability analysis and testing approach holds significant value in elevating the reliability standards of agricultural equipment as a whole and can be explored and implemented in other agricultural machinery contexts.

Keywords: natural rubber; intelligent profiling progressive automatic glue cutter; FMECA method; fuzzy comprehensive evaluation; reliability analysis

Citation: Zhang, H.; Chen, Y.; Cong, J.; Liu, J.; Zhang, Z.; Zhang, X. Reliability Study of an Intelligent Profiling Progressive Automatic Glue Cutter Based on the Improved FMECA Method. *Agriculture* **2023**, *13*, 1475. https://doi.org/10.3390/agriculture13081475

Academic Editor: Massimiliano Varani

Received: 10 July 2023
Revised: 23 July 2023
Accepted: 25 July 2023
Published: 26 July 2023

1. Introduction

The intelligent profiling step-by-step automatic rubber cutting machine offers several advantages, including addressing the shortage of rubber workers, revolutionizing the rubber work process, high automation levels, and independence from environmental constraints. These benefits effectively reduce the reliance on manual labor for rubber cutting, lower labor costs, and increase natural rubber output. Moreover, the machine's reliability directly influences the yield and quality of natural rubber. Currently, rubber cutting machines experience a significant failure rate, which greatly diminishes the quality of the cut rubber and can even damage the rubber trees, thereby adversely affecting the income of rubber farmers. Therefore, it is imperative to conduct a reliability analysis of rubber cutting machines. Failure is a disruptive event that can cause production delays and compromise the overall reliability of a system [1–3]. In order to cope with various failure modes that may occur, appropriate decisions based on the multicriteria decision-making (MCDM) process are made at different stages such as design, manufacturing, and operation to improve system reliability [4].

FMECA (failure modes, effect, and criticality analysis) mainly consists of two parts: failure mode and effect analysis (FMEA) and criticality analysis (CA). It is commonly used to find and solve various known or potential failures in equipment systems, which plays a crucial role in improving the reliability and service life of equipment. However, when using the traditional FMECA method to analyze the reliability of equipment, there is too much subjectivity, when determining the order of hazard level only qualitative analysis can be performed (not quantitative analysis) and it is difficult to find out the weak links in the system accurately by calculating the objective results, and it cannot provide technical support for the daily maintenance of various equipment systems. In recent decades, significant efforts have been made by scholars and researchers to enhance the FMECA methodology [5]. Several improved methods have emerged, focusing on the following areas. Bozdag et al. proposed a novel fuzzy FMECA method based on fuzzy sets [6]. This approach considered the optimal weights of risk factors and integrated them using an ordered weighted average operator based on the cut concept. Liu et al. [7] introduced fuzzy directed graph and matrix methods into FMECA, developing a new FMECA model that considered the relative weights of risk factors expressed linguistically. These weights were transformed into fuzzy numbers and risk priority indices for failure modes were calculated using corresponding fuzzy risk matrices. Zhou et al. [8] proposed a generalized evidence FMECA model (GEFMECA) to handle uncertain risk factors encompassing both conventional and incomplete risk factors. By utilizing the generalized evidence theory, the issue of relative weights among risk factors was effectively resolved. Additionally, Liu et al. [9] introduced an integrated FMECA method based on interval intuitionistic fuzzy sets (IVIFS) and multi-attribute boundary approximation region comparison (MABAC) methods. This approach established a linear programming model for obtaining weight information among risk factors when complete weight information was not available, thereby determining the optimal weights for these factors. Yang et al. [10] proposed a fuzzy rule-based Bayesian inference method for prioritizing failure modes. Jee et al. [11] introduced a new fuzzy inference system (FIS)-based risk assessment model for FMECA, employing a two-stage approach to reduce the collection of fuzzy rules. Gajanand et al. [12] combined FMECA with a fuzzy linguistic scaling method, presenting a novel reliability-centered maintenance strategy that utilizes weighted Euclidean distance and fuzzy logic-based center-of-mass defuzzification to rank failure modes. Sayyadi et al. [13] developed a new FMECA model based on an intuitionistic fuzzy approach, enabling the evaluation of failure modes in the presence of fuzzy concepts and limited data. Jian et al. [14] combined intuitionistic fuzzy sets (IFSS) with evidence theory, proposing a novel FMECA failure mode risk assessment method. Jiang et al. [15] utilized fuzzy affiliation in the proposed FMECA fuzzy evidence method to evaluate the risk factors of failure modes, ranking them using the Dempster–Shafer evidence theory to consolidate characteristic information. Aydogan [16] proposed a method that combines the rough hierarchical analysis and fuzzy TOPSIS methods for organizational performance analysis in a fuzzy environment. Liu et al. [17] introduced an intuitive fuzzy hybrid TOPSIS method to determine the risk priority of failure modes in FMECA. Silvia et al. [18] proposed an optimization method for the maintenance activities of complex systems by integrating reliability analysis and MCDM methods. They used *AHP* for weight assessment and fuzzy TOPSIS for risk prioritization. Zhou et al. [19] introduced gray theory and fuzzy theory into FMECA for tanker equipment failure prediction. They determined the risk priority of failure modes using the fuzzy risk priority number (FRPN) obtained from fuzzy set theory and the gray correlation coefficient from gray theory. Liu et al. [20] developed an FMECA framework by integrating the cloud model and PROMETHEE method to handle diverse risk assessments from FMECA team members and prioritize failure mode risks. Mandal et al. [21] proposed a method to rank human errors using the VIKOR method. Baloch et al. [22] integrated the fuzzy VIKOR method and data envelopment analysis into FMECA to determine the ranking of potential modalities and select the most important damage modalities. Liu et al. [23] proposed an improved approach for FMECA using fuzzy evidential reasoning (FER) theory. This approach addresses two limitations of traditional FMECA: the acquisition and aggregation of evaluations from different experts and the determination of the risk priorities for failure modes. Liu et al. [24] further proposed a new FMECA failure mode prioritization risk assessment model based on FER and belief rule base (BRB) methods. In this model, the diverse and uncertain evaluations provided by experts are captured and aggregated using FER, whereas the nonlinear and uncertain relationships between risk factors and corresponding risk levels are modeled using the BRB. Du et al. [25] presented a fuzzy FMECA method based on evidential reasoning (ER) and TOPSIS to accurately identify and aggregate risk factors. Su et al. [26] proposed an improved method for dealing with conflicting evidence combinations by employing an uncertain inference method based on the Gaussian distribution to reconstruct the basic belief assignment (BBA) while considering the weight of each expert. Jiang et al. [27] proposed an improved application of the theory of evidence to FMECA, which redistributes the underlying belief assignments by considering the credibility coefficients obtained based on the distance of the evidence to minimize conflicts between expert opinions. The applied research conducted by these experts and scholars, focusing on enhancing the FMECA method in the reliability analysis of different equipment systems, establishes a theoretical foundation for analyzing the reliability of the intelligent profiling progressive automatic rubber cutter discussed in this paper.

This paper focuses on conducting reliability research on an intelligent profiling progressive automatic rubber cutter. The research utilizes an improved FMECA method as the research tool. The primary objective is to assess and enhance the reliability level of the rubber cutter. The application of the improved FMECA methodology innovatively combines expert knowledge and engineering experience to provide a quantitative analysis solution for the reliability of rubber cutting equipment through improved fuzzy theory. The study aims to identify key areas for improving the reliability of the rubber cutter. The findings of this research will provide theoretical guidance for subsequent improvement designs and the daily maintenance of rubber cutters.

2. Materials and Methods

2.1. Basic Theory of the Traditional FMECA Method

The fundamental principle of the traditional FMECA is to systematically examine the structure of the system, identify potential failure modes, analyze the underlying causes of failures, and use statistical methods to estimate the severity (S), occurrence (O), and detection (D) of the failure consequences based on technical specifications, historical data, and user requirements. Subsequently, the risk priority number (RPN) is calculated, and the failure modes are prioritized according to their RPN values. Appropriate improvement or maintenance measures can then implemented to reduce the RPN and ensure the reliability of the system.

The typical steps involved in conducting FMECA analysis are as follows, as illustrated in Figure 1:

(1) Product Definition: Provide a description of the product's composition, environmental conditions during operation, functional aspects, and operational procedures;

(2) FMECA Method Selection: Choose the appropriate FMECA method based on the analysis objective and the product's development stage and develop a corresponding FMECA analysis table;

(3) FMEA Analysis Implementation: Identify potential failure modes, describe their effects, investigate the contributing causes for each mode, assess failure detection methods, and analyze potential compensatory measures;

(4) Hazard Analysis: Evaluate the hazards associated with the identified failure modes.

In traditional FMECA, each failure mode identified within a system is evaluated using three risk factors: severity (S), occurrence (O), and detectability (D). The RPN is calculated by multiplying the values of S, O, and D, providing a ranking for the failure modes [28]. Typically, experts assign scores ranging from 1 to 10 to the risk factors S, O, and D, with higher values indicating more severe situations. The RPN value is utilized to determine the risk priority of each failure mode, allowing analysts to identify inherent vulnerabilities in the system. A higher RPN indicates greater importance [29], suggesting a more significant impact on the system and a higher risk priority. To ensure safety and reliability, preventive and improvement measures should give priority to higher-risk failure modes in order to prevent their occurrence.

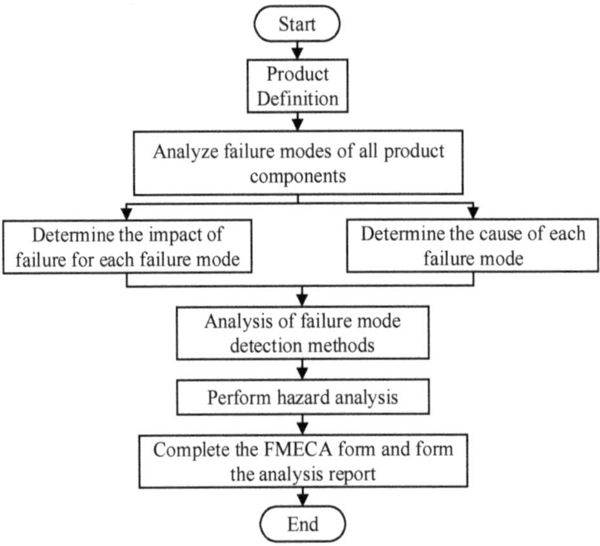

Figure 1. The basic analysis steps of the traditional FMECA method.

However, the conventional crisp value of the RPN has faced criticism for several reasons, [30–35] as explained below:

1. The relative importance of the three risk factors is often not considered or assumed to be equal, which may not reflect the actual situation in many cases;

2. Multiplying the values of S, O, and D across different groups may result in the same RPN values but the risk hazards in each group could be completely different. This can lead to an unreasonable allocation of limited resources and time, or worse, the neglect of high-risk failure modes;

3. Calculating the mean RPN is sensitive and contentious when the values of the risk factors change. Even a slight change in the value of a risk factor can have a significantly different effect on the RPN, particularly when other risk factor values are large;

4. The evaluation of risk factors S, O, and D usually utilizes discrete ordered metric scales. As a result, the multiplicative calculations lack meaningfulness since the resulting RPN may show discontinuity, with multiple gaps and a wide range from 1 to 1000. In such situations, the ranking results of failure modes lose their significance and can potentially mislead;

5. The accurate determination of these three risk factors is often challenging. FMECA team members express their evaluations using linguistic terms such as high, medium, or low;

6. Due to variations in expertise and backgrounds, FMECA team members may evaluate the same risk factors differently and some evaluations may be ambiguous and uncertain. The traditional FMECA approach lacks comprehensive methods to describe group judgments and explore the inherent connections between different evaluations. [36].

Although the FMECA method facilitates the timely identification of structural design flaws, comparison of alternative solutions, and decision-making support for improving design and maintenance strategies, the analysis process poses challenges due to multiple evaluation factors, qualitative assessments, and the incommensurability of failure consequences and impacts. As a result, analysts often encounter difficulties in producing precise and effective analysis results.

2.2. Enhancing the Fundamentals of the FMECA Method

The fuzzy comprehensive evaluation method is a quantitative approach based on fuzzy mathematics that converts qualitative assessments into quantitative evaluations using fuzzy principles. It facilitates the overall evaluation of objects or phenomena influenced by multiple factors, providing clear and systematic results. This method is especially valuable for addressing problems involving fuzziness and difficulties in quantification, making it suitable for a range of non-deterministic scenarios.

The enhanced FMECA method integrates fuzzy theory with the traditional FMECA method to analyze the reliability of equipment systems. By leveraging the strengths of both reliability analysis methods, it effectively addresses the limitations of the traditional FMECA approach. The improved method allows for the quantification of analysis results and enhances the accuracy and reliability of comprehensive hazard-level rankings for each failure mode. Figure 2 illustrates the fundamental evaluation process of the enhanced FMECA method.

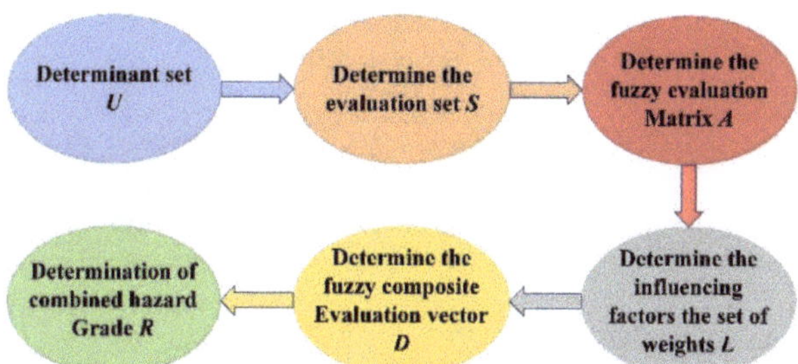

Figure 2. The fundamental evaluation process of the enhanced FMECA method.

2.2.1. Defining the Set of Factors

The factor set, represented by U, encompasses the collection of factors that exert influence on the evaluation object.

$$U = \{u_1, u_2, u_3, \ldots, u_i, \ldots, u_m\} \tag{1}$$

where u_i denotes the ith influencing factor, $i = 1, 2, 3, \ldots, m$.

2.2.2. Determining the Evaluation Set

Defining the evaluation set, represented by S, includes all the evaluation results provided by experts for the evaluated object.

$$S = \{s_1, s_2, s_3, \ldots, s_j, \ldots, s_n\} \tag{2}$$

where, s_j denotes the jth evaluation result made by the judging expert, $j = 1, 2, 3, \ldots, n$.

2.2.3. Establishing the Fuzzy Evaluation Matrix

A fuzzy mapping $f: U \to S$ is needed to establish the relationship between the evaluation results and the influence factors. This mapping is denoted as $f: U \to F(S)$, where each influence factor u_i is mapped to its respective fuzzy evaluation result $f(u_i)$. Applying this mapping allows us to determine the degree of affiliation a_{ij} between an influence factor u_i and an evaluation result s_j.

$$f(u_i) = \mathbf{A}_i = (a_{i1}, a_{i2}, a_{i3}, \ldots, a_{ij}, \ldots, a_{in}) \tag{3}$$

The set $\mathbf{A}_i$ represents the evaluation results for the individual influence factor, u_i. These individual evaluation results are organized as rows in the evaluation matrix and then transposed to form an *m*-by-*n* matrix. The resulting matrix, denoted as $\mathbf{A}$, represents the fuzzy evaluation of the influence factors.

$$\mathbf{A} = [\mathbf{A}_1, \mathbf{A}_2, \cdots, \mathbf{A}_n]^{\mathrm{T}} = \begin{bmatrix} a_{11} & a_{12} & \cdots & a_{1n} \\ a_{21} & a_{22} & \cdots & a_{2n} \\ \vdots & \vdots & \vdots & \vdots \\ a_{m1} & a_{m2} & \cdots & a_{mn} \end{bmatrix} \tag{4}$$

An evaluation team, comprising r experts, is formed, and each expert is responsible for providing their evaluation result s_j for each influence factor u_i. If there are r_{ij} experts who evaluate u_i as s_j, the evaluation set for u_i can be obtained using the following procedure:

$$\begin{aligned} \mathbf{A}_i &= \left\{ \frac{r_{i1}}{r}, \frac{r_{i2}}{r}, \frac{r_{i3}}{r}, \cdots, \frac{r_{ij}}{r}, \cdots, \frac{r_{in}}{r} \right\} \\ &= \left\{ a_{i1}, a_{i2}, a_{i3}, \cdots, a_{ij}, \cdots, a_{in} \right\} \end{aligned} \tag{5}$$

where $\sum\limits_{j=1}^{n} a_{ij} = 1$.

2.2.4. Determine the Weights for the Set of Influence Factors

Given the significant variations in the degree of harm caused by different influencing factors on each failure mode, it is crucial to prioritize and rank the factors according to their impact. Prior to conducting the comprehensive evaluation by experts, the weights are determined in a specific order. These weights collectively constitute the weight set of influencing factors, denoted as $\mathbf{L}$.

$$\mathbf{L} = \{l_1, l_2, l_3, \cdots, l_i, \cdots, l_n\} \tag{6}$$

where $0 < l_i < 1, i \in [1, n]$, and $\sum\limits_{i=1}^{n} l_i = 1$.

Several methods are available for determining the weight set of influencing factors, including survey statistics, expert scoring methods, and the analytic hierarchy process (AHP). Among these methods, the survey statistics method involves analyzing a significant amount of data through empirical research to determine the weights. However, this method is labor intensive and time consuming, making it impractical in many situations. The expert scoring method determines the weights through expert evaluations. Although this method saves time and effort, it is subjective and prone to errors, posing challenges in obtaining the ideal and precise weights. In contrast, the analytic hierarchy process method can address the limitations of the expert scoring method by mitigating human biases in the weight determination process. This method ensures the attainment of objective and effective weights. In this study, the 1–9 scale model of the analytic hierarchy process is used to determine the weight set of influencing factors [8]. The following steps outline the procedure:

(1) The relative significance of the influencing factors u_i and u_j is conveyed by using b_{ij}, resulting in the formation of a judgment matrix $\mathbf{B}$.

$$\mathbf{B} = \begin{bmatrix} b_{11} & b_{12} & \cdots & b_{1n} \\ b_{21} & b_{22} & \cdots & b_{2n} \\ \vdots & \vdots & \vdots & \vdots \\ b_{n1} & b_{n2} & \cdots & b_{nn} \end{bmatrix} \tag{7}$$

where the values of b_i are referred to in Table 1. It is evident that $b_{ii} = 1$ and $b_{ij} = 1/b_{ji}$;

Table 1. AHP analysis of the relative importance degree of the influencing factors.

Implications	b_{ij}
u_i is as important as u_j	1
u_i is slightly more important than u_j	3
u_i is significantly more important than u_j	5
u_i is strongly more important than u_j	7
u_i is definitely more important than u_j	9
The importance of u_i over u_j is between the above two scale values	2, 4, 6, 8

(2) A consistency test is conducted on the judgment matrix **B** by determining the consistency ratio K_C. To begin, the maximum characteristic root $\lambda_{\max}$ associated with the judgment matrix **B** and the consistency index I_C are calculated as follows:

$$I_C = \frac{\lambda_{\max} - n}{n - 1} \tag{8}$$

Subsequently, the value of I_T, representing the average random consistency index of the judgment matrix, is determined. The specific values of I_T for the judgment matrices of order 1–13 can be obtained from Table 2.

Table 2. Values 1–13 of I_T judgment matrix.

n	1	2	3	4	5	6	7	8	9	10	11	12	13
I_T	0	0	0.58	0.90	1.12	1.24	1.32	1.41	1.45	1.49	1.52	1.54	1.56

Based on the calculated values of I_C and I_T, the consistency ratio K_C is computed, which is defined as follows:

$$K_C = \frac{I_C}{I_T} \tag{9}$$

when $K_C < 0.1$, it can be concluded that the consistency of judgment matrix **B** meets the required criteria. However, if K_C exceeds 0.1, adjustments to the judgment matrix **B** are necessary;

(3) The weights of each factor are determined using the square root method. The weighting term for the ith factor of the analyzed failure mode is calculated as follows:

$$l_i = \frac{\sqrt[n]{\prod_{j=1}^{n} b_{ij}}}{\sum_{i=1}^{n} \sqrt[n]{\prod_{j=1}^{n} b_{ij}}} \tag{10}$$

Subsequently, the set of factor weights for this specific failure mode is obtained as $L_1 = \{l_1, l_2, \ldots, l_i, \ldots, l_n\}$ where $0 < l_i < 1$, ensuring the fulfillment of the normalization condition:

$$\sum_{i=1}^{n} l_i = 1 \tag{11}$$

2.2.5. The Calculation of the Fuzzy Comprehensive Evaluation Vector

By multiplying the previously derived fuzzy evaluation matrix **A** of influence factors with the set of influence factor weights **L**, we can obtain the fuzzy comprehensive evaluation vector, denoted as **D**, which can be expressed as follows:

$$\mathbf{D} = \mathbf{L} \times \mathbf{A} = [l_1, l_2, \cdots, l_n] \times \begin{bmatrix} a_{11} & a_{12} & \cdots & a_{1n} \\ a_{21} & a_{22} & \cdots & a_{2n} \\ \vdots & \vdots & \vdots & \vdots \\ a_{m1} & a_{m2} & \cdots & a_{mn} \end{bmatrix} \tag{12}$$

2.2.6. Determining the Comprehensive Hazard Level

To enable a clearer comparison of the hazard levels associated with each failure mode in relation to the evaluation object, it is essential to calculate a specific value by applying weighted averaging to the fuzzy comprehensive evaluation vector **D**. This value, represented as **R**, indicates the ranking of the hazard levels for each failure mode with respect to the evaluation object.

$$\mathbf{R} = \mathbf{D} \cdot \mathbf{S}^{\mathrm{T}} \tag{13}$$

where **S** in denotes the evaluation result matrix.

3. Results and Analysis

The quality of rubber cutting plays a crucial role in achieving high yields from natural rubber trees in agricultural production. Therefore, the performance and reliability of rubber cutters directly affect the quality of rubber cutting, necessitating a reliability analysis. This study aims to analyze the reliability of a specific type of rubber cutter, specifically an intelligent profiling progressive automatic rubber cutter, as an illustrative example.

3.1. Intelligent Profiling Progressive Automatic Gum Cutter

The structure of the intelligent profiling progressive automatic rubber cutter, as depicted in Figure 3, mainly comprises several key components, including the adjustable flexible tooth-belt fixing device, adaptive tree profiling cutting device, circumferential motion device, vertical motion device, reduction drive device, and a complete set of screws, motors, and other auxiliary elements.

Before operating the intelligent profiling automatic rubber cutter, careful considerations are taken into account for the variations in the upper and lower bark sizes of the natural rubber tree, as well as the irregularities on its surface. To address these factors, the rubber cutter is first secured to the rubber tree using the adjustable flexible tooth-belt fixing device. This ensures that the center axis of the upper and lower drive teeth stays aligned with the center axis of the rubber tree during the cutting process, enabling precise and stable autonomous execution of the rubber cutting task. When the rubber cutter receives the rubber cutting command, it starts by extending the push rod, which causes the adaptive tree imitation device to stick to the surface of the rubber bark, thereby entering the cutting state. Afterwards, the rubber cutting knife performs a spiral movement along the natural rubber trunk, facilitated by the compound motion transmission device. This movement imitates the cutting trajectory of humans, facilitating efficient rubber cutting operations. After the completion of rubber cutting, the stepping motor drives the rubber cutting device to descend along the screw by a specified distance (tare consumption). This ensures the appropriate consumption of bark for the subsequent cutting cycle. Ultimately, a complete rubber cutting operation is achieved. The entire cutting process can be remotely controlled using an intelligent control module, allowing for the smooth execution of automated cutting operations.

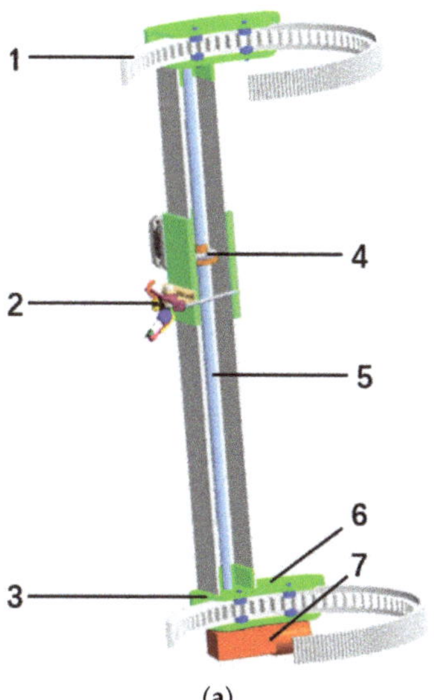

Figure 3. Intelligent profiling step-by-step automatic glue cutter. (**a**) Three-dimensional structure picture. (**b**) Pictures of Gumlin field work. (1) Adjustable flexible belt fixing device; (2) Adaptive tree-adaptive miter devices; (3) Circular motion device; (4) Vertical motion device; (5) Complete set of screws; (6) Reduction gearing; (7) Motor.

3.2. Analysis of the FMEA Method for an Intelligent Profiling Progressive Automatic Glue Cutter

A comprehensive analysis of the failure modes of the intelligent profiling progressive automatic rubber cutter was carried out, as depicted in Table 3.

Table 3. Analysis table of FMEA method for intelligent profiling progressive automatic glue cutter.

Code	Failure Mode	Failure Analysis	Fault Impact	Fault Checking Method	Troubleshooting Measures
1	Motor shaft damage; fracture or deformation	Excessive torque due to overload	Decreased functionality	Regular inspection; instrument testing	Motor replacement
2	Tooth-belt slippage	Excessive load causes the transmitted force to be greater than the limit of the sum of the frictional forces between the belt and the gear	Decreased functionality	Visual inspection	Increase the belt width or replace the belt

Table 3. *Cont.*

Code	Failure Mode	Failure Analysis	Fault Impact	Fault Checking Method	Troubleshooting Measures
3	Blade deformation or breakage	Insufficient blade strength; improper cutting depth and blade installation angle resulting in excessive load causing blade breakage	Loss of function	Visual inspection	Replacement of high strength blades; correction of cutting depth and blade installation angle
4	Circumferential and vertical movement device rotation is not flexible, there is a jamming phenomenon	The upper and lower tooth-belt gap has foreign matter and installation is not parallel to cause the center axis of the transmission teeth and rubber tree center axis offset; poor lubrication	Decreased functionality	Regular inspections; instrument testing	Removal of foreign objects; enhance lubrication
5	Unstable amount of skin consumption	The height of the descending screw and the distance from the cutting knife to the guiding depth limiting wheel are not consistent when cutting rubber	Decreased functionality	Regular inspections; instrument testing	Correction of the height of the lowering screw and the distance from the cutting knife to the guiding depth limit wheel
6	Unstable cutting depth	Improper installation and cutting angle of the blade; spring tension failure, etc., led to jumping of the cutting knife	Decreased functionality	Instrument inspection; visual inspection	Correct blade mounting and cutting angle; replace spring

3.3. Analysis of the Traditional FMECA Method for the Intelligent Profiling Progressive Automatic Glue Cutter

The reliability analysis of the intelligent profiling progressive automatic rubber cutter was conducted using the conventional FMECA method based on the FMEA table.

The RPN is used as a quantitative measure of the hazard, evaluating the potential severity associated with each failure mode. By evaluating the levels of fault occurrence probability, impact severity, and detection difficulty, a comprehensive analysis of the impact is conducted. A higher RPN value indicates a higher hazard associated with the corresponding failure mode, which can be expressed as follows:

$$RPN = ESR \times OPR \times DDR \tag{14}$$

where ESR represents the level of impact severity, OPR denotes the level of probability of occurrence, and DDR signifies the level of detection difficulty. The ESR is established based on the failure mode, impact, and hazard analysis guide, as well as the maintenance experience of cutter maintenance engineers. The hazard degree values are assigned as follows: I is assigned a value of 5, II is assigned a value of 4, III is assigned a value of 3, IV is assigned a value of 2, and V is assigned a value of 1. The value of the OPR is determined based on the failure mode, impact, hazard analysis guide, and the available cutter failure data. The possibility of occurrence for each type of hardware failure of the cutter is categorized into five intervals. The corresponding probabilities of fault occurrence P are $P > 10^{-1}$, $10^{-2} < P \leq 10^{-1}$, $10^{-4} < P \leq 10^{-2}$, $10^{-6} < P \leq 10^{-4}$, and $P \leq 10^{-6}$, and the corresponding OPR values are $5, 4, 3, 2$, and 1, respectively. The DDR is defined based on various factors, including the precision of the current testing equipment, the expertise of the testing personnel, and the employed testing methods. The assessment of inspection complexity for each component of the cutter involves categorizing them into five types: undetectable, difficult to detect, detectable, easy to detect, and directly identifiable. Correspondingly, the assigned DDR values for these categories are $5, 4, 3, 2$, and 1, respectively.

The RPN values for each failure mode of the intelligent profiling progressive automatic rubber cutter are obtained based on the value criteria of ESR, OPR, and DDR. Equation (15) was applied to derive these values and experts from the rubber cutter maintenance industry were invited to provide scores, as illustrated in Table 4.

Table 4. Table of RPN values for each failure mode of the intelligent profiling progressive automatic glue cutter.

Projects	ESR	OPR	DDR	RPN
Failure Mode 1	4	2	4	32
Failure Mode 2	3	3	2	18
Failure Mode 3	4	3	2	24
Failure Mode 4	2	3	2	12
Failure Mode 5	3	3	3	27
Failure Mode 6	4	3	2	24

Based on Table 4, the hazard levels for the six failure modes can be determined in descending order as follows: Failure Mode 1, Failure Mode 5, Failure Mode 3, Failure Mode 6, Failure Mode 2, and Failure Mode 4. Failure Mode 1 and Failure Mode 5 exhibit the highest RPN values and pose the greatest hazards. Failure Mode 6 and Failure Mode 3 have identical RPN values, indicating the same level of hazard for both failures.

The preparation of the FMEA worksheet, as shown in Table 4, is found to be highly beneficial during the analysis process. By comparing the RPN values before and after implementing maintenance measures, the evaluation of maintenance policies is facilitated. Moreover, a more comprehensive understanding of the failure causes and effects contributes to a better assessment of risk factors.

However, the RPN approach has several shortcomings that have led to the adoption of alternative methods in FMEA. The main criticisms include the following:

(1) Potential result unification: The RPN values may be the same for two different failure modes, despite their values for factors such as ESR, OPR, and DDR being different;

(2) Overlooking significant aspects of failures: In certain cases, failures with high severity may garner inadequate attention due to the low values of other risk factors, resulting in a low RPN value;

(3) Lack of clear distribution pattern: RPNs are distributed from 1 to 125, and the relationship between neighboring numbers at intervals of 5 or 10 relate to each other.

3.4. Analysis of the Improved FMECA Method for the Intelligent Profiling Progressive Automatic Glue Cutter

The improved FMECA method is employed to quantitatively analyze the FMEA results and assess the reliability of the intelligent profiling progressive automatic glue cutter. This analysis provides a comprehensive ranking of the hazard level for each failure mode, enabling the implementation of appropriate measures to promptly address failures and ensure the efficient and reliable operation of the cutter.

The analysis of the intelligent profiling progressive automatic glue cutter within the framework of the FMEA method focuses on evaluating common failure modes through fuzzy synthesis. The evaluation process is outlined in the following steps:

(1) Determination of the factor set: The assessment of fault hazard level in the intelligent profiling progressive automatic glue cutter involves the consideration of the following factors: fault occurrence probability (u_1), degree of fault impact (u_2), testing difficulty (u_3), and fault repair difficulty (u_4). These factors collectively form the set U;

(2) Establishment of the evaluation set: The primary objective of this study is to identify weaknesses in the intelligent profiling progressive automatic glue cutter and provide guidance for its improvement and maintenance. Since there is no absolute "good" or "bad" state for each influencing factor, an intermediate scale comprising three levels—"better", "average", and "worse"—is introduced between the two extremes. The evaluation levels for the impact factors are categorized into five levels represented by the set $S = \{1, 2, 3, 4, 5\}$. The specific evaluation levels for each influencing factor are presented in Table 5.

Table 5. Evaluation grade table of each influence factor.

Influencing Factors	Evaluation Level				
	1	**2**	**3**	**4**	**5**
Fault occurrence probability u_1	Almost never happens	Rarely happens	Occasional	Sometimes it happens	Frequent
Degree of fault impact u_2	Almost no effect	Mild faults	Moderate failure	Critical Failure	Fatal Failure
Difficulty of testing u_3	Can be found directly	Easy to detect	Not easy to detect	Hard to detect	Undetectable
Difficulty of repairing faults u_4	Simple debugging	Reinstallation	Replacement Parts	Replace the whole machine	Unrepairable

(3) Determining the fuzzy evaluation matrix involves assigning fuzzy sets to the failure probability, failure impact degree, detection difficulty, and maintenance difficulty of Failure Mode 1. Let the fuzzy set for the failure probability be denoted as $\mathbf{a}_1 = \{0, 0.2, 0.5, 0.3, 0\}$, the fuzzy set for the failure impact degree as $\mathbf{a}_2 = \{0.1, 0.45, 0.3, 0.15, 0\}$, the fuzzy set for the detection difficulty as $\mathbf{a}_3 = \{0, 0.15, 0.7, 0.1, 0.05\}$, and the fuzzy set for the maintenance difficulty as $\mathbf{a}_4 = \{0, 0, 0.5, 0.4, 0.1\}$. Consequently, the fuzzy evaluation matrix for Failure Mode 1 can be derived as follows:

$$\mathbf{A}_1 = \begin{bmatrix} \mathbf{a}_1 & \mathbf{a}_2 & \mathbf{a}_3 & \mathbf{a}_4 \end{bmatrix}^{\mathrm{T}} = \begin{bmatrix} 0 & 0.2 & 0.5 & 0.3 & 0 \\ 0.1 & 0.45 & 0.3 & 0.15 & 0 \\ 0 & 0.15 & 0.7 & 0.1 & 0.05 \\ 0 & 0 & 0.5 & 0.4 & 0.1 \end{bmatrix} \tag{15}$$

(4) Determining the weight set of influencing factors for Failure Mode 1 involves evaluating the relative importance of these factors in the decision model based on their impact on the decision outcomes. By performing the corresponding calculations, the weight values for each influencing factor are obtained, as presented in Table 6.

Table 6. Judgment matrix and weight value of each influencing factor of the plugging device.

Influencing Factors	u_1	u_2	u_3	u_4	Weighting Value l_i
u_1	1	5	1/3	5	0.2804
u_2	1/5	1	1/7	1	0.0678
u_3	3	7	1	7	0.5747
u_4	1/3	1	1/7	1	0.0771

The judgment matrix can be obtained from the data provided in Table 6 as follows:

$$\mathbf{B} = \begin{bmatrix} 1 & 5 & 1/3 & 5 \\ 1/5 & 1 & 1/7 & 1 \\ 3 & 7 & 1 & 7 \\ 1/3 & 1 & 1/7 & 1 \end{bmatrix} \tag{16}$$

The consistency test of the judgment matrix $\mathbf{B}$ is conducted based on the value of the consistency ratio K_C. Firstly, the corresponding maximum characteristic root $\lambda_{\max} = 4.13$ is calculated from matrix $\mathbf{B}$. Substituting it into Equation (8), the consistency index $I_C = 0.043$ is obtained, and from Table 2, $I_T = 0.90$. By substituting these values into Equation (9), $K_C = 0.048 < 0.1$, indicating that the consistency of judgment matrix $\mathbf{B}$ satisfies the requirements. Therefore, the set of factor weights corresponding to Failure Mode 1 can be obtained as follows:

$$\mathbf{L}_1 = \{0.2804, 0.0678, 0.5747, 0.0771\} \tag{17}$$

(5) Fuzzy integrated evaluation of the cavity rate overload fault. According to Equation (12), the fuzzy comprehensive evaluation vector for fault mode 1 is $\mathbf{D}_1 = \mathbf{L}_1 - \mathbf{A}_1 = [0.3604\ 0.2381\ 0.2962\ 0.1053]$. This indicates that the membership of the cavity rate overload fault to hazard level 1, 2, 3, and 4 is 0.4410, 0.2425, 0.2672, and 0.0493, respectively;

(6) Determining the comprehensive hazard level of excessive cavitation failure. According to Equation (13), the comprehensive hazard level of the cavity rate fault can be obtained as follows:

$$\mathbf{R}_1 = \mathbf{D}_1 \times [1, 2, 3, 4, 5]^{\mathrm{T}} = 3.069 \tag{18}$$

(7) Similarly, the fuzzy evaluation matrices for Failure Modes 2 to 6 are determined as follows:

$$\mathbf{A}_2 = \begin{bmatrix} 0 & 0.6 & 0.3 & 0.1 & 0 \\ 0 & 0.8 & 0.2 & 0 & 0 \\ 0.7 & 0.3 & 0 & 0 & 0 \\ 0.1 & 0.4 & 0.3 & 0.1 & 0.1 \end{bmatrix}$$

$$\mathbf{A}_3 = \begin{bmatrix} 0.1 & 0.3 & 0.3 & 0.2 & 0.1 \\ 0 & 0.1 & 0.2 & 0.3 & 0.4 \\ 0.6 & 0.3 & 0.1 & 0 & 0 \\ 0.1 & 0.2 & 0.7 & 0 & 0 \end{bmatrix}$$

$$\mathbf{A}_4 = \begin{bmatrix} 0 & 0.3 & 0.4 & 0.2 & 0.1 \\ 0.2 & 0.5 & 0.3 & 0 & 0 \\ 0.6 & 0.3 & 0.1 & 0 & 0 \\ 0.5 & 0.2 & 0.3 & 0 & 0 \end{bmatrix}$$

$$\mathbf{A}_5 = \begin{bmatrix} 0.3 & 0.2 & 0.4 & 0.1 & 0 \\ 0 & 0.1 & 0.5 & 0.3 & 0.1 \\ 0 & 0.1 & 0.4 & 0.5 & 0 \\ 0.4 & 0.3 & 0.3 & 0 & 0 \end{bmatrix}$$

$$\mathbf{A}_6 = \begin{bmatrix} 0.2 & 0.1 & 0.5 & 0.2 & 0 \\ 0 & 0.1 & 0.2 & 0.3 & 0.4 \\ 0.3 & 0.6 & 0.1 & 0 & 0 \\ 0.2 & 0.3 & 0.4 & 0.1 & 0 \end{bmatrix}$$

Given that the relative importance of the influencing factors for each failure mode remains consistent, as indicated in Table 5, the same weight set is employed, which is denoted as $\mathbf{L}_1 = \mathbf{L}_2 = \mathbf{L}_3 = \mathbf{L}_4 = \mathbf{L}_5 = \mathbf{L}_6$. Utilizing Equation (12), we can obtain the fuzzy comprehensive evaluation vector for each failure mode as follows:

$$\mathbf{D}_2 = \mathbf{L}_2 \times \mathbf{A}_2 = [0.41, 0.4257, 0.1208, 0.0358, 0.0077]$$

$$\mathbf{D}_3 = \mathbf{L}_3 \times \mathbf{A}_3 = [0.3806, 0.2787, 0.2091, 0.0764, 0.0552]$$

$$\mathbf{D}_4 = \mathbf{L}_4 \times \mathbf{A}_4 = [0.3969, 0.3059, 0.0982, 0.0561, 0.0280]$$

$$\mathbf{D}_5 = \mathbf{L}_5 \times \mathbf{A}_5 = [0.1150, 0.1435, 0.3991, 0.3357, 0.0068]$$

$$\mathbf{D}_6 = \mathbf{L}_6 \times \mathbf{A}_6 = [0.2439, 0.4028, 0.2421, 0.0841, 0.0271]$$

The combined hazard level for each failure mode can be determined using Equation (13):

$$\mathbf{R}_2 = \mathbf{D}_2 \times [1, 2, 3, 4, 5]^{\mathrm{T}} = 1.8055$$

$$\mathbf{R}_3 = \mathbf{D}_3 \times [1, 2, 3, 4, 5]^{\mathrm{T}} = 2.1469$$

$$\mathbf{R}_4 = \mathbf{D}_4 \times [1, 2, 3, 4, 5]^{\mathrm{T}} = 1.6677$$

$$\mathbf{R}_5 = \mathbf{D}_5 \times [1, 2, 3, 4, 5]^{\mathrm{T}} = 2.9761$$

$$\mathbf{R}_6 = \mathbf{D}_6 \times [1, 2, 3, 4, 5]^{\mathrm{T}} = 2.2477$$

Based on the calculated integrated hazard level for each fault, the descending order of hazard levels for the six failure modes is as follows: Failure Mode 1, Failure Mode 5, Failure Mode 6, Failure Mode 3, Failure Mode 2, and Failure Mode 4. After conducting numerous statistical analyses on data from field cutting tests in the forest, the findings demonstrate that the evaluation results are consistent with the real-world scenario, with an accuracy exceeding 95%. Additionally, this method can be applied to analyze the reliability of other agricultural equipment. Additionally, this method enables the analysis of reliability for other agricultural equipment, identification of the most critical potential faults, and implementation of timely measures to mitigate and enhance the design, thus improving the overall reliability of agricultural equipment.

4. Discussion

Natural rubber, as one of the four major industrial materials in modern society, exhibits excellent abrasion resistance, impact resistance, elasticity, and heat dissipation. In particular, at low temperatures it demonstrates ductility and resilience that are incomparable to synthetic rubber. It has extensive applications in industrial production, national defense equipment, transportation, medicine, health care, and other fields. Therefore, natural rubber plays a critical role in the national economy and people's livelihoods. Natural rubber is primarily obtained by the semi-spiral ring cutting of rubber trees. In order to reduce the intensity of artificial rubber cutting and alleviate labor difficulties, the integration of agro-mechanical and agronomic approaches has led to the research and development of intelligent rubber cutting equipment that is replacing the traditional manual rubber cutting. This transition is an inevitable trend. Intelligent rubber cutting equipment significantly reduces the dependence on manual labor, lowers the cost of rubber cutting, and improves the output rate of natural rubber. The reliability of this equipment directly affects the production and quality of natural rubber. The reliability of rubber cutting equipment directly impacts the output and quality of natural rubber. Currently, the failure rate of rubber cutting machines remains significantly high. Once a failure occurs, it greatly reduces the quality of the rubber cuts and may even lead to serious damage to the rubber trees, thereby affecting the income of rubber farmers. Currently, no relevant scientific literature exists on the reliability of intelligent rubber cutters. This paper introduces an innovative method to analyze the reliability of intelligent profiling step-by-step automatic rubber cutters. The original contributions of this paper can be summarized as follows:

(1) Research on the FMECA analysis method based on fuzzy comprehensive evaluation was carried out, and a fuzzy comprehensive evaluation model was established. The model provides a theoretical basis for the reliability design of agricultural equipment and lays the foundation for the application of reliability analysis in the field of agricultural machinery;

(2) A qualitative and quantitative analysis of the reliability of the intelligent profiling progressive automatic glue cutter was carried out by using the improved FMECA method, and the comprehensive hazard ranking of each failure was obtained. We propose improvement measures and formulate a preventive maintenance outline, which can provide a theoretical basis for the improved design of the cutter and check the possible failures of the cutter beforehand, thus reducing the failure rate of the cutter and greatly improving the reliability level of the intelligent profiling progressive automatic cutter;

(3) Taking the intelligent profiling progressive automatic rubber cutter as an example, the above model was used to verify and analyze its frequent failure modes; the results showed that this evaluation result was consistent with the actual situation, indicating that the evaluation model is an effective method that can make full use of the system's fuzzy information for reliability analysis. Furthermore, this innovative agricultural equipment reliability analysis and testing approach holds significant value in elevating the reliability standards of agricultural equipment as a whole and can be explored and implemented in other agricultural machinery contexts.

5. Conclusions

This study applies an improved FMECA method to analyze the reliability of an intelligent profiling step-type automatic rubber cutter. The traditional FMECA method suffers from subjectivity and a limited ability for quantitative analysis. To address these issues, we introduced a fuzzy theoretical method in combination with the traditional FMECA method, thus proposing an improved FMECA method based on fuzzy comprehensive judgment. Through this approach, we quantified the qualitative analysis problems, calculated the qualitative analysis results, and analyzed the reliability of the rubber cutter. The study quantifies the problems, calculates the hazard rankings of each failure mode of the rubber cutter, and identifies the areas requiring reliability improvement. It addresses the shortcomings of the original hazard ranking and optimizes the cycle for replacing spare parts, along with the frequency of maintenance and inspection in the field cutting work within the natural rubber forest. This method offers a theoretical basis for the subsequent improvement design of the rubber cutter. Additionally, it enables pre-emptive detection of the potential failures of the rubber cutter, thereby reducing the incidence of failures and enhancing the reliability of the intelligent profiling step-by-step automatic rubber cutter. Moreover, this type of agricultural equipment reliability analysis and detection method exhibits significant technical innovation and positive implications for enhancing the reliability level of agricultural equipment. Furthermore, its applicability can be explored for other agricultural equipment as well.

Author Contributions: H.Z.: conceptualization, methodology, investigation, and writing—original draft. Y.C.: investigation and writing—review and editing. J.C.: writing—review and editing. J.L.: writing– review and editing. Z.Z.: writing—review and editing. X.Z.: conceptualization, writing—review and editing, project administration, and funding acquisition. All authors have read and agreed to the published version of the manuscript.

Funding: This work was supported by the Key Research and Development Projects of Hainan Province (Grant No.ZDYF2021XDNY198), the Innovation Platform for Academicians of Hainan Province (Grant No.YSPTZX202008, YSPTZX202109), and the National Modern Agricultural Industry Technology System Project (Grant No.CARS-33-JX2).

Institutional Review Board Statement: Not applicable.

Informed Consent Statement: Not applicable.

Data Availability Statement: Not applicable.

Conflicts of Interest: The authors declare that there are no conflict of interest.

References

1. Linton, J.D. Facing the challenges of service automation: An enabler for e-commerce and productivity gain in traditional services. *IEEE T Eng. Manage* **2003**, *50*, 478–484. [CrossRef]
2. Scheu, M.N.; Tremps, L.; Smolka, U.; Kolios, A.; Brennan, F. A systematic Failure Mode Effects and Criticality Analysis for offshore wind turbine systems towards integrated condition based maintenance strategies. *Ocean. Eng.* **2019**, *176*, 118–133. [CrossRef]
3. Arabsheybani, A.; Paydar, M.M.; Safaei, A.S. An integrated fuzzy MOORA method and FMEA technique for sustainable supplier selection considering quantity discounts and supplier's risk. *J. Clean. Prod.* **2018**, *190*, 577–591. [CrossRef]
4. Stamatis, D.H. *Failure Mode and Effect Analysis—FMEA from Theory to Execution*; ASQC Qual Press: New York, NY, USA, 1995.
5. Aswin, K.R.; Renjith, V.R.; Akshay, K.R. FMECA using fuzzy logic and grey theory: A comparitve case study applied to ammonia storage facility. *Int. J. Syst. Assur. Eng.* **2022**, *13*, 2084–2103. [CrossRef]
6. Bozdag, E.; Asan, U.; Soyer, A.; Serdarasan, S. Risk prioritization in Failure Mode and Effects Analysis using interval type-2 fuzzy sets. *Expert. Syst. Appl.* **2015**, *42*, 4000–4015. [CrossRef]
7. Liu, H.; Chen, Y.; You, J.; Li, H. Risk evaluation in failure mode and effects analysis using fuzzy digraph and matrix approach. *J. Intell. Manuf.* **2016**, *27*, 805–816. [CrossRef]
8. Zhou, D.; Tang, Y.; Jiang, W. A Modified Model of Failure Mode and Effects Analysis Based on Generalized Evidence Theory. *Math. Probl. Eng.* **2016**, *2016*, 4512383. [CrossRef]
9. Liu, H.; You, J.; Duan, C. An integrated approach for failure mode and effect analysis under interval-valued intuitionistic fuzzy environment. *Int. J. Prod. Econ.* **2019**, *207*, 163–172. [CrossRef]
10. Yang, Z.; Bonsall, S.; Wang, J. Fuzzy Rule-Based Bayesian Reasoning Approach for Prioritization of Failures in FMEA. *IEEE Trans. Reliab.* **2008**, *57*, 517–528. [CrossRef]
11. Jee, T.L.; Tay, K.M.; Lim, C.P. A New Two-Stage Fuzzy Inference System-Based Approach to Prioritize Failures in Failure Mode and Effect Analysis. *IEEE Trans. Reliab.* **2015**, *64*, 869–877. [CrossRef]
12. Gupta, G.; Mishra, R.P. A Failure Mode Effect and Criticality Analysis of Conventional Milling Machine Using Fuzzy Logic: Case Study of RCM. *Qual. Reliab. Eng. Int.* **2017**, *33*, 347–356. [CrossRef]
13. Sayyadi, T.H.; Ayatollah, A.S. A model for failure mode and effects analysis based on intuitionistic fuzzy approach. *Appl. Soft Comput.* **2016**, *49*, 238–247. [CrossRef]
14. Guo, J. A Risk Assessment Approach for Failure Mode and Effects Analysis Based on Intuitionistic Fuzzy Sets and Evidence Theory. *J. Intell. Fuzzy Syst.* **2016**, *30*, 869–881. [CrossRef]
15. Jiang, W.; Xie, C.; Zhuang, M.; Tang, Y. Failure mode and effects analysis based on a novel fuzzy evidential method. *Appl. Soft Comput.* **2017**, *57*, 672–683. [CrossRef]
16. Aydogan, E.K. Performance measurement model for Turkish aviation firms using the rough-AHP and TOPSIS methods under fuzzy environment. *Expert. Syst. Appl.* **2011**, *38*, 3992–3998. [CrossRef]
17. Liu, H.; You, J.; Shan, M.; Shao, L. Failure mode and effects analysis using intuitionistic fuzzy hybrid TOPSIS approach. *Soft Comput.* **2015**, *19*, 1085–1098. [CrossRef]
18. Carpitella, S.; Certa, A.; Izquierdo, J.; La Fata, C.M. A combined multi-criteria approach to support FMECA analyses: A real-world case. *Reliab. Eng. Syst. Safe* **2018**, *169*, 394–402. [CrossRef]
19. Zhou, Q.; Thai, V.V. Fuzzy and grey theories in failure mode and effect analysis for tanker equipment failure prediction. *Safety Sci.* **2016**, *83*, 74–79. [CrossRef]
20. Liu, H.; Li, Z.; Song, W.; Su, Q. Failure Mode and Effect Analysis Using Cloud Model Theory and PROMETHEE Method. *IEEE Trans. Reliab.* **2017**, *66*, 1058–1072. [CrossRef]
21. Mandal, S.; Singh, K.; Behera, R.K.; Sahu, S.K.; Raj, N.; Maiti, J. Human error identification and risk prioritization in overhead crane operations using HTA, SHERPA and fuzzy VIKOR method. *Expert. Syst. Appl.* **2015**, *42*, 7195–7206. [CrossRef]

22. Baloch, A.U.; Mohammadian, H. Fuzzy failure modes and effects analysis by using fuzzy Vikor and Data Envelopment Analysis-based fuzzy AHP. *Int. J. Adv. Appl. Sci.* **2016**, *3*, 23–30. [CrossRef]
23. Liu, H.; Liu, L.; Bian, Q.; Lin, Q.; Dong, N.; Xu, P. Failure mode and effects analysis using fuzzy evidential reasoning approach and grey theory. *Expert. Syst. Appl.* **2011**, *38*, 4403–4415. [CrossRef]
24. Liu, H.; Liu, L.; Lin, Q. Fuzzy Failure Mode and Effects Analysis Using Fuzzy Evidential Reasoning and Belief Rule-Based Methodology. *IEEE Trans. Reliab.* **2013**, *62*, 23–36. [CrossRef]
25. Du, Y.; Mo, H.; Deng, X.; Sadiq, R.; Deng, Y. A new method in failure mode and effects analysis based on evidential reasoning. *Int. J. Syst. Assur. Eng.* **2014**, *5*, 1–10. [CrossRef]
26. Su, X.; Deng, Y.; Mahadevan, S.; Bao, Q. An improved method for risk evaluation in failure modes and effects analysis of aircraft engine rotor blades. *Eng. Fail. Anal.* **2012**, *26*, 164–174. [CrossRef]
27. Jiang, W.; Xie, C.; Wei, B.; Zhou, D. A modified method for risk evaluation in failure modes and effects analysis of aircraft turbine rotor blades. *Adv. Mech. Eng.* **2016**, *8*, 2071833545. [CrossRef]
28. Liu, H.; Fan, X.; Li, P.; Chen, Y. Evaluating the risk of failure modes with extended MULTIMOORA method under fuzzy environment. *Eng. Appl. Artif. Intel.* **2014**, *34*, 168–177. [CrossRef]
29. Liu, H.; You, J.; You, X.; Shan, M. A novel approach for failure mode and effects analysis using combination weighting and fuzzy VIKOR method. *Appl. Soft Comput.* **2015**, *28*, 579–588. [CrossRef]
30. Liu, H.; Liu, L.; Liu, N. Risk evaluation approaches in failure mode and effects analysis: A literature review. *Expert. Syst. Appl.* **2013**, *40*, 828–838. [CrossRef]
31. Certa, A.; Hopps, F.; Inghilleri, R.; La Fata, C.M. A Dempster-Shafer Theory-based approach to the Failure Mode, Effects and Criticality Analysis (FMECA) under epistemic uncertainty: Application to the propulsion system of a fishing vessel. *Reliab. Eng. Syst. Safe* **2017**, *159*, 69–79. [CrossRef]
32. Mohsen, O.; Fereshteh, N. An extended VIKOR method based on entropy measure for the failure modes risk assessment—A case study of the geothermal power plant (GPP). *Safety Sci.* **2017**, *92*, 160–172. [CrossRef]
33. Huang, J.; Li, Z.S.; Liu, H. New approach for failure mode and effect analysis using linguistic distribution assessments and TODIM method. *Reliab. Eng. Syst. Safe* **2017**, *167*, 302–309. [CrossRef]
34. Kerk, Y.W.; Tay, K.M.; Lim, C.P. An Analytical Interval Fuzzy Inference System for Risk Evaluation and Prioritization in Failure Mode and Effect Analysis. *IEEE Syst. J.* **2017**, *11*, 1589–1600. [CrossRef]
35. Certa, A.; Enea, M.; Galante, G.M.; La Fata, C.M. ELECTRE TRI-based approach to the failure modes classification on the basis of risk parameters: An alternative to the risk priority number. *Comput. Ind. Eng.* **2017**, *108*, 100–110. [CrossRef]
36. Song, W.; Ming, X.; Wu, Z.; Zhu, B. A rough TOPSIS Approach for Failure Mode and Effects Analysis in Uncertain Environments. *Qual. Reliab. Eng. Int.* **2014**, *30*, 473–486. [CrossRef]

 *agriculture*

Article

Design and Testing of a Directional Clamping and Reverse Breaking Device for Corn Straw

Xun He [1,2], Xudong Fan [1], Wenhe Wei [1], Zhe Qu [1], Jingzhao Shi [1], Hongmei Zhang [1,2,*] and Bo Chen [1]

[1] College of Mechanical and Electrical Engineering, Henan Agricultural University, Zhengzhou 450002, China; hexun@henau.edu.cn (X.H.); 13525977568@163.com (X.F.); wenhe@stu.henau.edu.cn (W.W.); quzhe071171@henau.edu.cn (Z.Q.); haujingzhao@126.com (J.S.); chenbozz@stu.henau.edu.cn (B.C.)

[2] Research Center of Low-Carbon Agricultural Intelligent Equipment Engineering Technology in Henan Provence, Zhengzhou 450002, China

[*] Correspondence: zhanghongmei0905@henau.edu.cn

Abstract: Realizing high-quality and increased production of fresh corn and promoting diversified development of the corn industry structure not only can effectively promote the development of agricultural economy, but also can enrich people's dietary culture. However, existing fresh corn machinery has a high rate of ear damage during the harvesting process, and the overall harvesting efficiency is not ideal. To reduce damage during the harvesting of fresh corn, a device for breaking ears of fresh corn was designed based on the directional clamping of corn straw reverse breaking method. Based on the physico-mechanical characteristics parameters of fresh corn ears, the main structural parameters of the directional clamping and conveying mechanism and the ear-breaking mechanism were determined. The overall inclination angle of the device was 15°, and the effective conveying length of the directional clamping mechanism was 550 mm; the ear-snapping mechanism was a snapping roll composed of a pair of six radial distribution function fingers, with an effective operating radius of 320 mm. By simulating and analyzing the reverse breaking movement of directional clamping corn straw, the key motion parameter ranges of the directional clamping conveying mechanism and breaking mechanism were obtained. The results of the bench test showed that under the optimal conditions of a directional clamping feeding speed of 1.67 m/s, a breaking wheel speed of 80 rpm, and a travel speed of 1.06 m/s, the lowest ear damage rate was 0.57%, and the lowest impurity rate was 1.87%. In addition, it was observed that flexible harvesting can improve harvest efficiency and quality. The study also found that actively applying force to the device can effectively avoid the problem of machine blockage and reduce the damage rate of ears (the following text uses ears instead of fresh corn ears).

Keywords: fresh corn; directional clamping; reverse ear breaking; bench test; optimal design

Citation: He, X.; Fan, X.; Wei, W.; Qu, Z.; Shi, J.; Zhang, H.; Chen, B. Design and Testing of a Directional Clamping and Reverse Breaking Device for Corn Straw. *Agriculture* **2023**, *13*, 1506. https://doi.org/10.3390/agriculture13081506

Academic Editor: Massimiliano Varani

Received: 21 June 2023
Revised: 20 July 2023
Accepted: 24 July 2023
Published: 27 July 2023

1. Introduction

Fresh corn, also known as fruit corn, is usually picked during the milk ripening period and is a nutritious and delicious vegetable suitable for different populations. With the support of relevant policies, the fresh corn industry has developed rapidly [1,2]. Fresh corn has characteristics such as high moisture content, thin skin, poor grain compression resistance, and susceptibility to damage. During the process of picking and transporting ears, it is necessary to maintain the integrity of the ears to the greatest extent possible to ensure the quality of the crops after harvest [3]. The traditional process of harvesting fresh corn is manual harvesting, which not only has low efficiency but also leads to a shortage of labor. The mechanized harvesting of fresh corn is an effective means to solve the current problem of rural labor shortage, which can increase farmers' enthusiasm for planting, reduce labor intensity, and promote yield increase and harvest [4]. However, due to significant climate differences in different regions, corn planting modes and harvesting methods are also divided into various types. In practice, specialized fresh corn harvesters are often used for harvesting operations [5,6], and the market penetration rate of specialized harvesting machinery is relatively low [7–9].

At present, fresh corn harvesters that simulate manual picking methods have become a potential solution to the problem of a high rate of ear damage during harvesting [10,11]. So far, relevant scholars have conducted research on harvesting machinery for fresh corn and reduced the damage rate of ears. According to the literature review, the structural improvement of the ear-picking device may have a significant impact on the optimization of the above issues [12,13]. Regarding the research on fresh corn ear-picking devices, existing device types inevitably encounter phenomena such as corn ear biting, broken stems, and machine blockage, which affects the application of corn harvesters [14].

The ear-picking device is the most important working component of a corn harvester, which directly affects the quality of harvested ears [13]. The ear-picking device is divided into the horizontal roller type, the stem-pulling roller and ear-picking plate combination type, the biomimetic ear-picking type, etc. according to its structure [15]. The horizontal roller type ear-picking device can achieve effective harvesting of fresh corn through the difference in torque force and friction force. For example, Zhansheng Zhang et al. [16] (2021) used a horizontal roller type ear-picking mechanism longitudinally configured with high- and low-torsion rollers to simulate the operation process of manually breaking fresh corn. The combined ear-picking device of stem-pulling roller and ear-picking plate reduces the power loss during the ear-picking process by adding an ear-picking plate above the stem-pulling roller [17] (Tianyu Li, 2019). Guanqiang Zhu et al. [11] (2023) designed a corresponding ear-picking device using a biomimetic ear-picking method. This scheme adopted a new ear-harvesting posture adjustment method, and the average power consumption of the entire device was reduced to over 3.0 kW. Xirui Zhang et al. [18] (2019) used an oblique roller to break the ears of fresh corn, separating them from the stems and achieving the harvest requirements for crispy and tender corn.

Because most of the existing corn-harvesting machines on the market are horizontal roller harvesting mechanisms and have a large inventory, it is possible to improve and optimize the existing harvesting devices, which can save energy and protect the environment [19]. This study aims to design an ear-breaking device for harvesting fresh corn and verify its feasibility, especially to determine how the flexible ear-picking mechanism affects the breaking rate of ears.

This article proposes to design a new ear-breaking device for harvesting fresh corn based on the method of directional clamping of corn straw and reverse ear breaking. Based on clarifying the physico-mechanical properties of fresh corn, the key mechanisms of the ear-breaking device were optimized and designed; aimed at the goal of reducing the rate of ear damage and impurity content, a bench test was conducted using the orthogonal rotation center combination experimental design method to obtain and verify the optimal combination of key mechanism motion parameters, thereby providing theoretical and technical bases for the optimization design of fresh corn combine harvesters.

2. Design of Directional Clamping Corn Ear-Breaking Device

2.1. Design Requirement

The optimal harvest period for fresh corn is in the late stage of milk ripening and the early stage of wax ripening. At this time, the moisture content of the stems of fresh corn plants is higher, and compared to ordinary grain corn, the stems are thin and brittle. Completing the fresh corn ears picking operation can easily cause a higher ear impurity rate. Meanwhile, the moisture content of fresh corn kernels ranges from 65% to 75%, which can easily cause serious mechanical damage during the picking process, thereby affecting the quality and storage time of fresh corn and reducing its economic value. The design requirements for fresh corn ear picking equipment [20,21] are:

I.   Reduce or avoid the possibility of mechanical collision between fresh corn ears and operating components, reduce the probability of mechanical damage to ears, and ensure that the crushing rate of fresh corn seeds is less than 1%.

II.   Avoid cutting off the stems of fresh corn plants by the ear-picking device, ensure that the impurity content of fresh corn ears is less than 1.5%, and avoid blockage of the ear-picking device.

2.2. Structure and Working Principle

The directional clamping corn ear-breaking device is the header part of a corn combine harvester, mainly composed of a frame, a disc cutter device, a directional clamping mechanism, an ear-breaking mechanism, a stem-pulling mechanism, an ear collection mechanism, a transmission mechanism, etc. (Figure 1). When harvesting fresh corn, as the combine moves in opposite directions, the corn plant is continuously fed to the directional clamping conveying mechanism, which clamps the stem above the ears and delivers it to the breaking mechanism, and the ear breaking is completed under the action of ear-breaking rollers composed of a pair of six radial distribution function fingers. During the breaking process, the corn stalks were not cut, similar to the manual breaking process. The directional clamping mechanism played a role in supporting the breaking process. After the breaking operation was completed, the cutting knife cut the stalks, and the pulling roller squeezed the corn stalks to achieve continuous, stable production.

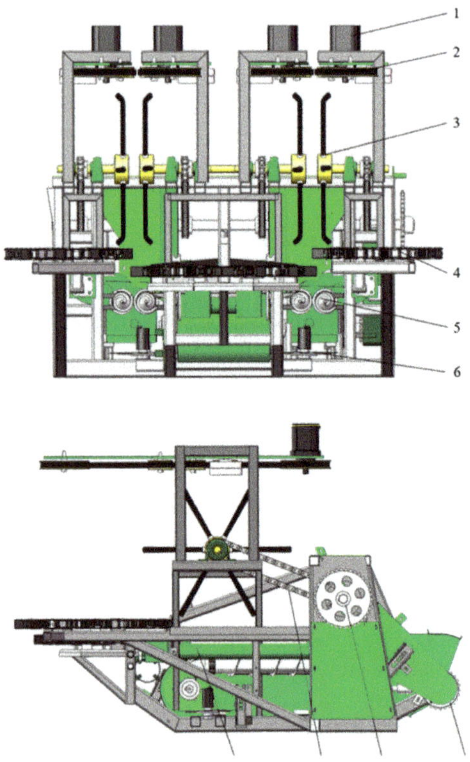

Figure 1. Directional clamping corn straw breaking header. 1. Hydraulic motor. 2. Directional clamping mechanism. 3. Breaking mechanism. 4. Reel chain. 5. Stem pulling roller. 6. Stem cutting mechanism. 7. Header power input shaft. 8. Breaking mechanism power spindle. 9. Breaking transmission chain. 10. Pulling chain transmission shaft.

2.3. Directional Clamping Mechanism

As illustrated in Figure 2, the directional clamping mechanism imitates the manual breaking of corn stalks. Before butting the corn roots, the cutter is transported to the breaking mechanism under the action of the clamping belt and supports the corn stalks to ensure that the stalks are not broken. After the breaking of the ears is completed under the action of the breaking wheel fingers, the directional clamping and conveying process can be completed.

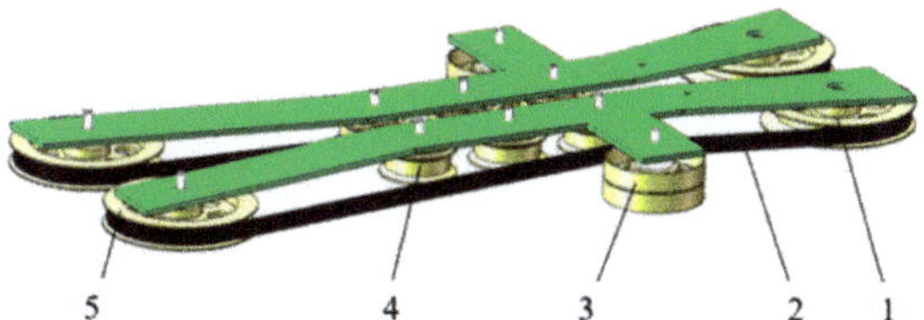

Figure 2. Structural diagram of directional clamping and conveying mechanism. 1. Active pulley. 2. Clamping belt. 3. Tensioning pulley. 4. Guide pulley. 5. Passive pulley.

To ensure effective clamping of the stems during the clamping and conveying process, it is advisable to keep the clamping position within the same section of the stems as much as possible. Therefore, the inclination angle of the directional clamping and conveying mechanism should not be too large. According to the measurement of fresh corn plants, the diameter of the stems above the ears decreases as the height increases; therefore, the clamping position of the directional clamping and conveying system should be controlled, and the inclination angle should not be too large so ζ should satisfy:

$$\sin \zeta \leq \frac{d_m}{l_f} \tag{1}$$

In the formula, d_m is the single stem length, mm; l_f is the length of directional clamping and conveying mechanism, mm; ζ is the angle of device inclination, °.

Considering the clamping stability and structural layout during harvesting operations, an effective clamping conveying length of 550 mm was designed, and the calculated inclination range is $0° \leq \zeta \leq 21°$. Based on practical considerations, the inclination angle of the clamping and conveying mechanism is taken as $\zeta = 15°$, and due to the integrity of the header, the overall tilt angle of the header is 15°.

To achieve effective clamping of the stems during the harvesting operation, the clamping and conveying speed exerts a significant influence on the breaking effect. To ensure the stability and efficiency of the breaking operation, the changes in the inclination angle and speed of the corn stems during the directional clamping and conveying process were estimated. As shown in Figure 3, during the process of ears breaking, if the forward speed of the machine is equal to the clamping and feeding speed of the stem, the rapid feeding of adjacent plants may affect the ear-breaking effect of a single plant. Therefore, to reduce the influence of the front plant on subsequent plants, the directional clamping and conveying speed should not be less than the forward speed of the machine.

In the analysis of the process of gripping corn plants, and during the ear-breaking operation, it is necessary to make the ears setting position L_j in the ear-breaking mechanism.

Above the minimum effective ear-breaking height H_1, the tilt angle of the corn plant should be less than α_1.

If the operating speed of the machine during field operations is v_j and the spacing between plants is S_1, then at time t_1 when advancing by one pre-set unit spacing between plants:

$$t_1 = \frac{S_1}{v_j} \tag{2}$$

In the formula, S_1 is the single row spacing, mm; v_j is the operation speed of machine, m/s.

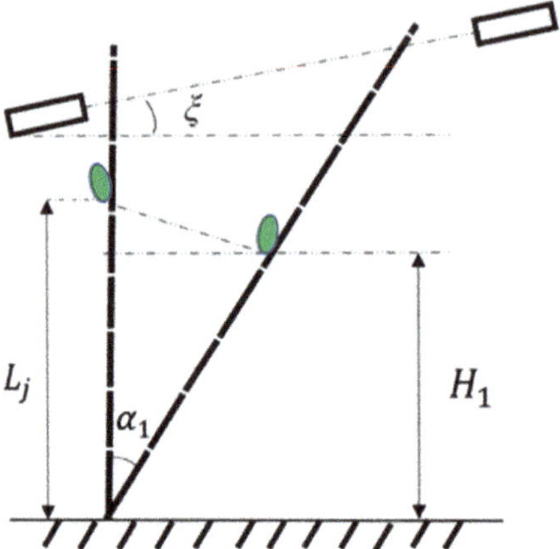

Figure 3. Schematic diagram of plant inclination angle under clamping and conveying.

During the directional clamping and conveying process, the upper end of the corn plant tilts to a certain extent under the action of the clamping and conveying mechanism to change the feeding posture of the fresh corn plant. When the critical condition is reached, the ears setting position tilts to the lowest effective ear-breaking height H_1. At this point, adjacent corn plants also enter the feed inlet of the directional clamping conveyor mechanism, and the clamping conveyor time t_2 is:

$$t_2 = \frac{L_j \times \tan \alpha}{v_f \times \cos \xi} \tag{3}$$

In the formula, v_f is the speed of directional clamping and conveying mechanism, m/s. If $t_1 = t_2$, then:

$$v_f = \frac{L_j \times \tan \alpha_1}{S_1 \times \cos \xi} v_j \tag{4}$$

According to the measurement of the characteristics of fresh corn plants, the range of ears height is 0.73–1.08 m, and the spacing between fresh corn plants is 0.25–0.35 m. The mechanical operation speed is taken as 0.85–1.35 m/s based on the existing harvester operation speed. The clamping conveyor belt speed v_f is calculated from Equations (1) and (2) to be 1.83–2.92 m/s. According to the ratio between the machine forward speed and the clamping feeding speed, the maximum speed of the clamping conveyor mechanism is 1.83 m/s. In summary, the speed range of the directional clamping conveying mechanism is 1.35–1.83 m/s.

2.4. Design and Analysis of the Ear-Breaking Mechanism

The ear-breaking device is the core component of the corn harvester, and the harvesting efficiency and ear damage rate of fresh corn are determined mainly by the ear-breaking device. At present, there are two main types of corn ear-breaking devices: the roller type and the combination of ear-breaking plate and stem-pulling roller. When the roller-type ear-breaking mechanism is in harvesting operation, the direct contact between the ears and the ear-breaking roller can cause severe ear biting due to the high moisture content of fresh corn [22]. The combined ear-breaking device mainly uses the stem-pulling roller to pull the corn stem downward, and when the ears come into contact with the ear-breaking plate, the ears cannot pass through. When the separation occurs between the ears and stem, the instantaneous impact on the same part of the corn ear is greater, resulting in a higher rate of damage. At the same time, different collision conditions from different angles also cause different amounts of damage to the impact surface of fresh corn ears.

2.4.1. Stress Analysis of Fresh Corn Ears

The fresh corn-clamping and breaking device utilizes the breaking finger and the gravity of the ears itself to complete the breaking operation during the breaking operation, which can reduce the rate of damage to ears [23]. As shown in Figure 4, during the growth of fresh corn plants, the balance equation between the weight torque of corn ears and the torque T of stems under stable conditions is as follows:

$$T = GL_1 \sin \theta \tag{5}$$

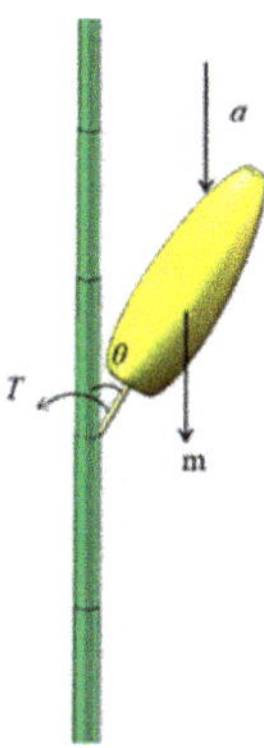

Figure 4. Stress analysis of ears.

Assuming that during the process of ears breaking, ears breaking refers to applying a downward acceleration a to the ears, the inertial force on the ears is given by:

$$mgL_1 \sin \theta + maL_2 \sin \theta = m(gL_1 + aL_2) \sin \theta \tag{6}$$

In the formula, m is the mass of an ear, kg; L_1 is the distance from the center of gravity of the ear to the growth point, mm; L_2 is the distance from the point of force application to the growth point, mm; θ is the angle between the ear and the stem, °.

According to Equation (6), the combined moment of gravity and force of the ears during the breaking process is greater than the torque of the stem when it is stable, so the breaking operation can be completed.

Therefore, compared to traditional ear-breaking methods, the reverse ear-breaking method requires less force to complete the ear-breaking operation under the self-weight of the ears.

2.4.2. Design of the Ear-Breaking Mechanism

To reduce the damage rate of fresh corn ears, an active ear-breaking wheel was designed based on the principle of manual ears breaking, which mainly consists of a rotating wheel hub, ear-breaking fingers, and flexible rubber sleeves, as shown in Figure 5.

When performing ear-breaking operations, ears breaking refers to the process of gradually increasing the force on the corn plant. Therefore, the relationship between the installation height of the rotation axis of the ear-breaking wheel and the average ear height should meet the following requirements:

$$H \leq \bar{h} \tag{7}$$

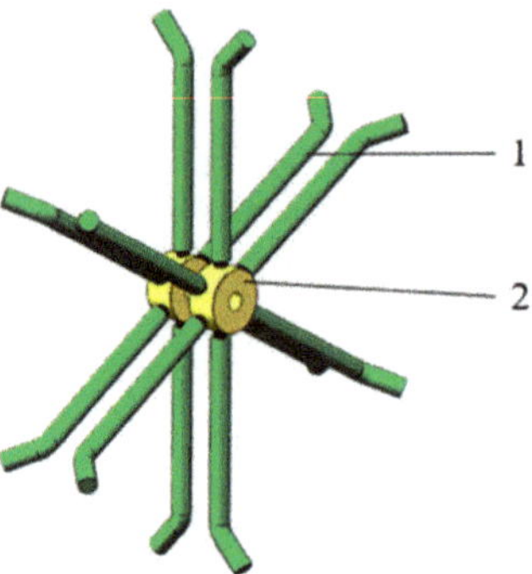

Figure 5. Schematic diagram of ears breaking mechanism. 1. Flexible breaking finger. 2. Rotating wheel hub.

By measuring the physical parameters of fresh corn, 50 plants of Wannuo 2018 and Starnuo 41 varieties were randomly selected during the harvest period. The physical characteristics [24] (including plant growth height, stem diameter, ear diameter, and ear height) of fresh corn plants were measured (Table 1).

Table 1. The physical characteristics of fresh corn.

Breed	Position	Average Value	Breed	Position	Average Value
	Plant height	2250 mm		Plant height	2310 mm
	Ear height	937 mm		Ear height	1005 mm
	Lower diameter of spike position	28.5 mm		Lower diameter of spike position	30.1 mm
Wannuo 2018	Diameter at ear position	19.3 mm	Stanuo 41	Diameter at ear position	21.7 mm
	Diameter of the four nodes on the ear	16.7 mm		Diameter of the four nodes on the ear	18.3 mm
	Ear diameter at the large end	52.3 mm		Ear diameter at the large end	48.7 mm

Using a tensile tester to perform reverse breaking force testing on ears, the measured breaking force ranged from 36.8 N to 72.4 N. Mechanical properties were measured using a ZQ-890A mechanical (ZhiQu Precision Instrument Co., Ltd., Dongguan, China.) testing machine (with a force range of 100 kg/1 kN). Using the three-point bending measurement method, the average yield strength of the stem was found to be 9.26 MPa, the average compressive strength of the ears with bracts was 0.205 MPa, and the average compressive strength of the ears without bracts was 0.138 MPa.

The average ear height difference of a single variety is 296.5 mm, and the effective ear height difference is reduced under the action of the directional clamping conveyor mechanism, that is, $dx \leq 300$ mm. Therefore, to ensure that all corn plants of the same variety can be within the effective ear-breaking range, the installation of the ear-breaking wheel rotation axis does not exceed the average ear-breaking height (a static analysis thereof is represented in Figure 6).

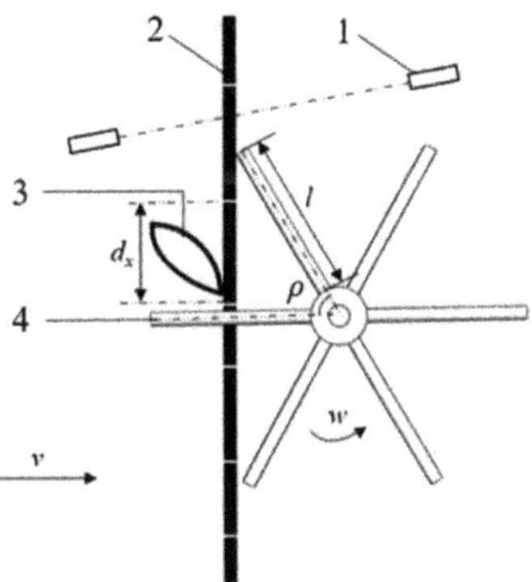

Figure 6. Static ear breaking interval. 1. Directional clamping mechanism. 2. Corn plants. 3. Fresh corn ears. 4. Breaking fingers.

To ensure that fresh corn ears can enter the ear-breaking fingers, the distance between adjacent ear-breaking fingers should not be too small. If it is too small, it will prevent the ear-breaking fingers from acting on the ears, affecting the efficiency of the operation; therefore, the distance between adjacent fingers is such that:

$$d_z = 2l_z \times \sin \frac{\rho}{2} \tag{8}$$

In the formula, d_z is the distance between the fingers of two adjacent broken ears, mm; l_z is the breaking finger length, mm; ρ is the angle between adjacent ear fingers, °.

Due to the differences in growth among different corn varieties, the angle between ears and stems is also different. Therefore, the effective length of ears in the vertical direction is:

$$d_g = l_g \times \cos \beta \tag{9}$$

In the formula, d_g is the distance between the fingers of two adjacent broken ears, mm; l_g is the breaking finger length, mm; β is the angle between ear and stem, °.

At the same time, the effective length d_g of the ear in the vertical direction and the distance d_z between adjacent ear fingers should meet the following requirements:

$$d_g \leq d_z \tag{10}$$

According to the measurement of planting parameters for fresh corn plants in the field, the average plant spacing of fresh corn is 325 mm, and the machine operating speed is 1.2 m/s. During pre-testing, when the rate of rotation of the ear-breaking wheel is 90 rpm, the ear-breaking operation is good, so 90 rpm is used as a benchmark in subsequent calculations. The relationship between ear-breaking time interval and ear-breaking index is determined using Equation (11):

$$t = \frac{60}{n \times z} \tag{11}$$

In the equation, t is the interval time, s; n is the speed of spinning wheel, rpm; z is the number of broken ear fingers.

From Equation (11), the rate of rotation of the ear-breaking wheel is inversely proportional to the index of the ear-breaking finger. When the rotational speed of the ear-breaking wheel is a fixed value, the time interval between ear-breaking fingers decreases as the number of roots of the ear-breaking fingers increases. To ensure that there is no missed picking during the ear-breaking operation, Equation (12) must be satisfied:

$$t \leq \frac{d}{v} \tag{12}$$

In the equation, d is the plant spacing of fresh corn, mm; v is the forward speed of machine, m/s.

From this, it can be concluded that $t \leq 0.25$ s. Considering the rotational balance and structural dimensions of the ear-breaking wheel, when $z = 6$, $t = 0.11$ s satisfies Equation (12).

In summary, the length of the ear-breaking fingers is 300 mm, and the number of ear-breaking fingers is 6. The inner part is an iron roller, and the outer part is a flexible rubber sleeve. The material used for the outer rubber sleeve is polyurethane, with a Shore hardness of 65A and a thickness of 7 mm.

2.4.3. Analysis of the Effect of Breaking Ears on the Growth Angle of Fresh Corn Ears

When the ear-breaking wheel is used for the ear-breaking operation, the growth direction of fresh corn ears has a significant impact on the ear-breaking effect of the ear-breaking finger. Therefore, the gap between the ear-breaking fingers must meet the requirement of producing ear-breaking force for fresh corn ears growing in different directions. During the ear-breaking operation, the corn stem is directionally clamped and fed into the gap between the ear-breaking fingers. As the diameter of the ear is smaller than the gap, the ear-breaking finger can apply the ear-breaking force to separate the ear from the stem from top to bottom. Due to the randomness of the direction between the ears and the gap between the ear-breaking wheel during the operation of the machine, there are several states where the angle between the ears axis and the ear-breaking finger axis appears, as illustrated in Figure 7.

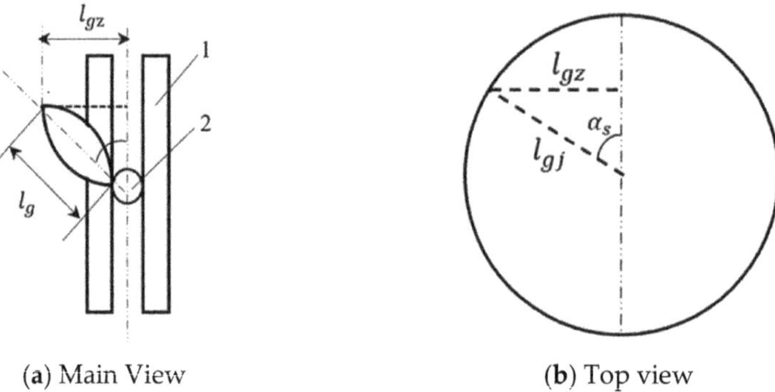

(**a**) Main View   (**b**) Top view

Figure 7. The relative position of the ear-breaking finger and the ears. 1. Ear-breaking roller. 2. Corn plants.

Based on the measurement data of the length of fresh corn ears and the angle between ear and stem, the horizontal projection l_{gz} of the top of the ear to the axis of the stem can be calculated as:

$$l_{gj} = l_g \times \sin \beta \tag{13}$$

$$l_{gz} = l_{gj} \times \sin \alpha_s \tag{14}$$

From the above equation, in the unilateral case:

I.   Fresh corn ears grow at a certain angle range in the opposite direction to the direction of motion of the harvesting machinery, and the angle between the ears and stems is: $0° \leq \alpha_s < 31.2°$. The ear-breaking force is applied to the top one-third of the ears, and as the ear-breaking finger rotates, the point of action of the resultant ear-breaking force slides down to the bottom of the ears;

II.   Fresh corn ears grow at a certain angle range compared to harvesting machinery, and the angle between the ears and corn stems is $31.2° \leq \alpha_S < 148.8°$; the ear-breaking force is applied to the middle one-third of the ears by the ear-breaking finger. As the ear-breaking finger rotates, the point of action of the resultant ear-breaking force slides down to the bottom of the ears or even to the stem itself;

III.   Fresh corn ears grow in the same direction as harvesting machinery, and the angle between the ears and stems is $148.8° \leq \alpha_S \leq 180°$; the ear-breaking force is applied to the lower one-third of the ears by the ear-breaking finger. As the ear-breaking finger rotates, the point of action of the resultant ear-breaking force slides down to the top of the ears.

During this process, the stressed parts of fresh corn vary, so the impact force on the ears with bracts is a sliding impact. This breaking method can reduce the proportion of damaged ears.

2.4.4. Finite Element Analysis of Fruit Stem Bending Fracture

The solution idea of finite element method is to discretize a global solution domain into multiple small subdomains and connect them into a whole through adjacent nodes [25,26]. Xiaodong Guan et al. [23] investigated collisions of ears at different angles using ANSYS. This article aims to determine the impact of bending load on corn plants. The plant model was imported into ANSYS, and corresponding loads were applied to observe the effect. By adding forces, the effects of applied loads on the bending fracture trend of fruit stems were examined, and the theoretical analysis of ear picking was conducted.

In ANSYS Workbench, the designed three-dimensional model of fresh corn plants was imported and the mechanical properties parameters of fresh corn input [17] (as shown in Table 2):

Table 2. Model parameters of fresh corn plants.

Young's Modulus (N·mm^{-2})	Density (kg·mm^{-3})	Poisson's Ratio
3.67×10^3	4.5×10^7	0.33

The fresh corn plant was divided into grids and Boolean operations were used to connect the components. To ensure the independence of the three components, the Glue command in Boolean operations was used to connect the common boundaries of each component. To observe the stress analysis of the fracture site, the area where the stem is connected to the upper and lower sections of the fruit stem and the area where the fruit stem is connected to the ear was also meshed (Figure 8).

(a) Overall grid   (b) Local mesh

Figure 8. Grid division of fresh corn plants.

The fresh corn-clamping and breaking device designed in this article is fixed at both ends during the breaking operation, exerting a force on the ears. When bent to a certain extent, the ears are picked off; therefore, in finite element analysis, fresh corn plants are fixed and supported in both an up and down motion, and loads are added to the ears. According to the measurement of fresh corn ear-breaking force described in Section 2, the range of ear-breaking force is 36.8 to 72.4 N. Due to errors in experimental measurement, and based on the compressive strength data measured on the mechanical properties of corn ears, under the condition of meeting the compressive strength of corn ears with bracts, a force of 120 N is applied to the ears in a vertical downward direction. For the convenience of observation, the model style during deformation is retained in finite element analysis (by zooming in and out on the observed area). As shown in Figure 9a, there is a certain amount of deviation due to the thin upper stem of the spike. The stress cloud diagram of the connection between the fruit stem and the stem and ear after adding the load is shown in Figure 9b.

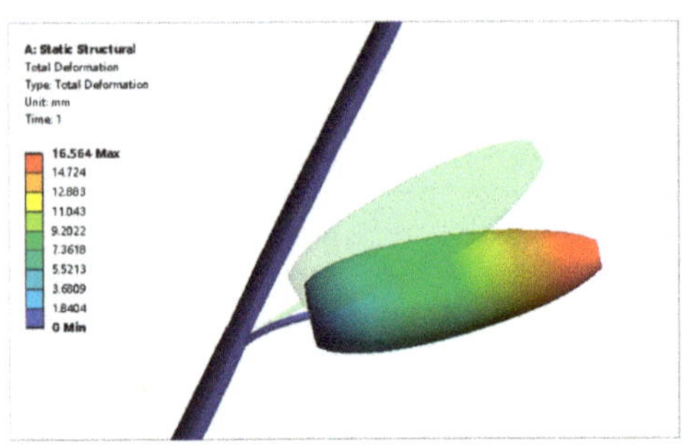

(**a**) Total deformation diagram

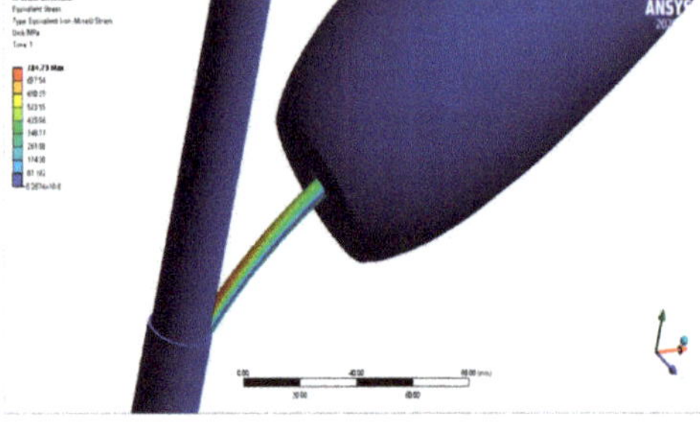

(**b**) Equivalent force cloud diagram

Figure 9. Strain cloud diagram under vertical force.

Through observation and research, it has been found that under uniform load applied to the surface of ears, force is transmitted to the fruit handle and straw, causing them to undergo bending deformation under load. The maximum stress concentration is found to be close to the point of connection between the fruit handle and straw; therefore, it can be determined that, when bending load is applied to the surface of the ears, the fruit handle will first bend and then break, and the breaking position is easy to occur near the connection position between the fruit handle and straw. Therefore, the reverse ear-breaking operation designed can be deemed reasonable.

3. Bench Test

3.1. Fresh Corn Clamping and Breaking Device Test Bench

The directional clamping device in the fresh corn clamping and breaking device needs to have good stable clamping ability of the corn stem, to avoid excessive changes in the posture of the corn plant that may affect the breaking effect; the ear-breaking device requires a stable ear-breaking operation path, a certain degree of wear resistance and flexibility, and an efficient harvesting effect. According to the operational requirements of the fresh corn clamping and breaking device, a test bench for the fresh corn clamping and breaking device was designed, which mainly consists of a directional clamping and conveying mechanism, a breaking mechanism, and a stem pretreatment mechanism, as shown in Figure 10. The entire test bench is powered by electricity, and its specific structural parameters are listed in Table 3.

Table 3. Test bench size parameters.

Position	Size Parameter
Overall dimensions of the test bench	1800 mm × 1300 mm × 2125 mm
Dimensions of lifting inner frame	1120 mm × 1090 mm × 1180 mm
Directional clamping and conveying mechanism	1000 mm × 400 mm
Rotation radius of ear-breaking mechanism	$r = 320$ mm
Operation length of stem-pulling roller	580 mm
Range of platform lifting angle	0~40°

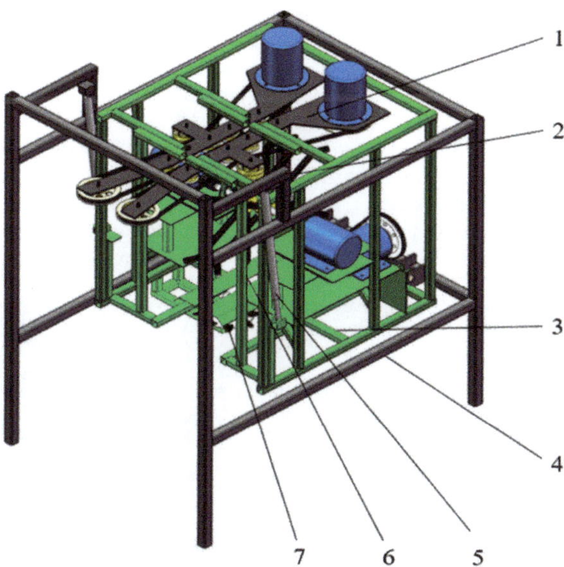

Figure 10. Structure diagram of the test bench for the fresh corn clamping and breaking device. 1. Directional clamping mechanism. 2. Ear picking mechanism. 3. Inner frame. 4. Outer frame. 5. Lifting push rod. 6. Stem pretreatment mechanism. 7. Cutting mechanism.

3.2. Tracking of Ears Movement Trajectory during Breaking

The process was theoretically verified using the directional clamping corn ear-breaking device test bench. High-speed photography technology was used to collect images of the trajectory changes of the ears being picked during the ear-breaking process. Later, motion-tracking software (Adobe After Effects 2021) was used to mark the position of the ears and track its motion trajectory [23], determining the contact situation of ears falling onto the collection board (Figure 11).

Figure 11. Image collection during ear-breaking process. 1. Test bench. 2. Control box. 3. Display. 4. Feeding device. 5. High-speed camera.

We tracked the motion trajectory through the ear-breaking process and used Adobe After Effects 2021 software to mark the bottom and top of the ear, as shown in Figure 12:

the ear-breaking finger acts on the middle of the ear, and the stem has a certain bending strength. During the application of the ear-breaking force, the ear-breaking force did not reach the maximum bending force of the stem, resulting in a small degree of bending of the stem. Therefore, in the actual ear-breaking process, breaking the ear refers to sliding from the middle of the ear to the connection between the stem and the stem. During this process, the ear is affected by its own self-weight and breaking forces, and the displacement of the top changes quickly. It can achieve effective breaking of the ear using the combined force of its own gravity and breaking force. The actual operation and virtual simulation results are not significantly different, meeting the finite element and virtual simulation analysis of fresh corn breaking. The device design can thus be deemed reasonable.

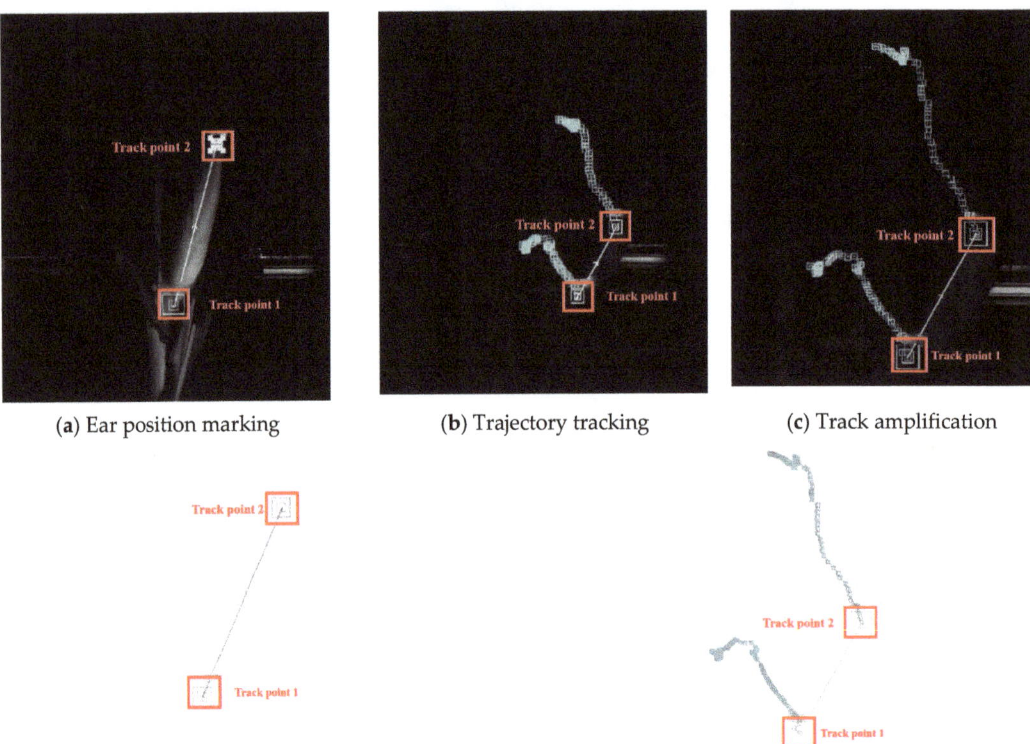

(**a**) Ear position marking   (**b**) Trajectory tracking   (**c**) Track amplification

(**d**) Trajectory extraction before ear breaking   (**e**) Trajectory extraction after ear breaking

Figure 12. Tracking and extraction of ear trajectory during ear-breaking process.

3.3. Bench Test

3.3.1. Protocol

The experiments of gripping and breaking ears of fresh corn were conducted on a self-designed and constructed breaking ears test bench. As shown in Figure 13, the overall system adopts three-phase motor control (model: YX3-100L-6, manufacturer: Shanghai Dashu Motor Co., Ltd., Shanghai, China). By adjusting the output power of the frequency converter (model: CDI-EM60G1R5S2B, manufacturer: Hangzhou Delixi Frequency Converter Co., Ltd., Hangzhou, China), the operating parameters of each mechanism were controlled. After each mechanism had stabilized and was running smoothly, the test was started. The directional movement of fresh corn plants fixed on the feeding device simulated the relative movement of machinery and plants during harvesting operations.

Figure 13. Physical image of the test bench. 1. Test bench. 2. Control box. 3. Feeding device.

The experimental material is fresh corn from the Wannuo 2018 variety, and the moisture content of fresh corn ears needs to be measured before the experiment. Moisture is used mainly to measure the moisture content of materials [27]. The measurement method is the constant temperature drying method. Five fresh corn plants are randomly selected from each point in the measurement area according to the five-point sampling method, and the mass q of the seeds is measured. The weight of the seeds is not less than 50 g. The temperature of the moisture meter is adjusted to $105 \pm 2\,°C$, and each part of the material is placed in the test box. The material is dried in the tray for 5 h until the mass is constant, as shown in Figure 14. The mass q' of each part is measured at this time, and the formula for calculating water content is shown in Equation (15):

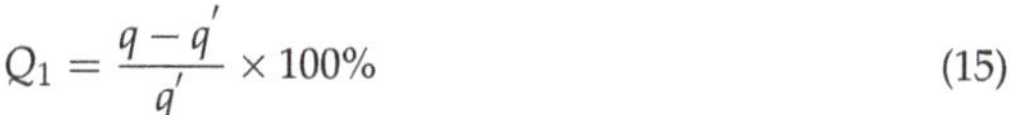

$$Q_1 = \frac{q - q'}{q'} \times 100\% \tag{15}$$

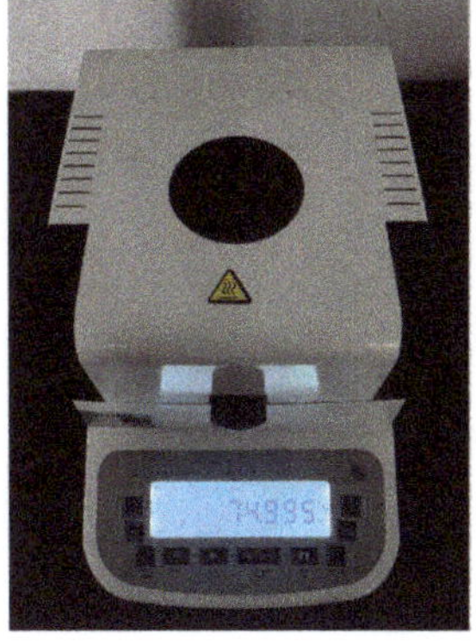

Figure 14. Measurement of grain moisture content.

In the equation, Q_1 is the moisture content of fresh corn ears, %; q is the quality before drying, g; q' is the quality after drying, g.

The average moisture content of the ears measured in the experiment was 67.23%, which meets the harvest period standard. Corn plants with intact appearance, no sagging ears, and good growth conditions were selected as the experimental material. The experimental process used the quadratic regression orthogonal rotation combination method and Design Expert 8.0.6 software to determine the orthogonal coding values of three factors and five levels, as shown in Table 4.

Table 4. Encoding of test factors.

Encoding Values	Factor		
	Clamping Feeding Speed (m·s⁻¹)	Spinning Finger Rotation Speed (rpm)	Forward Speed (m·s⁻¹)
2	1.83	110.00	1.35
1	1.71	100.00	1.25
0	1.59	90.00	1.10
−1	1.47	80.00	0.95
−2	1.35	70.00	0.85

To ensure the accuracy of the test results, each group of experiments was conducted three times, and 10 plants were harvested at a time. The overall operation of the device was stable, and there were no blockages. The test results of measuring the target factors using the constructed bench test are summarized in Table 5.

Table 5. Test plan and results.

Number	Factor			Objective Function	
	Clamping Feeding Speed X_1	Spinning Finger Rotation Speed X_2	Forward Speed X_3	Ear Damage Rate R_1	Harvest Impurity Content R_2
1	1.47	80.00	0.95	0.47	1.92
2	1.71	80.00	0.95	0.37	1.75
3	1.47	100.00	0.95	0.82	1.73
4	1.71	100.00	0.95	0.85	1.93
5	1.47	80.00	1.25	0.52	2.14
6	1.71	80.00	1.25	0.65	1.52
7	1.47	100.00	1.25	0.23	2.30
8	1.71	100.00	1.25	0.74	2.17
9	1.39	90.00	1.10	0.48	2.29
10	1.83	90.00	1.10	0.83	1.54
11	1.59	70.00	1.10	0.31	1.73
12	1.59	110.00	1.10	0.85	1.98
13	1.59	90.00	0.85	0.64	1.72
14	1.59	90.00	1.35	0.31	2.14
15	1.59	90.00	1.10	0.37	1.76
16	1.59	90.00	1.10	0.32	1.71
17	1.59	90.00	1.10	0.24	1.68
18	1.59	90.00	1.10	0.47	1.66
19	1.59	90.00	1.10	0.56	1.49
20	1.59	90.00	1.10	0.41	1.77

3.3.2. Test Indicators

The target factors for selecting experimental indicators include the ear damage rate and the ear impurity rate. To ensure the accuracy of measurement data, fresh corn planting parameters were measured in the field, and the number of plants per mu (the mu is a Chinese measure of land area, with 15 mu being equivalent to one hectare) was calculated. The target factors for testing 30 plants were converted into the rate of damage and harvest impurity rate per mu of fresh corn plants. The formula for calculating the number of ears per mu is as follows:

$$S = 667 \div (l_h \cdot l_z) \cdot j_s \tag{16}$$

In the formula, S is the yield per mu of the ears, piece; l_h is the average row spacing, cm; l_z is the average plant spacing, cm; j_s is the ears setting rate, %.

Referring to the agricultural machinery promotion appraisal [20] (the specified indicators for ears harvesting in the technical conditions of fresh corn harvesters), it is required that the impurity content of fresh corn ears should be $\leq 2\%$, and the damage rate of fresh corn should be $\leq 4\%$ for sweet corn; for glutinous corn and sweet glutinous corn, a value of $\leq 2\%$ is acceptable.

I. Due to the high moisture content of fresh corn, threshing measurement is not applicable. Therefore, after collecting all ears in the grain box and removing the bracts, the condition of each ear being damaged by the machine was checked, and the total number of grains and the number of damaged grains (with obvious cracks and peels) were calculated according to the following formula:

$$Z_s = \frac{W_s}{W_q} \times 100\% \tag{17}$$

In the formula, Z_s is the grain damage rate, %; W_s is the number of damaged grains, piece; W_q is the total number of ears and grains, piece.

II. In the area of measurement, this research collected the harvested grain from the grain bin, weighed the total mass and the mass of impurities (including stems and leaves), and calculated the impurity content as follows:

$$G_n = \frac{W_n}{W_p} \times 100\% \tag{18}$$

In the formula, G_n is the ears impurity content, %; W_n is the quality of debris, g; W_p is the total mass of grain harvested in the grain tank, g.

4. Results

Based on bench test data, regression analysis was conducted using Design Expert 8.0.6 data analysis software [28] to obtain regression equations for ear damage rate R_1 and harvest impurity rate R_2, and their significance was tested [29].

4.1. Analysis of the Impact of Ear Damage Rate

4.1.1. Multiple Regression Analysis of Ear Damage Rate

Based on the analysis of experimental data and multiple regression analysis, an analysis of variance was conducted on the rate of ear damage (Table 6). The primary and secondary order of the interaction between factors is as follows: X_2, $X_{2\times3}$, X_1^2, X_1, $X_{1\times3}$, X_3, X_2^2, $X_{1\times2}$, X_3^2.

Table 6. Analysis of variance for ear damage rate.

Project	Source	Sum of Squares	Freedom	Mean Square	F Value	p Value
	model	0.73/0.69	9/7	0.082/0.099	7.70/7.88	0.018/0.0011
	X_1	0.098/0.098	1/1	0.098/0.098	9.29/7.87	0.0123/0.0159
	X_2	0.17/0.17	1/1	0.17/0.17	16.37/13.86	0.0023/0.0029
	X_3	0.063/0.063	1/1	0.063/0.063	5.92/5.01	0.0353/0.0449
	$X_1 X_2$	0.033	1	0.033	3.07	0.1102
	$X_1 X_3$	0.063/0.063	1/1	0.063/0.063	5.96/5.04	0.0348/0.0444
Damage rate	$X_2 X_3$	0.13/0.13	1/1	0.13/0.13	12.53/10.61	0.0054/0.0069
	X_1^2	0.12/0.12	1/1	0.12/0.12	11.54/9.27	0.0068/0.0102
	X_2^2	0.062/0.057	1/1	0.062/0.057	5.85/4.58	0.0361/0.0535
	X_3^2	0.012	1	0.012	1.10	0.3190
	residual	0.11/0.15	10/12	0.011/0.012		
	misfitting term	0.042/0.087	5/7	8.492×10^{-3}/0.012	0.67/0.98	0.6643/0.5296
	pure error	0.063/0.063	5/5	0.013/0.013		
	total	0.84	19			

Note: The diagonal line represents the analysis of variance for R_1 after removing insignificant factors. $p < 0.01$ represents a highly significant impact, $0.01 \leq p \leq 0.05$ represents a significant impact, and $0.05 \leq p \leq 0.1$ represents a more significant impact.

Among them, the rate of rotation of the ear-breaking wheel X_2, the rotational speed of the ear-breaking wheel $X_{2\times3}$, and the machine forward speed $X_{2\times3}$. The secondary term X_1^2 of the clamping feeding speed has a very significant impact on the ear damage rate R_1 ($p < 0.01$); the clamping feeding speed X_1, the interaction term $X_{1\times3}$ between the clamping feeding speed and the machine forward speed, the machine forward speed X_3, and the secondary term X_2^2 of the ear-breaking wheel speed have a significant impact on the ear damage rate R_1 ($0.01 < p < 0.05$). The clamping feeding speed and the rate of rotation of the ear-breaking wheel $X_{1\times2}$, and the quadratic term X_3^2 of the machine forward speed have no significant impact on the ear damage rate R_1 ($0.1 < p$). The interaction terms of the insignificant terms are regressed to the sum of squares and degrees of freedom, and the residual term is included. Continuing with the analysis of variance, the regression equation for the influence of various factors on the ear damage rate R_1 is:

$$R_1 = +2.940 - 5.063X_1 - 0.007X_2 + 3.779X_3$$
$$+1.849X_1X_3 - 0.086X_2X_3 + 0.871X_1^2 + 0.0006X_2^2 \tag{19}$$

According to the mismatch in the above equation, $p = 0.5296$, so it is insignificant, indicating that there are no other main factors affecting the experimental indicators. There is a significant quadratic relationship between the experimental indicators and the experimental factors [27].

4.1.2. Response Surface Analysis of Ear Damage Rate

Response surface analysis of the effects of interaction factors on the rate of ear damage was conducted using Design Expert 8.0.6 data analysis software. The experimental data were processed to obtain support for the significant and significant interaction between the feeding speed X_1, the rate of rotation X_2 of the ear-breaking wheel, and the machine forward speed X_3 on the rate of damage to the ears, as shown in Figure 15.

I. For the ear breakage rate R_1, when the speed of the ear-breaking wheel $X_2 = 90$ rpm, the interaction between the clamping feeding speed and the machine forward speed is shown in Figure 15a.

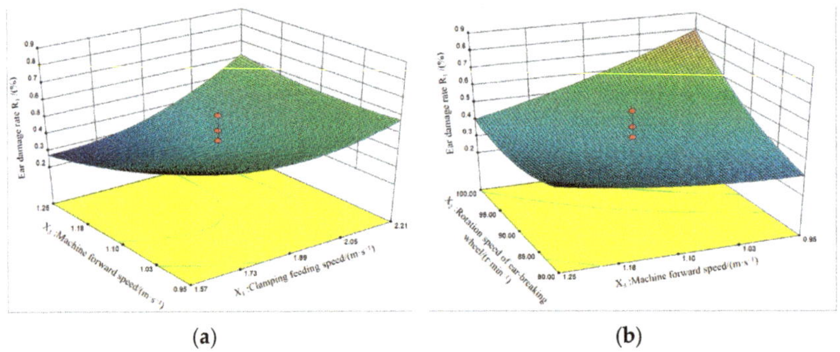

(a)   (b)

Figure 15. Effect of the interaction of experimental factors on the rate of ear damage. (**a**) The interactive effect of the clamping feeding speed and machine forward speed on R_1; (**b**) The interactive effect of the rotation speed of ear-breaking wheel and machine forward speed on R_1.

When the clamping feeding speed is constant, the ear breakage rate R_1 gradually decreases with the increase in the machine's forward speed.

As the machine's forward speed increases, the overall tilt of the fresh corn plant decreases. The ear-breaking force generated by the rotation of the ear-breaking finger is a fixed value, and the effective breaking force acting on the ear of the fruit increases. The number of effective breaking operations increases, resulting in a decrease in the value of the ear damage rate R_1.

When the machine advances at a certain speed, the rate of damage R_1 gradually increases with the increase in the clamping feeding speed.

As the clamping feeding speed increases, the overall tilt angle of the fresh corn plant gradually increases. The increase in tilt angle reduces the effective number of times the ear-breaking wheel is applied to the ear, and the decrease in ear breaking force is likely to increase the bending force on the plant, leading to the inability to effectively pick the ear. This phenomenon may be attributed to the short clamping time actually applied to the ear.

That is to say, compared to the influence of the forward speed of the machine, when the speed of the ear-breaking wheel is constant, the clamping feeding speed is the main factor affecting the ear breakage rate R_1.

II.   For the ear breakage rate R_1, when the directional clamping feeding speed $X_1 = 1.59\,\text{m/s}$, the interaction between the ear-breaking wheel speed and the machine forward speed is as shown in Figure 15b.

When the machine advances at a certain speed, the ear damage rate R_1 gradually increases with the increase in the rate of rotation of the ear-breaking wheel.

As the rate of rotation of the ear-breaking wheel increases, the force exerted by the ear-breaking wheel on the corn ear gradually increases. Due to the high moisture content of the grains, the smaller compressive strength is not sufficient to withstand the impact force exerted by the ear-breaking finger per unit area, resulting in an increase in the ear damage rate R_1.

When the rate of rotation of the ear-breaking wheel is constant, the ear breakage rate R_1 tends to decrease with the increase in the machine's forward speed.

This may be because as the front speed increases, the overall tilt angle of the fresh corn plant gradually decreases during this process, and the effective number of ear breaking times caused by the ear-breaking wheel on the corn ears increases, resulting in an increase in the number of ear breaking times of the device and a decrease in the ear damage rate R_1.

This means that compared to the influence of the forward speed of the machine, the rotational speed of the ear-breaking wheel is the main factor affecting the ear breakage rate R_1 when the clamping feeding speed is constant.

4.2. Analysis of the Impact of Impurity Rate in Harvest

4.2.1. Multiple Regression Analysis of Impurity Rate in Harvest

Based on the analysis of experimental data and multiple regression analysis, the variance of the impurity content in the harvest was analyzed (Table 7). The primary and secondary order of the interaction between factors is: X_1, X_3, X_3^2, X_1^2, X_2, $X_{1\times2}$, $X_{2\times3}$, $X_{1\times3}$, X_2^2.

Among them, the clamping feeding speed X_1, the machine forward speed X_3, and the machine forward speed secondary term X_3^2 have a significant impact on the harvest impurity content R_2 ($p < 0.01$); the secondary term of clamping feeding speed X_1^2, the rate of rotation of the ear-breaking wheel X_2, the interaction term of clamping feeding speed and ear-breaking wheel speed $X_{1\times2}$, the interaction term of ear-breaking wheel speed and machine forward speed $X_{2\times3}$, the interaction term of clamping feeding speed and machine forward speed $X_{1\times3}$, and the secondary term of ear-breaking wheel speed X_2^2 have a significant impact on the yield impurity rate R_2 ($0.01 < p < 0.05$). There are no insignificant factors in the yield impurity rate, so an analysis of variance was conducted on the overall factors, the regression equation for the impact of each factor on the yield impurity rate R_2 can be obtained as follows:

$$R_2 = +23.6232 - 4.4796X_1 - 0.2494X_2 - 10.6688X_3 + 0.0336X_1X_2 \\ -2.0313X_1X_3 + 0.0683X_2X_3 + 0.8564X_1^2 + 0.0007X_2^2 + 4.1333X_3^2 \tag{20}$$

Table 7. Analysis of variance of the impurity content in harvest.

Project	Source	Sum of Squares	Freedom	Mean Square	F Value	p Value
	model	1.07	9	0.12	10.07	0.0006
	X_1	0.29	1	0.29	24.43	0.0006
	X_2	0.11	1	0.11	9.27	0.0124
	X_3	0.17	1	0.17	14.12	0.0037
	$X_1 X_2$	0.092	1	0.092	7.86	0.0187
	$X_1 X_3$	0.076	1	0.076	6.46	0.0293
Impurity	$X_2 X_3$	0.084	1	0.084	7.14	0.0234
content	$X_1{}^2$	0.11	1	0.11	9.42	0.0119
	$X_2{}^2$	0.064	1	0.064	5.41	0.0423
	$X_3{}^2$	0.12	1	0.12	10.59	0.0087
	residual	0.12	10	0.012		
	misfitting term	0.066	5	0.013	1.27	0.4004
	pure error	0.052	5	0.010		
	total	1.18	19			

Note: $p < 0.01$ represents a highly significant impact, $0.01 \leq p \leq 0.05$ represents a significant impact, and $0.05 \leq p \leq 0.1$ represents a relatively significant impact.

From the mismatch of the above regression equation, $p = 0.4004$, so it is insignificant, indicating that there are no other main factors affecting the experimental indicators. There is a significant quadratic relationship between the experimental indicators and the experimental factors.

4.2.2. Response Surface Analysis of Impurity Rate in Harvest

The experimental data were processed using Design Expert 8.0.6 data analysis software to obtain the significant and significant interaction between the clamping feeding speed X_1, the rate of rotation X_2 of the ear-breaking wheel, and the machine forward speed X_3 on the yield impurity rate. The response surfaces of the experimental factor indicators are shown in Figure 16, respectively.

I. For the impurity content R_2 of the harvest, when the machine advances at a speed X_3 of 1.10 m/s, the interaction between the clamping feeding speed and the speed of the breaking wheel is shown in Figure 16a.

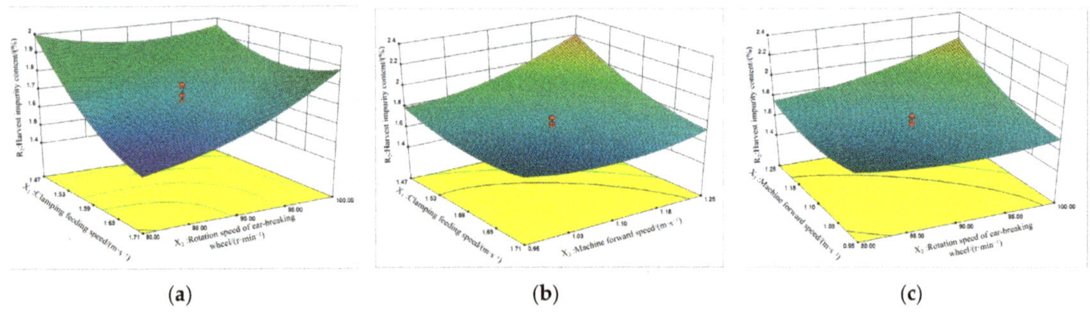

(a)   (b)   (c)

Figure 16. Response surface of experimental factor interaction on the impurity content in harvest. (a) The interactive effect of clamping feeding speed and the rotation speed of ear-breaking wheel on R_2; (b) The interactive effect of clamping feeding speed and machine forward speed on R_2; (c) The interactive effect of the rotation speed of ear-breaking wheel and machine forward speed on R_2.

When the clamping feeding speed is constant, the impurity content R_2 in the harvest gradually increases with the increase in the rate of rotation of the ear-breaking wheel.

This may be because as the rotational speed of the ear-breaking wheel increases, the rotational speed of the ear-breaking finger also correspondingly increases. The number of times the stems and leaves of fresh corn plants are subjected to the action of the ear-breaking finger increases, and the likelihood of the stems and leaves breaking increases. The proportion of broken stems and leaves to the harvest impurity rate increases, leading to an increase in the harvest impurity rate R_2.

When the rate of rotation of the ear-breaking wheel is constant, the impurity content R_2 in the harvest shows a decreasing trend with the increase in the clamping feeding speed.

This may be due to the increase in clamping feeding speed and the increase in stem inclination, resulting in a decrease in the contact time between the plant stem and the force of the ear-breaking finger and a decrease in the amount of stem and leaf falling, resulting in a decrease in the harvest impurity content R_2.

That is to say, compared to the influence of clamping feeding speed, the rotational speed of the ear-breaking wheel is the main factor affecting the harvest impurity rate R_2 when the machine's forward speed is constant.

II.   For the harvest impurity content R_2, when the rate of rotation of the ear-breaking wheel X_2 is 90 rpm, the interaction between the clamping feeding speed and the machine forward speed is as shown in Figure 16b.

When the machine advances at a certain speed, the impurity content R_2 in the harvest gradually decreases with the increase in the clamping feeding speed.

This may be due to the increase in clamping feeding speed and the inclination of the stem, which reduces the impact force on the stem after the breaking of the ear by the breaking finger. As a result, the degree of damage to the stem and leaves decreases, resulting in a decrease in the impurity rate R_2 of the harvest.

When the clamping feeding speed is constant, the impurity content R_2 in the harvest gradually increases with the increase in the forward speed of the machine.

This may be due to the increase in the forward speed of the machine, which increases the feeding amount of fresh corn plants and the number of interactions between fresh corn plants. The ear-breaking finger acts on the intertwined stems and leaves, and the area of action applied by the ear-breaking finger on the stems and leaves increases, increasing the possibility of stem and leaf breakage or damage, thereby increasing the impurity content R_2 in the harvest.

That is to say, compared to the influence of clamping feeding speed, the forward speed of the machine is the main factor affecting the impurity rate R_2 of the harvest when the rotational speed of the ear-breaking wheel is constant.

III.   For the impurity content R_2 of the harvest, when the clamping feeding speed X_1 is 1.59 m/s, the interaction between the speed of the breaking wheel and the forward speed of the machine is as shown in Figure 16c.

When the forward speed of the machine is constant, the impurity content R_2 in the harvest gradually increases with the increase in the rate of rotation of the ear-breaking wheel.

This may be due to the increase in rotational speed of the ear-breaking wheel, which increases the number of times and impact force that the ear-breaking finger acts on fresh corn plants. The instantaneous impulse acting on the stems and leaves is larger, increasing the degree of damage to the stems and leaves, and thus increasing the impurity rate R_2 in the harvest.

When the rate of rotation of the ear-breaking wheel is constant, the impurity content R_2 in the harvest gradually increases with the increase in the machine's forward speed.

This phenomenon is similar to the clamping feeding speed and machine forward speed as influencing factors when the speed of the ear-breaking wheel is constant.That is, when the clamping feeding speed and the rotating speed of the ear-breaking wheel are constant, the higher the forward speed of the machine, the more the feeding amount will cause the stems to intertwine with each other. When the ear-breaking finger applies force to it, the possibility of the stems and leaves breaking or damaging increases, resulting in an increase in the impurity content R_2 of the harvest.

That is to say, compared to the influence of the rotational speed of the ear-breaking wheel, the forward speed of the machine is the main factor affecting the impurity content R_2 of the harvest when the clamping feeding speed is constant.

4.3. Selection of Optimal Operation Parameters

Based on the response surface analysis of the ear damage rate and harvest impurity rate mentioned above, three regression models were solved using the optimization function in Design Expert 8.0.6 analysis software. According to the actual operating environment, performance requirements, and above analysis results of the fresh corn gripping and breaking device, the optimization and constraint conditions were selected as follows:

$$\begin{cases} 1.35\,\mathrm{m/s} \leq X_1 \leq 1.83\,\mathrm{m/s} \\ 70.00\,\mathrm{rpm} \leq X_2 \leq 110.00\,\mathrm{rpm} \\ 0.85\,\mathrm{m/s} \leq X_3 \leq 1.35\,\mathrm{m/s} \\ \min R_1(X_1, X_2, X_3) \\ \min R_2(X_1, X_2, X_3) \\ 0.23\% \leq R_1 \leq 0.85\% \\ 1.49\% \leq R_2 \leq 2.00\% \end{cases} \tag{21}$$

Based on the output values of certain parameters, various optimized combinations of mechanism operating parameters were obtained. Considering the stability of the overall operation of the harvesting device while ensuring the efficiency of the ear-breaking operation, the optimal combination of operation parameters was selected from multiple optimization results: a directional clamping feeding speed of 1.67 m/s, an ear-breaking wheel speed of 80 rpm, and a machine forward speed of 1.06 m/s were used: the corresponding average ear damage rate was 0.37%, and the average harvest impurity rate was 1.55%. Compared to the experimental data of the biomimetic ear picking mechanism designed by Guangqiang Zhu's team [11], this parameter combination reduces the average ear damage rate by 0.51%.

Based on the theoretical operation parameters obtained from orthogonal experiments, the optimization results were experimentally verified to ensure the reliability of the parameter optimization results and the actual operation situation. The bench test validation results were taken as the average value, as displayed in Table 8. The actual experimental values and predicted values of ear damage rate and harvest impurity rate had errors of 0.2% and 0.32%, respectively. The error between the bench test measurement results and the orthogonal software analysis prediction results was relatively small. From this, the regression equation established in this study could be deemed reliable.

Table 8. Statistical results: experimental validation.

Comparative Test Verification	Ear Damage Rate	Harvest Impurity Content
Theoretical data	0.37%	1.55%
Actual test mean	0.57%	1.87%

4.4. Discussion

From the analysis of the influencing factors of the two experimental indicators (ear damage rate R_1 and harvest impurity rate R_2) in Sections 4.1 and 4.2 of this chapter, we found that the experiment still has some limitations.

In Section 4.1, we found in the analysis of R_1 that the clamping feeding speed (X_1) and the breaking wheel speed (X_2) may be the two main factors (both will increase the value of R_1), and the order of influence is X_2, X_1, X_3. As shown in Table 6, the p-value of X_2 is ≤ 0.01, which is a very significant impact, while the p-value of X_3 and X_1 is greater than 0.01, which is a significant impact. An increase in the value of X_3 will result in a decrease in the value of the R_1 test indicator, which is the result we hope to achieve. However, within the range of its influence, the forward speed of the machine (X_3) is only taken as a certain value within the range of 0.85–1.35 m/s, which has certain limitations.

In addition, X_2 accounts for a significant proportion of the impact on R_1, which means that the value of R_1 increases. This may be because the impact force applied by the ear-breaking finger is more likely to cause damage to the ear compared to the bending force on the ear when the clamping device is in contact. Because the moisture content of fruit ears is high, their compressive strength is low. However, this experiment did not investigate the range within which the compressive strength of the ears and the impact force generated by the breaking fingers were generated in the experiment.

In Section 4.2, where R_2 may have more influencing factors, but its main influencing factors are similar to Section 4.1, namely X_1, X_2, and X_3. The order of influence of these three is X_1, X_3, X_2. (As shown in Table 7, p-values of X_1 and $X_3 \leq 0.01$ are extremely significant effects; p-values of $X_2 > 0.01$ are significant effects.)

X_1 has a higher proportion of influence in multiple regression analysis, but the influence value of X_3 is not significantly different from X_1, and the feedback categories generated by the two on experimental indicators are also not the same. X_1 will reduce the value of R_2, while X_3 will increase the value of R_2. What we hope to see is that X_1 can reduce the value of the experimental indicator R_2.

Both X_2 and X_3 increase the value of R_2, but the influence of X_3 is relatively large. This may be because as the value of X_3 increases, excessive feeding leads to the entanglement of stems and leaves, increasing the likelihood of stem and leaf breakage or damage. In contrast, when the value of X_2 increases, the number of times the ear-breaking value acts on the stems and leaves increases, and the breakage of the stems and leaves is not as obvious as the phenomenon of intertwining between the stems. The specific value of stem and leaf breakage was not studied in this experiment.

Finally, the operation of the ear-picking device in this article is in a flexible ear-picking state. For the case of a rigid ear-picking state, no comparative experiment was conducted.

5. Conclusions

This article verifies the feasibility of the fresh corn ear-picking device through bench tests and analyzes the effects of various factors on the yield of ear damage and impurity content through orthogonal experiments of three factors and five levels.

I. Based on theoretical calculations and single-factor experiments, the effective operating speed range of each mechanism to ensure efficient harvesting operations was determined. Second, gradient partitioning of its parameters was performed using Design Expert 8.0.6 data analysis software, and orthogonal experimental parameter combinations were designed for testing and measurement.

After analyzing the experimental data and multiple regression analysis, it was found that the main and secondary order of the impact on the damage rate of fresh corn ears is as follows: the rotation speed of the ear-breaking wheel, the interaction term between the rotation speed of the ear-breaking wheel and the machine forward speed, the secondary term of the clamping feeding speed, the clamping feeding speed, the interaction term between the clamping feeding speed and the machine forward speed, the machine forward speed, and the secondary term of the ear-breaking wheel speed. The interaction term between the clamping feeding speed and the rotational speed of the ear-breaking wheel, as well as the quadratic term of the machine forward speed, have no significant impact on the ear breakage rate R_1. The most obvious factor is the rotational speed of the ear-breaking wheel, which is the same as other types of ear-picking mechanisms for the rate of ear damage. However, due to the buffering effect of the flexible ear-breaking finger, the force change of the ear is relatively gentle.

After analyzing the experimental data and multiple regression analysis, it was found that the primary and secondary order of the impact of harvest impurity rate on the damage rate of fresh corn ears was as follows: clamping feeding speed, machine forward speed, machine forward speed quadratic term, clamping feeding speed quadratic term, ear-breaking wheel speed, interaction term between clamping feeding speed and ear-breaking wheel speed, interaction term between ear-breaking wheel speed and machine forward speed, and interaction term between ear-breaking wheel speed and machine forward speed The interaction term between the support feeding speed and the forward speed of the machine, as well as the quadratic term of the rotation speed of the ear-breaking wheel, do not have insignificant factors in the impurity content of the harvest. The most obvious factor is the clamping feeding speed, mainly due to the frequency of posture changes of fresh corn plants, which offsets the impact of the ear-breaking wheel on the plants.

The optimal operating parameter ratio obtained from the bench test is as follows: clamping feeding speed 1.67 m/s, breaking wheel speed 80 r/min, and machine forward speed 1.06 m/s. In the bench test, it was found that actively applying the breaking force through the flexible picking mechanism can effectively reduce the crushing rate of fruit ears and also solve the problem of blockage during mechanical harvesting.

II. In order to ensure the reliability and stability of the fresh corn harvester, it is necessary to optimize the structural parameters of the harvester after conducting corresponding stem collection device design experiments. At the same time, it is necessary to optimize the material of the outer rubber of the ear-breaking finger to ensure the reliability of the ear-breaking mechanism. At the same time, to ensure the rationality of the header structure design, structural design and testing of the stem collection device will be carried out in the later stage based on this foundation.

III. During the design and bench test of the fresh corn clamping and breaking device, the main objective was to study and test the damage rate and impurity content of corn harvest ears. The theoretical calculation and optimization of the device's power consumption were not conducted. In the later stage, optimization design and improvement will be carried out based on the power consumption requirements of the harvesting device to meet the operational requirements of the fresh corn harvesting device.

IV. During the experiment of breaking ears of fresh corn, data statistics were conducted on the damage rate and impurity content of the fresh corn harvest. The breaking device designed in this article effectively reduced the damage rate and impurity content compared to other types of fresh corn picking devices. Among them, during the picking experiment, the damage rate of each group of operating parameters was less than 1%, which was 1% lower than the industry indicator's 2% damage rate. Due to the ear-picking device, the impurity content is mainly manifested as the breakage of fresh corn plant leaves. Therefore, the impurity content is selected within 2% to seek the optimal parameter ratio to meet the standard of mechanized corn harvesting operation.

Author Contributions: Conceptualization, X.H. and X.F.; methodology, H.Z. and X.H.; investigation, X.F., H.Z. and X.H.; writing—original draft preparation, X.H., X.F. and H.Z.; writing—review and editing, W.W., Z.Q. and J.S.; visualization, B.C. and W.W.; funding acquisition, X.H. and H.Z. All authors have read and agreed to the published version of the manuscript.

Funding: This research was funded by Henan Province Modern Agricultural Industry Technology System Corn Whole-Process Mechanization Special Project (HARS-22-02-G4).

Institutional Review Board Statement: Not applicable.

Informed Consent Statement: Not applicable.

Data Availability Statement: The data used to support the findings of this study are available from the corresponding author upon request.

Acknowledgments: The authors would like to appreciate the reviewers who provided helpful suggestions for this article.

Conflicts of Interest: The authors declare no conflict of interest.

References

1. Kamonporn, S.D.; Ketthaisong, K.L. Bioactive, antioxidant and enzyme activity changes in frozen, cooked, mini, super-sweet corn (*Zea mays* L. saccharata 'Naulthong'). *J. Food Compos. Anal.* **2015**, *44*, 1–9.
2. Wang, Y.; Zhang, C. Current status and suggestions for the development of corn harvesting machinery. *Agric. Eng.* **2013**, *3*, 34–36.
3. Shinners, K.J.; Adsit, G.S.; Binversie, B.N. Single–pass, split–stream harvest of corn and stover. *Trans. ASABE* **2007**, *50*, 355–363. [CrossRef]
4. Hou, C.; Dong, Y.; Liu, C.G. Current situation and quality analysis of China's corn harvester industry. *Agric. Eng.* **2019**, *9*, 20–25.
5. Xin, S.; Zhao, W.; Dai, F. Design of crawler type corn combine harvester with full film double ridge and trench sowing in arid areas. *Trans. Chin. Soc. Agric. Eng.* **2019**, *35*, 1–11.
6. Cui, T.; Liu, J.; Zhang, D. Design and test of a hybrid operating mechanism for harvesting and straw crushing. *Trans. Chin. Soc. Agric. Mach.* **2012**, *43*, 95–100.
7. Hou, S.; Wang, X.; Ji, Z. Experiments on the influence of corn straw morphological combinations on timely no-tillage sowing soil temperature and moisture in cold regions. *Agriculture* **2022**, *12*, 1425. [CrossRef]
8. Rovira-Más, F.; Han, S.; Wei, J. Autonomous guidance of a corn harvester using stereo vision. *Agric. Eng. Int. CIGR J.* **2007**. Available online: https://cigrjournal.org/index.php/Ejounral/article/view/944 (accessed on 23 July 2023).
9. Liu, Y.; Yang, R.; Wu, X. Design of ear picking device for corn harvesters in residential areas. *J. Agric. Mech. Res.* **2021**, *43*, 103–106+138.
10. Wang, X.; Geng, L.; Li, X. Design and experiment of low-injury picking test table for fresh-eating maize. *Agric. Eng.* **2017**, *7*, 68–71.
11. Zhu, G.; Li, T.; Zhou, F. Design and test of a biomimetic ear picking device for fresh corn. *J. JiLin Univ.* **2023**, *53*, 1231–1244.
12. Zhang, L.; Li, Q. Test on ear breaking speed and power consumption of bionic corn ear breaking device. *Trans. Chin. Soc. Agric. Eng.* **2015**, *31*, 9–14.
13. Chen, M.; Cheng, X.; Jia, X.; Zhang, L.; Li, Q. Optimization of structure and operation parameters of bionic hand ear breaking corn harvesting device. *Trans. Chin. Soc. Agric. Eng.* **2018**, *34*, 15–22.
14. Wang, Y.; Zhang, Q.; Yu, L. The application status and prospects of corn ear picking equipment. *J. Agric. Mech. Res.* **2011**, *33*, 228–231.
15. Rogovskii, I.L.; Liubarets, B.S.; Voinash, S.A. Research of diagnostic of combine harvesters at levels of hierarchical structure of systems and units of hydraulic system. *J. Phys. Conf. Ser.* **2020**, *1679*, 2–5. [CrossRef]
16. Zhang, Z. Research and design of key components of fresh corn harvester. *J. Farm Mach. Using Maint.* **2021**, *5*, 14–16.
17. Li, T.; Zhou, F.; Guan, X. Design and experiment on flexible low–loss fresh corn picking device. *J. Int. Agric. Eng. J.* **2019**, *28*, 2–4.
18. Zhang, X.; Wu, P.; Wang, K. Design and test of 4YZT–2 self–propelled fresh corn harvester. *Trans. Chin. Soc. Agric. Eng.* **2019**, *35*, 1–9.
19. Shahzad, M.W.; Burhan, M.; Ang, L. Energy–water–environment nexus underpinning future desalination sustainability. *Desalination* **2017**, *413*, 52–64. [CrossRef]
20. GB/T21962–2020; S. Corn Harvesting Machinery. China Machinery Industry Federation: Beijing, China, 2020; pp. 4–5.
21. Hao, W.; Liu, J.; Min, W. Changes of moistuer distribution and migration in fresh ear corn during storage. *J. Integr. Agric.* **2019**, *18*, 2644–2651.
22. Xu, W.; Zhao, J.; Cui, X. The bionics design and analysis on device of corn picking with dragging and cutting stem. *J. Agric. Mech. Res.* **2018**, *40*, 81–86.
23. Guan, X.; Li, T.; Zhou, F. Determination of bruise susceptibility of fresh corn to impact load by means of finite element method simulation. *Postharvest Biol. Technol.* **2023**, *198*, 2–9. [CrossRef]
24. Lisowski, A.; Swiatek, K.; Klonowski, J. Movement of chopped material in the dischargespout of forage harvester with a flywheel chopping unit: Measurements using maize and numerical simulation. *Biosyst. Eng.* **2012**, *111*, 381–391. [CrossRef]
25. Zhang, H.; Hu, R.; Kang, S. *ANSYS 12.0 Finite Element Analysis from Introduction to Proficiency*; Beijing Machinery Industury Press: Beijing, China, 2010.
26. Liu, W.; Gao, W.; Yu, G. *ANSYS 12.0 Classic*; Beijing Electronic Industry Press: Beijing, China, 2010.
27. Gorecki, J.; Lykowski, W. Influence of die land length on the maximum extrusion force and dry ice pellets density in ram extrusion process. *Materials* **2023**, *16*, 4281. [CrossRef] [PubMed]
28. Ferreira, S.L.; Bruns, R.E.; Ferreira, H.S. Box–Behnken design: An alternative for the optimiazation of analytical methods. *Anal. Chim. Acta* **2007**, *597*, 179–186. [CrossRef] [PubMed]
29. Geng, A.; Yang, J.; Zhang, J. Analysis of factors affecting mechanical damage of corn picking and harvesting. *Trans. Chin. Soc. Agric. Eng.* **2016**, *32*, 56–62.

 *agriculture*

Article

Optimal Design of and Experiment on a Dual-Spiral Ditcher for Orchards

Jianfei Liu, Ping Jiang, Jun Chen, Xiaocong Zhang, Minzi Xu, Defan Huang and Yixin Shi *

College of Mechanical and Electrical Engineering, Hunan Agricultural University, Changsha 410128, China; liujianfei@stu.hunau.edu.cn (J.L.); 1233032@hunau.edu.cn (P.J.); chenjun@stu.hunau.edu.cn (J.C.); cong@stu.hunau.edu.cn (X.Z.); xmzzrn@stu.hunau.edu.cn (M.X.); huangdefan@stu.hunau.edu.cn (D.H.)
* Correspondence: shiyixin@hunau.edu.cn

Abstract: This study aimed to design a counter-rotating dual-axis spiral ditcher to address the problems of the lack of ditching models, low ditching quality, and high power consumption in ditching operations in hilly and mountainous orchards. By establishing a discrete element simulation model of the interaction between the dual-spiral cutter and the soil, we analyzed the effects of different operating parameters on the ditching performance and power consumption to explore the mechanism of the interaction between the soil-engaging components and the soil; meanwhile, according to the simulation data, we took the forward speed, cutter speed, and ditching depth as the experimental influencing factors, and we took the ditching qualification rate and power consumption as the experimental indicators. Then, through a three-factor and three-level orthogonal experimental analysis, we obtained the optimal parameter combination. Based on the optimized simulation results, a prototype ditcher was fabricated, and the results of a soil trough test showed that the ditcher had a stable ditching performance, an average ditching qualification rate of 91.4%, an average soil crushing rate of 72%, and a ditching adjustment structure that could adjust the ditching depth and width within the ranges of 15–25 cm and 10–20 cm, respectively. The optimized ditcher had high efficiency, stability, and energy savings during its operation, and it was able to meet the agronomic requirements for ditching operations. This study provides a design basis and technical support for dual-spiral trial ditching operations.

Keywords: spiral cutter; discrete element simulation; parameter optimization

Citation: Liu, J.; Jiang, P.; Chen, J.; Zhang, X.; Xu, M.; Huang, D.; Shi, Y. Optimal Design of and Experiment on a Dual-Spiral Ditcher for Orchards. *Agriculture* **2023**, *13*, 1628. https://doi.org/10.3390/agriculture13081628

Academic Editor: Jacopo Bacenetti

Received: 2 August 2023
Revised: 15 August 2023
Accepted: 16 August 2023
Published: 18 August 2023

1. Introduction

Traditional ditching machines mainly include four types: the plow, disc, chain, and spiral types. They have the following characteristics and scopes of application: The plow ditcher is suitable for harsh farmland conditions but has low mechanical strength and low power output; the disc ditcher is suitable for southern farmlands, has high mechanical strength, and is more commonly used; the chain ditcher, which is suitable for digging deep and narrow ditches, has a good ditch shape and high efficiency and is often used for laying pipelines or cables; and the spiral ditcher has low resistance, has good operation quality, can throw soil and mix, and is often used for ditching and fertilizing in hilly and mountainous orchards [1–3]. It can also loosen soil and cut roots at the same time, promote the circulation of hot water and fertilizer, stimulate the growth of new fine roots, enhance the absorption capacity, and improve the subsequent fertilization effect [4,5]. Compared with that for other types, the research on the spiral ditcher is relatively undeveloped. With the standardization and normalization of orchard agriculture, as well as the increasing shortage in the labor force, the demand for mechanized ditching is increasing, which has promoted the innovation and development of various new types of ditching machines. To improve upon the conventional spiral ditcher, this study designed a counter-rotating dual-axis spiral ditcher.

In recent years, to improve the level of mechanization in orchards, relevant experts have designed a variety of new spiral ditchers. These ditchers are based on traditional operation methods, but they have been innovated and optimized by adopting different spiral structures and working parameters to adapt to different soil environments and agronomic needs. For example, the 1KS60-35X orchard-type dual-spiral fertilizer applicator developed by Xiao Hongru et al. 2017 can perform the integrated operations of soil breaking with the front axle and of soil stirring and fertilizing with the rear axle, thus effectively saving time and resources and improving the operation efficiency [6]. Ma Tie, Lin Jing, and Gao Wenying et al. designed a deep burial and returning machine for straw with a spiral ditching device and compared the advantages and disadvantages of single-spiral and dual-spiral structures; in addition, they drew on the principles of bionics to optimize the design of the spiral blade surface, enhanced the anti-blocking performance of the spiral blade, improved the adaptability of the spiral ditcher, and determined the optimal working parameters of the spiral ditcher through experiments [7,8]. Jiang Jianxin designed a dual-side spiral ditching device for small orchards that had two gears of operation speeds that could be flexibly adjusted according to the soil environment and agronomic needs, and they conducted a detailed design study on the structural parameters of the spiral cutter and the power parameters of the machine [9].

The simulation of the process of the interaction between a cutter and soil is an effective research method that can greatly shorten the research and development time and save research costs through analysis. This is of great significance for the optimization of agricultural machinery [10]. In recent years, discrete element simulation, which is a common simulation method, has been used by more and more scholars to study the interactions between soil and operation machinery. It provides a basis for the optimization of the structural and motion parameters of a cutter by analyzing the changes in the force state of the soil and cutter during operation [11,12]. For example, Wang Shaowei optimized a ditching cutter based on an inclined spiral ditcher that was developed by his research group for hilly orchards. He established a discrete element simulation model for analyzing the ditching power consumption and studied the law of the influence of the operation parameters on the ditching power consumption [13]. Zang Jiajun et al. designed a curved ditching and mixing device based on an orchard straw-covering machine to realize the integration of ditching, fertilizing, and mixing operations. They conducted a discrete element simulation of the ditching and fertilizing processes to determine the optimal pitch and operation speed of the cutter [14]. At present, there is still a lack of research on spiral ditchers. According to relevant studies, it was shown that an inclined shaft operation can effectively reduce the cutting power consumption of a cutter during operation [15]. Therefore, it is necessary to further study spiral ditchers and optimize the power consumption and effects of their cutters during operation.

This study was aimed at the demands of ditching operations in hilly and mountainous orchards, the design of a counter-rotating dual-axis spiral ditcher, and the establishment of a discrete element model of the interaction of a spiral cutter and soil by taking the ditching qualification rate and operation power consumption as the test indicators. We used EDEM software for the analysis of the simulated experiments, explored the effects of the cutter's structural and operation parameters on the ditching performance, and obtained the optimal combination of the structural and operation parameters for the cutter; finally, we verified the simulation results with a soil trough test, which proved the reliability of the simulation model and the rationality of the optimization scheme.

2. Materials and Methods

2.1. Design of the Overall Structure

The main components of the spiral ditcher were a three-point suspension frame, limit pull rod, movable hinge, spiral cutter, drive motor, reducer, etc. A diagram of the structure is shown in Figure 1. The machine was connected to a traction vehicle by a three-point suspension device. The power output from the motor was transmitted to the spiral cutter shaft through the reducer and a coupling to drive the cutter's rotation. The motion parameters of the cutter could be precisely controlled by adjusting the output speed through the motor control box. The device was also able to control the ditching depth and width by adjusting the length of the limit pull rod on the frame, which changed the operating angle of the cutter. The operation depth and width of the ditcher could be flexibly adjusted according to the agronomic needs and field conditions in different operation environments. The ditching cutter consisted of a pair of double-blade spiral cutters with opposite helixes, for which a coaxial double-blade structure that could alleviate the problem of unbalanced force on the cutter shaft during rotation was adopted. During operation, the ditching cutter first completed the soil entry process; from top to bottom in the forward direction, the left cutter shaft rotated clockwise, and the right cutter shaft rotated counterclockwise; that is, they were arranged in counter-rotation.

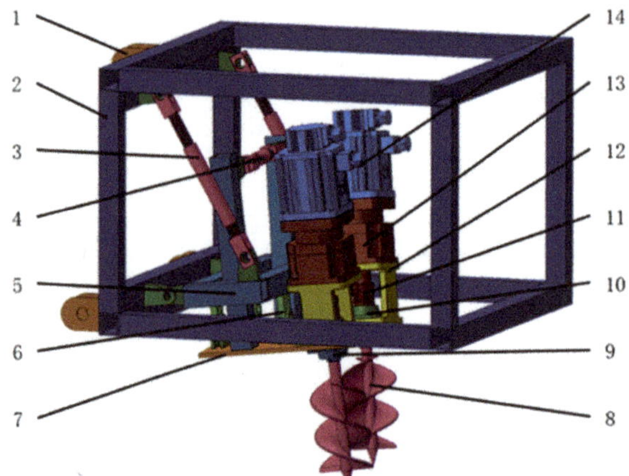

Figure 1. Virtual prototype of the dual-axis spiral trencher. 1. Three-point suspension device. 2. Installation frame of the cutter. 3. Limit rod A. 4. Limit rod B. 5. Motor mounting bracket. 6. Movable hinge. 7. Mounting plate. 8. Spiral cutter. 9. Bearing seat A. 10. Bearing seat B. 11. Coupling. 12. Reducer mounting seat. 13. Reducer. 14. Brushless DC motor.

The working principle was as follows: First, the output shaft of the motor drove the cutter shaft to rotate through the reducer, and it gradually lowered the cutter to cut into the soil. Then, when the cutter reached the predetermined trenching depth, the tractor moved forward and started trenching. The spiral blade continued to cut and throw out the soil at the rear end out of the trench. The spiral trencher was able to adjust the working angle of its trenching cutter according to the user's needs to achieve the desired effect of adjusting its trenching depth and width. The trenching depth could be adjusted from 150 mm to 250 mm, and the trenching width could be adjusted from 150 mm to 200 mm. When the three-point suspension position remained unchanged, the depth of the operation could be changed by adjusting the length of the limit rod to change the angle between the tool axis and the horizontal plane. The relationships between the angle of the cutter shaft, the horizontal plane, and the working depth are shown in Table 1.

Table 1. The relationship between angles and ditch depths.

Working depth/mm	150	170	190	210	230	250
Angle with horizontal plane/°	23.9	27.4	30.9	34.6	38.4	42.5

2.2. Design of the Trenching Components

The spiral ditching tool used a motor to drive the spiral blade to rotate at a high speed, thus cutting the soil particles and making them move along the spiral's direction. The soil particles were subjected to centrifugal force and moved to the outer edge of the spiral blade. At the same time, the ditch wall exerted friction on the soil particles, inhibiting them from rotating with the blade and causing them to rise along the spiral blade to the soil surface; then, they were thrown out of the ditch under the influence of centrifugal force. This method was able to effectively loosen the soil structure and improve the quality and efficiency of the operation [16]. The device mainly consisted of a keyway connecting shaft, an inner spiral cylinder, a spiral blade, and a soil-drilling cone. The connecting shaft, inner spiral cylinder, and drill tips were welded into one piece; the spiral blades were welded symmetrically in four sections on the inner cylinder. According to previous design experience, a blade shaft diameter of 20 mm, a blade outer diameter of 100 mm, and a blade thickness of 3 mm were chosen. In order to enhance the strength of the blade and break through the hard soil of the bottom plow layer, heat treatment was required [17].

The spiral blade is a key component of a soil-lifting machine, and its pitch type, number of threads, and installation method determine the quality of the soil-lifting effect [18]. By comprehensively considering the processing difficulty and operation effects, this study finally adopted a fixed pitch blade for trial production. In order to optimize the soil-lifting amount and operation efficiency, an inclined double-shaft spiral cutter was adopted, which was installed in a trapezoidal shape to adapt to the trapezoidal cross-section of ditches in hilly mountain orchards. The specific parameters of the fixed pitch blade and double-thread spiral blade are shown in Table 2, and their soil-lifting performance was tested and analyzed. Figure 2 shows a schematic diagram of the cross-section of the ditching process.

Table 2. Parameters of the spiral cutter.

Parameters	Main Shaft Outer Diameter/mm	Main Shaft Length/mm	Spiral Outer Diameter/mm	Spiral Angle/°	Blade Thickness/mm	Pitch/mm
Numerical value	20	30	100	15	3	70

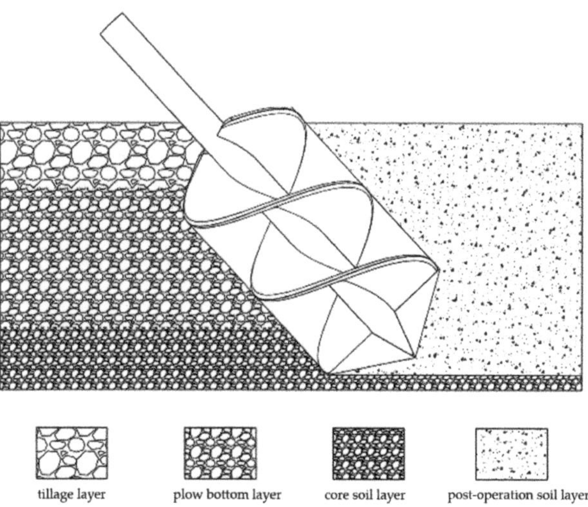

Figure 2. Schematic diagram of the cross-section of the ditching process.

The spiral blade is a core component of a spiral ditching machine, and the shape of its spiral line determines the effect of its interaction with soil particles. The spiral line of a double-blade spiral cutter can be expressed by a hyperbolic spiral; that is, when any point L on the cylinder moves along the generatrix at a constant speed from bottom to top while the generatrix rotates around the central axis of the cylinder, its motion trajectory is a spiral line, as shown in Figure 3. If the spiral line is flattened, it becomes a straight line on a plane. In the figure, the initial spiral angle corresponding to the starting point of the spiral line is α, and the height corresponding to the position of point L after the generatrix has turned n times is h. Such a spiral shape can allow a double-blade spiral cutter to more effectively cut soil particles during ditching and reduce the resistance of soil to the blade. The parameters of the spiral line can be optimized and adjusted according to different soil types and ditching operation requirements in order to achieve better ditching effects.

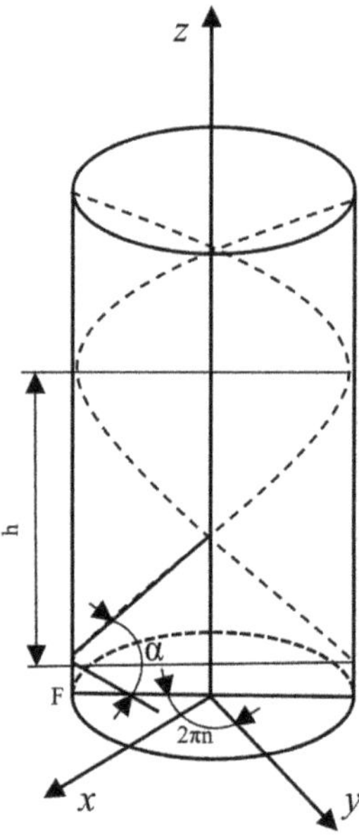

Figure 3. The principle of helix formation.

The spiral line equation of the double-spiral ditching tool is [19]:

$$x = r \cos 2\pi n \tag{1}$$

$$y = r \sin 2\pi n \tag{2}$$

$$z = al \tag{3}$$

$$l = 2\pi r n \tag{4}$$

α is the initial angle of the helix (°), n is the number of turns of the helix, and l is the arc length (m) of the busbar.

By taking the derivative of z, the tangent value of the helix angle at a certain point of the constant-pitch helix is:

$$z_2' = a \tag{5}$$

By substituting the starting position of the helix, $z' = \mathrm{tg}\alpha$, $z' = \mathrm{tg}\alpha$, $n = 0$, into Equation (4), the coefficient α of the quadratic term of the constant-pitch helix α is:

$$a_2 = \tan a \tag{6}$$

a is the initial angle (°) of the helix, and n is the number of turns of the helix.

2.3. Working Parameters of the Cutting Tool

During the trenching process, the effects of soil being cut and discharged out of the trench by the helical blade determine the quality and efficiency of trenching. The rotational speed of the cutting tool is an important factor affecting the trenching effect. If the trencher speed is too low, the soil will accumulate at the bottom of the trench, resulting in a reduction in the trench depth and width; if the speed is too high, the soil will be thrown too far, causing soil loss and backfilling difficulties. Therefore, according to the trenching conditions and requirements, an appropriate trencher speed should be selected. The calculation method for the rotational speed of the helical trenching tool is as follows [20–22]:

$$n = \frac{30}{\pi} F_r \sqrt{\frac{2g}{D}} \tag{7}$$

In the formula, F_r is the dimensionless similarity criterion (determined by the trenching helix angle; the range of values is 2.5–4.5), and D is the outer diameter of the helical blade (m).

The helix pitch angle affects the trenching efficiency, and the helix pitch angle in this machine's design was 15°. The relationship between the feed per revolution S of the helical trencher and the angle ζ (the value was 1°, and the range was 0.4°–1°) between the direction of motion of a point on the helical blade and the horizontal plane is given by the following formula:

$$S = \pi D \times 10^3 \mathrm{tg}\zeta \tag{8}$$

ζ is the angle between the direction of motion of a point on the helical blade and the horizontal plane.

In order to cause the soil to be smoothly thrown out of the trench, when the helix pitch angle was 15° and the angle ζ between the direction of motion of a point on the helical blade and the horizontal plane was 1°, the minimum value of the angle β between the soil's direction of motion and the horizontal plane and the value of F_r were obtained from the table; these were 15° and 3.86, respectively. By appropriately increasing β to 24°, the critical speed of the helical trencher could be calculated as 238 r·min^{-1} [23–25].

2.4. Establishment of the Simulation Model

This study used EDEM (DEM-Solutions, Edinburgh, UK, 2021) software to conduct the simulation experiments. By referencing the research of other scholars on the discrete element parameters of soil, a model that simulated the interaction between the tool and the soil was established [26]. In order to cause the soil simulation to be closer to the characteristics of the sticky soil in Hunan, China, a Hertz–Mindlin model with bonding was selected, and bond connections were added between the soil particles to form a whole that could withstand tangential and normal displacements [27,28]. When it was subjected to a force that exceeded the preset maximum shear stress, the bonds broke, and the particles were treated as separate individuals by the system in order to continue the simulation. By adding bonds, the physical properties of the soil could be more realistically reflected. The model was suitable for simulating a broken working state and could be more intuitively used to observe the breaking effect of the tool on the soil [29]. After the calculation was finished, the resistance of the tool in the forward process and the torque of the cutter roller in the soil could be directly analyzed by using the resulting data [30], thus providing data support for the improvement of the working parameters of the tool and the optimization of the structural design.

Based on the relevant literature, the physical parameters of both the soil particles and the tool were set in the preprocessing stage of the software. The soil had two types of density: the true density (the mass of soil per unit volume of solid particles) and the bulk density (the mass of soil per unit volume of the whole soil). For the same soil, the higher the compaction degree, the greater the bulk density. In order to simulate the actual soil, the soil model was divided into three layers: namely, the tillage layer, the plow bottom layer, and the core soil layer, with an increasing bulk density from top to bottom. The thickness of the tillage layer was about 10 cm, the thickness of the plow bottom layer was about 5 cm, and the thickness of the core soil layer was about 15 cm. In the modeling process, the setting of soil particles affected the quality and accuracy of the simulation calculations. Too large particles would cause distortion and an inability to reflect the actual physical process; too small particles would cause a sharp increase in the computer's calculation time. Therefore, the size of the soil particles needed to be appropriately selected. Considering the performance of the equipment used in the simulation platform, a particle diameter of 8 mm was selected, and the material's name was set as "soil" [31,32].

The final settings of the soil and tool materials are shown in Tables 3–5.

Table 3. Intrinsic parameters of the soil.

Parameters	Values
Poisson's ratio	0.38
Density	2670/2740/2820
Young's modulus	2.70×10^6

Table 4. Intrinsic parameters of the 65 Mn spring steel.

Parameters	Values
Poisson's ratio	0.30
Density	7860
Young's modulus	9.23×10^7

Table 5. Parameters of the soil particles.

Parameters	Values
Soil–soil restitution coefficient	0.50
Soil–soil rolling friction coefficient	0.55
Soil–soil static friction coefficient	0.10
Soil–65 Mn restitution coefficient	0.50
Soil–65 Mn rolling friction coefficient	0.40
Soil–65 Mn static friction coefficient	0.10
Normal stiffness coefficient (N/m)	5.00×10^7
Tangential stiffness coefficient (N/m)	6.00×10^7
Critical normal stress (Pa)	3.00×10^5
Critical shear stress (Pa)	3.00×10^5
Particle radius (mm)	8
Adhesion radius (mm)	9.2

According to the above parameters, a virtual soil tank with dimensions of 2000 mm × 800 mm × 300 mm (length × width × depth) was established. The soil particles were spherical particles of a uniform size, and to reflect the state of different sizes of particles in the soil, the particles were generated in a form with a normal distribution. An initial velocity of 0.5 m/s was applied to the particle Z-direction to promote falling compaction. The dynamic generation method was adopted for the creation of the particles, which took 3 s and generated 90,118 particles. Finally, by using the Analyst module, the simulation result file was output with a new file name, and this served as the virtual soil tank for the subsequent simulation experiments (see Figure 4).

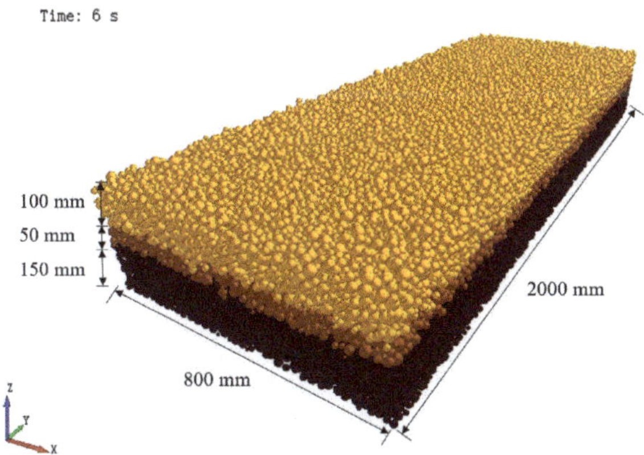

Figure 4. Discrete element soil tank model.

2.5. Experimental Scheme

This study adopted the discrete element simulation method to simulate the movement of the cutter through the soil tank and to explore the trenching performance of the spiral trenching tool. The simulation operation process is shown in Figure 5. Based on relevant design experience, the trenching performance was mainly affected by three factors—traction speed (A), cutter shaft speed (B), and trenching depth (C)—while keeping the field soil conditions constant. On this basis, a three-factor and three-level orthogonal experiment was designed, with the trenching depth qualification rate (Y1) and power consumption (Y2) as the experimental indicators; the L9(33) orthogonal table was used to arrange the experimental scheme. Table 6 presents the experimental factor levels. This study used SPSS 27 software to perform an analysis of variance and a range analysis on the experimental data and to obtain the effect size and the optimal combination of the factor levels for the experimental indicators.

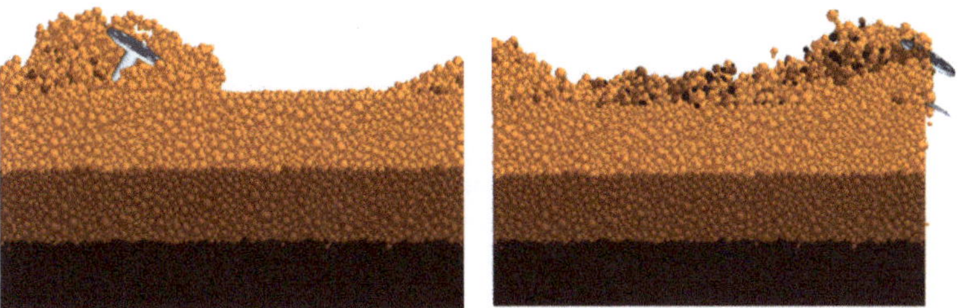

Figure 5. Simulation process diagram 7.1 s (**left**)–8.6 s (**right**).

Table 6. Table of the factor levels.

Level	Traction Speed (m/s)	Cutter Shaft Speed (rpm)	Trenching Depth (mm)
1	0.05	340	100
2	0.25	390	150
3	0.45	440	200

The calculation method for the trenching depth qualification rate is as follows:

$$S_1 = \frac{S_0}{S} \times 100\% \tag{9}$$

In the formula, S_1 is the trenching depth qualification rate (%), S_0 is the actual trenching depth (cm), and S is the theoretical trenching depth (cm).

The power consumption of the trenching machine mainly included two parts: One part was the torque power for the cutter's rotation, and the other part was the traction power for the advancement of the cutter. The sum of the two was the total power consumption of the trenching machine. In this study, the discrete element simulation method was used to numerically calculate and analyze the torque and resistance of the cutter while using different motion parameters.

A discrete element simulation of the torque and resistance of the trenching machine cutter was carried out to study its power consumption characteristics by using EDEM software. Because EDEM software cannot directly output the power consumption parameters, other methods were needed to calculate the power consumption. In this study, the postprocessing module Analyst Tree that was built into the software was used to export the torque and force data of the cutter during the simulation process, and the interval density of the data was set to 0.1 s to ensure their accuracy and continuity. Then, according to the set speed constant and the torque and force obtained from the simulation experiments, the power consumption formula was used to calculate the power consumption of the cutter. The motion trajectory of the cutter in the soil tank was simulated. The cutter rotated around the cutter shaft and moved at a constant speed along the direction of the y-axis. In order to eliminate the non-steady-state influence before and after the cutter entered the soil tank, only the time period from when the cutter completely contacted the soil to when it left the soil was analyzed, which was 7.1 s–8.6 s, in order to reflect the actual cutting state of the cutter. Figure 6 shows the effect of throwing soil on the border (left) and floating soil at the bottom of the ditch (right) after the operation.

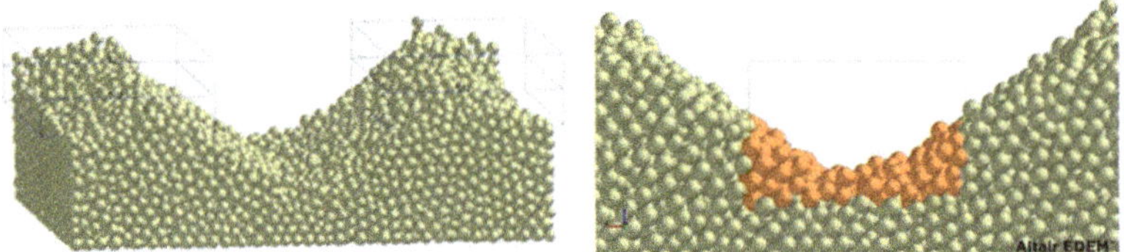

Figure 6. Effect diagram of throwing soil on the border (**left**) and floating soil at the bottom of the ditch (**right**).

Figure 7 shows examples of line charts of the torque and y-axis force on the cutter shaft over time under the following conditions: traction speed = 0.05 m/s, speed = 440 rpm, and working depth = 25 cm.

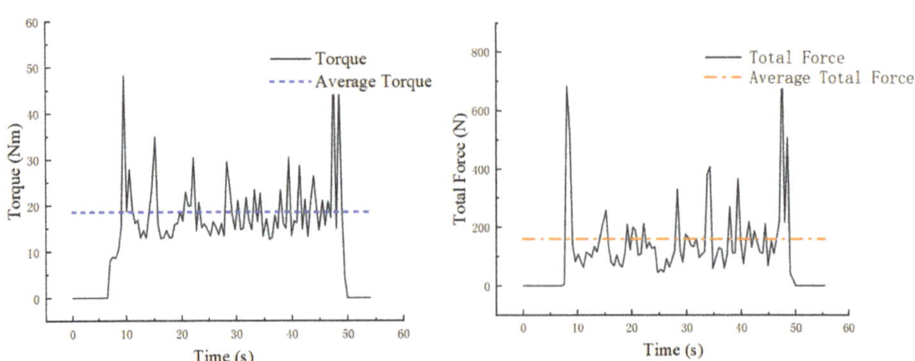

Figure 7. Time–torque relationship diagram (**left**) and time–force relationship diagram (**right**).

The torque power consumption and traction power consumption of the tool were added, and the total power consumption of the trenching process was calculated:

$$P_1 = \frac{T \cdot n}{9550} \tag{10}$$

In the formula, T is the torque (N·n) applied to the tool, and n is the rotational speed of the tool (r/min).

The calculation formula for the traction power consumption is:

$$P_2 = Fv \tag{11}$$

In the formula, F is the resistance (N) applied to the tool, and v is the forward speed of the tool (m/s).

That is,

$$P = P_1 + P_2 \tag{12}$$

3. Discussion

3.1. Analysis of the Simulation Results

SPSS data processing software was used to analyze the experimental results, as shown in Tables 7 and 8.

Table 7. Orthogonal test results.

Test Number	Traction Speed (m/s)	Cutter Shaft Speed (rpm)	Trenching Depth (mm)	Trench Depth Qualification Rate Y1 (%)	Power Consumption Y2 (w)
1	1 (0.05)	1 (340)	1 (15)	97.7	656
2	1	2 (390)	2 (20)	99.1	516
3	1	3 (440)	3 (25)	97.9	868
4	2 (0.25)	1	2	98.3	973
5	2	2	3	97.8	1009
6	2	3	1	98.5	893
7	3 (0.45)	1	3	96.2	1514
8	3	2	1	97.4	1031
9	3	3	2	98.0	1348
K1	680.03	1047.77	860.07		
K2	958.23	851.70	945.63		
K3	1297.57	1036.37	1130.13		
R	617.53	196.07	270.07		
Optimal level	A1	B2	C1		
Main and secondary factors		A > C > B			
Optimal combination		A1B2C1			

Table 8. Analysis of orthogonal test results.

Indicator	Source of Variation	Sum of Squares	Degrees of Freedom	Mean Square	F-Value	Significance
Trench depth qualification rate Y1	A	2.069	2	1.034	00	0.007
	B	1.029	2	0.514	133	0.015
	C	2.042	2	1.021	66.143	0.008
	Error	0.016	2	0.008	131.286	
	Total	5.156	2			
Power consumption Y2	A	573,889.8	2	286,944.9	168.103	0.06
	B	72,673.88	2	36,336.94	24.414	0.045
	C	114,297.9	2	57,148.95	33.679	0.029
	Error	3393.736	2	1696.868		
	Total	764,255.3	8			

According to the results of the data analysis shown in Tables 7 and 8, among the three experimental factors that affected the rate of the qualified trenching depth: traction speed, trenching depth, and cutter shaft speed, the traction speed and trenching depth had the most significant effects on the experimental results, while the effect of the cutter shaft speed was more significant. The importance of the experimental factors was in the following order: traction speed, trenching depth, and cutter shaft speed. The best combination was A1B2C1.

This study adopted an evaluation index based on the number of bond breakages between soil particles to analyze the soil fragmentation rate after the operation. Bonds were used as a measure of the force of interactions between soil particles. When the spiral trenching machine applied cutting force to the soil, the bonding bond was damaged, and the number of bonding bond connections showed a stable decreasing trend, leading to the separation and fragmentation of soil particles. Therefore, the change in bond number reflected the change in the soil fragmentation rate. A statistical function was set up in the postprocessing module of the software to count the number of bonds that were generated and broken, and graphs of the corresponding curves were drawn. By comparing the curve graphs generated with different combinations of the working parameters, the soil fragmentation by the spiral trencher could be analyzed and compared. The statistical results are shown in Figure 8.

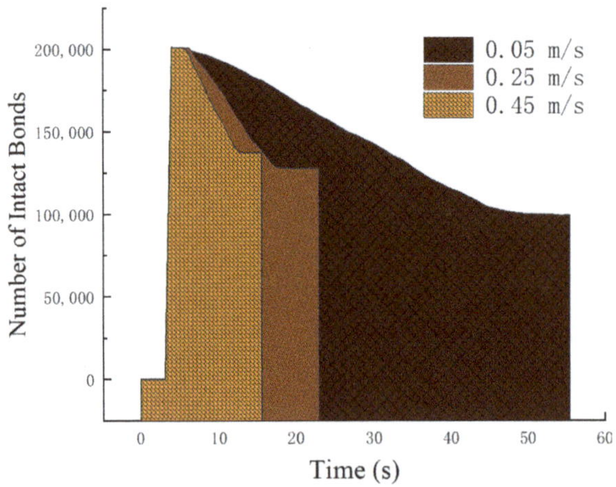

Figure 8. Changes in the bond number.

By comparing the line chart of bond changes under the combination of the three working parameters, combined with the working area of the tool in the soil tank, it could be calculated that the average crushing rate of the trenching tool in the effective operation range of the soil tank was 66%, which met the quality requirements of NYT 740-2003 field trenching machinery, and the number of bonding bonds did not change with time. Therefore, in order to improve the test efficiency, the number of bonding bonds reduced within 1 m of the tool was used as the operating effect index.

3.2. Soil Groove Test Conditions

The soil groove test bench of the Agricultural Training Center of Hunan Agricultural University was selected as the test site for the spiral trenching test. In the soil groove test bench, the spiral trencher was installed on a TCC electric four-wheel drive soil groove test car by using three-point suspension. The main parameters of the test car were as follows: a soil groove length of 30 m, a width of 2.95 m, a drive motor input voltage of 380 V, and a rated power of 30 kW; a stepless speed regulation could be achieved. The power output motor was a Taiwan Fuda frequency conversion motor with a base frequency of 50 Hz, and a stepless speed regulation was used by means of an ORS2000S 30kW frequency converter.

According to the simulation results, the initial motor selection was optimized, and the DC brushless motor model 110BL140-430 was chosen as the driving power source for the trenching rotary motion. Its rated power was 1500 W, its rated voltage was 48 V, its rated current was 42 A, its rated speed was 3000 rpm, and its rated torque was 14.4 Nm. In order to measure the influence of the spiral trencher on the soil properties, the following testing equipment was used: a M-85 soil density meter, a TTJSD-750-IV soil compaction meter, a WKT-M1 soil moisture meter, a WT-CF series high-precision electronic scale, a push–pull tester (HP-50k), a steel frame tape measure, and a meter ruler. The tester used in the experiment is shown in Figure 9.

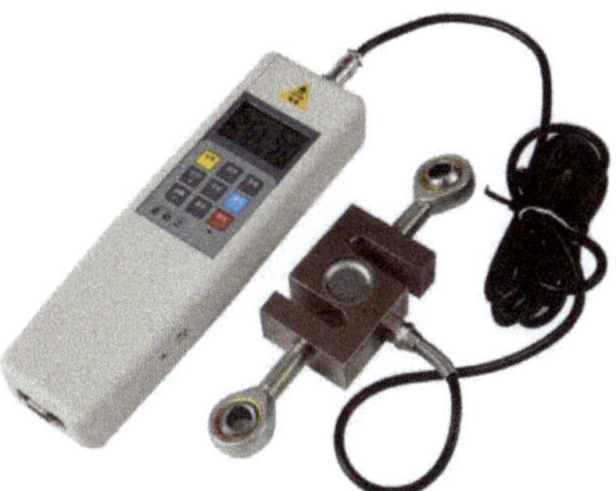

Figure 9. Push–pull tester (HP-50k).

3.3. Soil Tank Test Plan

In order to verify the correctness of the simulation model, a spiral trenching test was carried out in the experimental soil tank of Hunan Agricultural University. The size of the soil tank selected for the experiment was 6 m × 2 m × 0.6 m, and it was filled with compacted soil, as shown in Figure 10. Before the experiment, the density and moisture content of the soil were measured by using a WT-CF series high-precision electronic scale and WKT-M1 soil moisture meter, respectively, and the measurement results were 2712 kg/m^3 and 26.25%; the average hardness of 0–30 cm of soil was measured with an SC900 soil hardness tester, and this value was 0.61 MPa. According to a comparison, the parameters of the soil in the experimental field were basically consistent with those of the simulated soil.

Figure 10. Soil tank verification experiment.

We selected an inclination angle of 24° and an operation speed of 0.05 m/s as the optimal parameter combination according to the working conditions in the soil tank verification experiment. We attached the push–pull force tester between the tractor and the trenching tool to measure the traction resistance—that is, the magnitude of the force applied to the tool shaft in the simulation test. In each experiment, the machine advanced a distance of no less than 1 m to ensure the stability of the trenching. A total of five experiments were conducted for this study, and the average value was taken as the experimental result. The results measured by the tensile tester are shown in Figure 11 and Table 9. The error between the average value of the measured force and the simulation test results was 11%, which was close to the simulation test results.

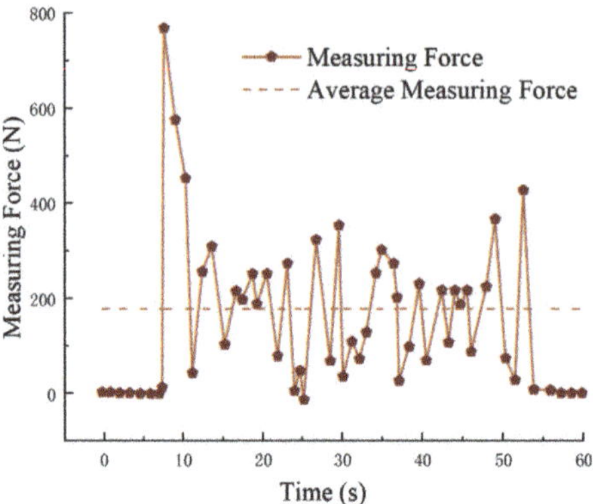

Figure 11. Change in measurements of traction in the soil tank.

Table 9. Dynamometer Data Analysis Table.

Indicator	Mean	Minimum Value	Median	Maximum Value
Measuring Force	237.57	78.14	266.05	416.75

In the soil tank test, priority was given to the qualified rate of trenching and the rate of soil fragmentation as the evaluation indicators. In order to evaluate the trenching quality, the trenching depth was measured every 200 mm along the machine's forward direction, and the trenching qualification rate was calculated. The trenching qualification rate was the proportion where the ratio of the actual trenching depth to the predetermined trenching depth reached more than 90%.

Five detection areas in the soil tank with an identical area were randomly selected after the operation, and the pulverization rate of the machine's operation was evaluated. The area of each detection area was 50 mm × 50 mm. In each detection area, the mass of all soil blocks with the longest side not exceeding 4 cm was measured, and the percentage of the total mass in all soil blocks in that detection area was calculated; this was used as the pulverization rate of that detection area. The higher the pulverization rate, the better the soil-breaking effect of the trencher. Figure 12 and Table 10 show a map of the distribution of the results of the soil tank verification test.

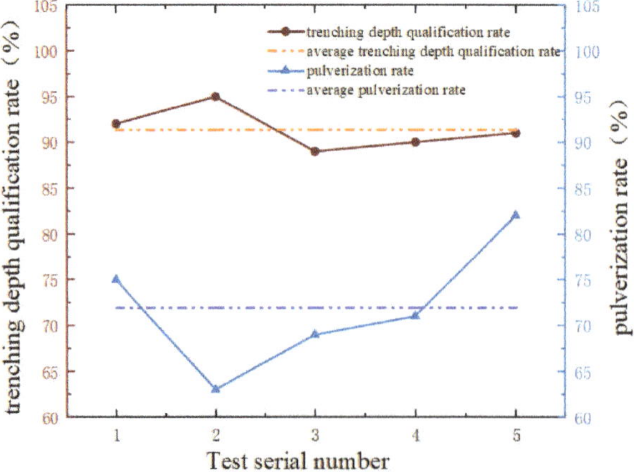

Figure 12. Chart of the soil tank test results.

Table 10. Analysis of the experimental data.

Indicator	Mean	Standard Deviation	Variance	Minimum Value	Median	Maximum Value
Trench depth qualification rate	91.4	2.28	5.2	89	91	95
Power consumption	72	6.99	48.8	63	71	82

Based on the measurements and calculations of the soil-crushing rate in the soil trough test field, this study also collected data on the trenching depth, trench shape, floating soil thickness at the bottom of the trench, and ridge width. By comparing the test results with the local agronomic requirements and relevant standards, the following conclusions were drawn: The average trenching depth in the soil trough test was 23.53 cm, which met the needs of agricultural production, and there was a small fluctuation in the trenching depth. The average trench width was 26.8 cm, the average trench bottom width was 18.4 cm, the average pass rate was 91.4%, and the trench shape was that of a ladder-shaped rectangle. The floating soil thickness at the bottom of the trench was 0.1 (trenching depth), the width consistency coefficient was 82%, and the trencher had good stability and accuracy during the walking process. The average soil fragmentation rate was 72%, which met the requirements of NY/T740-2003 and GB/T 5262-2008. The machine had a good soil-crushing effect, which was conducive to the improvement of soil aeration.

The results of the simulation experiment were compared to the results of the soil tank verification experiment. It was found that, under the same parameter conditions, they had high consistency, indicating that the actual trenching performance was close to the simulated trenching performance, thus proving the correctness of the simulation experiment. This showed that the motion analysis of the trenching tool and the simulation method used in this study were reliable and reasonable.

4. Conclusions

A new type of counter-rotating dual-axis spiral trenching tool was designed to meet the trenching demands for orchards in hilly and mountainous areas, and its working performance and influencing factors were analyzed and optimized in depth. The trenching tool was able to adapt to different agronomic requirements and operating environments, and it was able to achieve the effects of soil crushing, turning, and stirring while trenching, which was conducive to improving the interactions of water, fertilizer, and heat in the soil. We drew the following conclusions from the analysis of the simulation experiments in combination with the soil tank verification experiments.

1. A three-dimensional dynamic model of the soil particles and the tool was established by using the discrete element method, and the process of interaction between the soil and the tool during trenching was simulated. The motion trajectory of the soil particles and the laws of variation in the torque, force, and power of the tool were analyzed.
2. Based on the model created with the discrete element method, the parameters of the trencher, such as the spiral pitch angle; pitch; and motion parameters, including the forward speed, rotation speed, and working depth, were analyzed and optimized by using a sensitivity analysis and design optimization. We obtained an optimal parameter combination that improved the working efficiency and stability of the trencher.
3. By using an orthogonal test method, the effects of the forward speed, trenching depth, trencher rotation speed, and other parameters on the trenching depth qualification rate, total power consumption, and other performance parameters were studied. We found the optimal level and degree of influence. It was found that the forward speed was the most important factor that affected the working performance of the trencher, followed by the trenching rotation speed and depth.
4. While using the optimal parameter combination, a soil tank test was carried out to measure the actual working performance of the trenching tool, and the results were compared to those of the simulation test. The reliability and validity of the simulation test results were verified.

In this study, we analyzed and discussed the principle and method of the double-spiral trenching operation, which provides some reference and support for the design and technology involved so that related trenching operations can be used to better complete their objectives and tasks.

Author Contributions: Conceptualization, J.L. and P.J.; methodology, J.L.; software, J.L. and J.C.; validation, X.Z., M.X. and D.H.; formal analysis, J.L.; investigation, D.H.; resources, J.L.; data curation, J.L.; writing—original draft preparation, J.L.; writing—review and editing, J.L., Y.S. and P.J.; visualization, J.L.; supervision, Y.S. and P.J.; project administration, Y.S.; and funding acquisition, Y.S. and P.J. All authors have read and agreed to the published version of the manuscript.

Funding: This work was supported by the Scientific Research Project of the Hunan Provincial Education Department (grant number 21B0207), the Changsha City Natural Science Foundation (grant number kq2208069), Provincial-level special funding from Chenzhou National Sustainable Development Agenda Innovation Demonstration Zone (grant number 2022sfq20), and the subproject of the National Key R&D Plan (grant number 2022YFD2002001).

Data Availability Statement: Not applicable.

Conflicts of Interest: The authors declare no conflict of interest.

References

1. Wang, D.; Li, P.; Yi, X.; Liao, J.; He, Y.; Ran, J.; Zhang, F. Current status and development trend of orchard fertilization process and related machinery application. *J. Fruit Sci.* **2021**, *38*, 792–805.
2. Nie, Y. Research status and development trend of orchard trenching fertilizer applicator. *Agric. Sci. Technol. Equip.* **2020**, *5*, 55–56.
3. Song, Y.; Zhang, Z.; Fan, G.; Gao, D.; Zhang, H.; Zhang, X. Research status and development trend of orchard trenching fertilizer machinery in China. *J. Chin. Agric. Mech.* **2019**, *40*, 7–12.
4. Das, D.K.; Chaturvedi, O.P. Root biomass and distribution of five agroforestry tree species. *Agrofor. Syst.* **2008**, *74*, 223–230. [CrossRef]
5. Sundaram, S.R.; Tk, K. Redefining the logarithmic spiral trenching to understand root structure and distribution of trees. *Indian For.* **2021**, *147*, 202–204. [CrossRef]
6. Xiao, H.; Zhao, Y.; Ding, W.; Mei, S.; Han, Y.; Zhang, Y.; Yan, H.; Song, Z. Design and test of cutter shaft for 1KS60-35X type orchard double spiral trenching fertilizer applicator. *Trans. Chin. Soc. Agric. Eng.* **2017**, *33*, 32–39.
7. Gao, W.; Lin, J.; Li, B.; Ma, T. Parameter optimization and test of spiral trenching device for corn straw deep burial returning machine. *Trans. Chin. Soc. Agric. Mach.* **2018**, *49*, 45–54.
8. Lin, J.; Ma, T.; Gao, W.; Li, B.; Song, J.; Lv, Q. Design and test of spiral trenching device for straw deep burial returning machine. *J. Shenyang Agric. Univ.* **2018**, *49*, 41–48.
9. Jiang, J. Research on key technologies of small orchard double-sided spiral trenching machine. *Agric. Dev. Equip.* **2018**, *2*, 97–103.
10. Zeng, Z.; Ma, X.; Cao, X.; Li, Z.; Wang, X. Application status and prospect of discrete element method in agricultural engineering research. *Trans. Chin. Soc. Agric. Mach.* **2021**, *52*, 1–20.
11. Landry, H.; Lagu, C.; Roberge, M. Discrete element modeling of machine-manure interactions. *Comput. Electron. Agric.* **2006**, *52*, 90–106. [CrossRef]
12. Yang, W.; Wu, B.; Peng, Z.; Tan, S. Evaluation of trenching quality of vertical spiral ditcher based on discrete element method. *J. Southwest Univ. (Nat. Sci. Ed.)* **2019**, *41*, 1–7.
13. Wang, S.; Li, S.; Zhang, Y.; Zhang, C.; Chen, H.; Meng, L. Design and optimization of inclined spiral ditching component for orchard ditcher in mountainous area. *Trans. Chin. Soc. Agric. Eng.* **2018**, *34*, 11–22.
14. Zang, J. Design and Test of Organic Fertilizer Trenching Device for Dwarf Rootstock Orchard. Master's Thesis, Northwest A&F University, Xianyang, China, 2020.
15. Solhjou, A.; Fielke, J.M.; Desbiolles, J.M.A. Soil translocation by narrow openers with various rake angles. *Biosyst. Eng.* **2012**, *112*, 65–73. [CrossRef]
16. Li, Y.; Zhang, G.; Zhang, Z.; Zhang, Y.; Hu, T.; Cao, Q. Design and development of low-power small vertical-axis deep tillage machine segmented spiral tillage cutter. *Trans. Chin. Soc. Agric. Eng.* **2019**, *35*, 72–80.
17. Chen, G. Design and Optimization of Working Parts of Vertical Spiral Ditching Machine Based on Milling Force and BP Neural Network Model. *Mob. Inf. Syst.* **2022**, *2022*, 7971930. [CrossRef]
18. Li, Z. Design and Test of Spiral Toothed Stirrer Belt Leveling Device. Master's Thesis, Northeast Agricultural University, Harbin, China, 2021.
19. Xue, Z. Design and Optimization of Working Parts of Vertical Spiral Ditcher. Master's Thesis, Northwest A&F University, Xianyang, China, 2011.
20. Li, P.; Jiang, E.; Ding, Q. Critical speed analysis of vertical columnar variable pitch spiral ditcher. *Agric. Mech. Res.* **2005**, *1*, 129–130.

21. Lv, Z. Ways to reduce power consumption of vertical spiral ditcher. *Trans. Chin. Soc. Agric. Mach.* **1994**, *2*, 78–83.
22. Ma, A.; Liao, Q.; Tian, B.; Huang, H.; Zhou, S. Design and soil groove test of spiral orchard ditcher. *Hubei Agric. Sci.* **2009**, *48*, 1747–1750.
23. Huang, W. Design of vertical spiral ditcher. *Xinjiang Agric. Mech.* **1996**, *3*, 31–32.
24. *Chinese Academy of Agricultural Mechanization Sciences, Agricultural Machinery Design Manual Volume 1*; China Industrial Press: Beijing, China, 1971.
25. Song, K. Design of Electric-Driven Orchard Trenching Fertilizer Applicator. Master's Thesis, Nanjing Agricultural University, Nanjing, China, 2019.
26. Asaf, Z.; Rubinstein, D.; Shmulevich, I. Determination of discrete element model parameters required for soil tillage. *Soil Tillage Res.* **2007**, *92*, 227–242. [CrossRef]
27. Tamás, K.; Jóri, I.J.; Mouazen, A.M. Modelling soil–sweep interaction with discrete element method. *Soil Tillage Res.* **2013**, *134*, 223–231. [CrossRef]
28. Ucgul, M.; Fielke, J.M.; Saunders, C. Three-dimensional discrete element modelling (DEM) of tillage: Accounting for soil cohesion and adhesion. *Biosyst. Eng.* **2015**, *129*, 298–306. [CrossRef]
29. Shike, Z.; Yixin, S.; Junchi, Z.; Jianfei, L.; Defan, H.; Airu, Z.; Ping, J. Simulation Optimization and Experimental Study of the Working Performance of a Vertical Rotary Tiller Based on the Discrete Element Method. *Actuators* **2022**, *11*, 342.
30. Matin, M.A.; Fielke, J.M.; Desbiolles, J.M.A. Torque and energy characteristics for strip-tillage cultivation when cutting furrows using three designs of rotary blade. *Biosyst. Eng.* **2015**, *129*, 329–340. [CrossRef]
31. Xie, C. Research on Ring Trenching Fertilizer Applicator for Mountain Orchard. Master's Thesis, Southwest University, Chongqing, China, 2022.
32. Xiang, W.; Wu, M.; Lv, J.; Quan, W.; Ma, L.; Liu, J. Calibration of simulation physical parameters of sticky loam based on stacking test. *Trans. Chin. Soc. Agric. Eng.* **2019**, *35*, 116–123.

Article

Optimization of Shifting Quality for Hydrostatic Power-Split Transmission with Single Standard Planetary Gear Set

Zhaorui Xu [1,†]**, Jiabo Wang** [2,3,†]**, Yanqiang Yang** [1,†]**, Guangming Wang** [1,4,*] **and Shenghui Fu** [1,5]

1 College of Mechanical and Electronic Engineering, Shandong Agricultural University, Taian 271018, China; xzrsdau@163.com (Z.X.); yangyanqiang@sdau.edu.cn (Y.Y.); fush@sdau.edu.cn (S.F.)
2 College of Engineering, Nanjing Agricultural University, Nanjing 210031, China; gerab.wang@jsafc.edu.cn
3 School of Mechanical and Electrical Engineering, Jiangsu Vocational College of Agriculture and Forestry, Jurong 212400, China
4 Shandong Provincial Key Laboratory of Horticultural Machineries and Equipment, Taian 271018, China
5 College of Engineering, China Agricultural University, Beijing 100083, China
* Correspondence: gavinwang1986@163.com
† These authors contributed equally to this work.

Abstract: To improve the driving comfort of continuously variable transmission (CVT) tractors, the shifting quality of hydrostatic power-split transmission with a standard planetary gear set was optimized. Firstly, the powertrain of the CVT and two shift strategies, direct-shift and bridge-shift, were introduced; then, a dynamic model of tractor shifting was constructed, and the models of key components such as wet clutches and proportional pressure valves were experimentally verified. Finally, the control parameters of the above two shifting strategies were optimized, and the acceleration impact and sliding energy loss caused by them were compared. The results showed the following: the minimum peak acceleration of the bridge-shift method was 0.385807 m/s^2; the energy consumption of the bridge-shift method was significantly lower than that of the direct-shift method; the sliding friction work of clutches decreased by 14.92% and 75.84%, respectively, while their power loss decreased by 22.82% and 74.48%, respectively.

Keywords: tractor; power-split; continuously variable transmission; power shift

Citation: Xu, Z.; Wang, J.; Yang, Y.; Wang, G.; Fu, S. Optimization of Shifting Quality for Hydrostatic Power-Split Transmission with Single Standard Planetary Gear Set. *Agriculture* **2023**, *13*, 1685. https://doi.org/10.3390/agriculture13091685

Academic Editor: Kenshi Sakai

Received: 30 June 2023
Revised: 10 August 2023
Accepted: 21 August 2023
Published: 26 August 2023

1. Introduction

The working conditions of tractors are complex, as they require more gears to meet different operational needs, but this also leads to complex transmission structures and difficult gear selection. Tractors with continuously variable transmission (CVT) can effectively solve the above problems. At present, there are three common forms of tractor CVT [1–3]: hydrostatic transmission, steel belt transmission, and hydrostatic power-split transmission. Among them, the energy consumption of hydrostatic transmission is very high, and the torque transmitted by steel belt transmission is very limited, while hydrostatic power-split transmission has both a high efficiency and a large load driving capacity [4–6]. Since the release of the first CVT tractor "926 Vario" by Fent in 1996, hydrostatic power-split transmissions have gradually been applied to various pieces agricultural machinery [7]. Afterwards, transmission manufacturers began to introduce various concepts of tractor CVTs [8,9], such as the Eccom produced by ZF and the Autopowr produced by John Deere. The above-mentioned transmissions all adopt multi-range technology to achieve continuous speed adjustment [10–12], so it is necessary to conduct research to improve the quality of the shift [13–15], including changing the gear ratio, the clutch engagement time, and the displacement ratio of the swash plate axial piston units. For tractors with heavy load operations such as plowing as their main operating conditions, while paying attention to their riding comfort we also need to consider issues such as power interruption and clutch damage, all of which are important criteria for evaluating the shifting quality. Typically, cascading multiple planetary gears together to form a compound planetary gear set can improve the shifting quality through speed synchronization [10]. This is currently the mainstream strategy for CVT design, as its shifting logic is very simple. In this research field, Bao et al. [16] constructed a clutch control system based on solenoid directional valves and optimized the clutch pressure, the flow, and the displacement ratio of the pump to the motor to improve the shifting quality of the power-split CVT. Chen et al. [17] proposed a simulation model of a similar hydraulic control system, making it possible to further study the shifting dynamics of CVT tractors through computer simulation. Iqbal et al. [18] conducted similar work. Wang et al. [19] analyzed the reliability of clutch control systems based on on–off logic and discussed the possible influence of hydraulic system failures on the shifting quality. To further improve the shifting quality of this type of transmission, it is necessary to use proportional pressure valves to accurately control the clutch action. For example, Xiang et al. [20] proposed a control strategy for dual-clutch transmissions that can maintain the sliding in the torque phase to improve the shifting quality. Li and Görges [21] conducted similar work. Li et al. [22] used a PID controller to track the pressure of the clutch to ensure the repeatability of proportional pressure control. Although a compound planetary gear set can improve the shifting smoothness through speed synchronization, its structure needs to fully consider support and load balance issues when applied, which brings difficulties to its design, manufacturing, and assembly. In contrast, using a single standard planetary gear to merge the power is simple and cost effective. Currently, some companies such as Hofer have shown great interest in this new concept of transmission. However, the shift logic of the transmission is complex, requiring the simultaneous adjustment of multiple wet clutches and swash plate axial piston units during shifting, which is much more difficult to control than traditional power-split CVTs. To improve the shifting quality of this cost-effective power-split CVT with a single planetary gear set and promote its application in tractors, a new strategy called the bridge-shift method is proposed in this study.

2. Materials and Methods

2.1. Powertrain

The hydrostatic power-split transmission proposed in this study has two ranges, HM_1 and HM_2, in the forward direction, which can achieve a stepless speed regulation of the tractor within a range of 0–30 km/h. The principle of the transmission is shown in Figure 1. The engine power is divided into two parts on the input shaft, with part of the power being transferred to the sun gear of the planetary gear set through the swash plate axial piston units and the rest of the power entering the ring gear of the planetary gear set through the gear train. The above two parts of power are marked as the hydraulic circuit power and mechanical circuit power, respectively. The transmission ratio of the mechanical circuit is fixed, so the output speed of the transmission only depends on the displacement ratio of the pump to the motor, which is numerically equal to the actual displacement of the pump divided by the rated displacement of the motor. Since the displacement of the pump changes in two directions with the inclination angle of its swash plate, the displacement ratio ranges from −1 to +1 ("+" indicates that the speed direction of the pump and motor is the same, while "−" indicates the opposite). In each range, the displacement ratio of −1 corresponds to the lowest speed of the tractor, while the displacement ratio of +1 corresponds to the highest speed of the tractor.

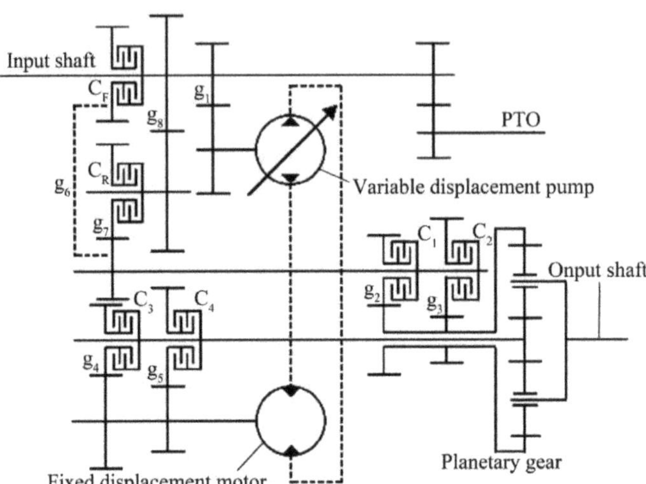

Figure 1. Transmission scheme of hydrostatic power-split CVT. Note: the symbol g represents the gear pair, and the symbol C represents the wet clutch.

Before starting the tractor, the transmission control unit (TCU) needs to adjust the displacement ratio of the pump to the motor to -1 (i.e., the displacement ratio corresponding to the minimum CVT output speed of the range HM_1), engage clutches C_1 and C_3, and separate clutches C_R, C_2, and C_4. Then, the TCU slowly engages the clutch C_F to bring the tractor to its minimum operating speed.

After starting, the transmission operates in the range HM_1. As the displacement ratio changes in the direction of "$-1 \rightarrow +1$", the tractor speed continuously increases. Once the tractor reaches its predetermined speed, the TCU separates clutches C_1 and C_3, engages clutches C_2 and C_4, and reversely adjusts the displacement ratio of the pump to the motor to achieve the equal-speed shifting of the transmission, thereby switching the working range of the transmission from HM_1 to HM_2. The speed adjustment process of the ranges HM_1 and HM_2 is completely the same and will not be repeated here.

When clutches C_R, C_1, and C_4 are engaged and clutches C_F, C_2, and C_3 are separated, the transmission operates in the reverse range HM_R. The speed of the tractor in this range covers two directions, and the displacement ratio corresponding to its zero speed is approximately -0.9. When the displacement ratio changes from -0.9 to $+1$, the tractor can achieve a stepless speed regulation within the range of 0–16 km/h in the reverse direction.

The clutch schedule of this transmission is shown in Table 1.

Table 1. Clutch schedule of the hydrostatic power-split transmission.

Working Range	Clutches						Displacement Ratio of Pump to Motor	Tractor Speed at Rated Engine Speed/(km/h)
	C_1	C_2	C_3	C_4	C_F	C_R		
HM_1	•		•		•		$-1 \rightarrow +1$	$2 \rightarrow 14$
HM_2		•		•	•		$-1 \rightarrow +1$	$12 \rightarrow 30$
HM_R	•			•		•	$-0.9 \rightarrow +1$	$0 \rightarrow -16$

2.2. Control Strategies

The transmission involves the separation or engagement of four clutches during shifting, and the action timing of each clutch will have a significant impact on the shifting process. For example, after the separation of clutches C_1 and C_3, the speed of the tractor will continuously decrease under the action of the load. If the engagement of clutches C_2 and C_4 is slow, the driving and driven plates of the clutch will be in a continuous sliding state, which will burden the cooling system and shorten the service life of the clutch. In severe cases, it can also directly cause power interruption. On the contrary, if the engagement of clutches C_2 and C_4 is very fast, the speed difference between the driving and driven plates of the clutch is quickly eliminated, which will cause severe speed oscillations and significantly reduce the riding comfort of the tractor. The actual shifting process is very complex, which requires precise control of the actions of each clutch in the time domain. To solve this problem, two shifting strategies are proposed: the direct-shift and bridge-shift strategies.

Taking the switching of range HM_1 to range HM_2 as an example, the former separates clutches C_1 and C_3 while directly engaging clutches C_2 and C_4, while the latter needs to insert a transitional state of C_2 and C_3 engagement during the aforementioned process, as shown in Figure 2.

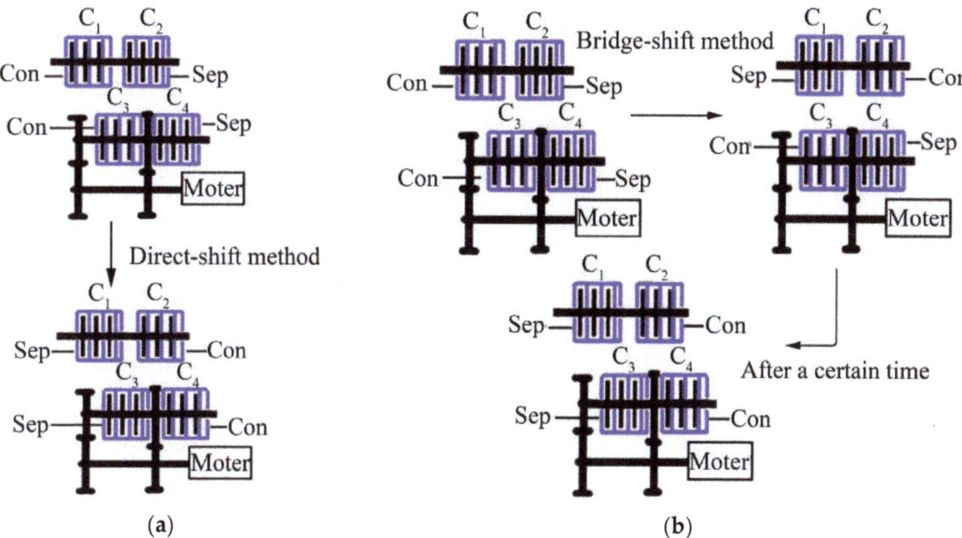

Figure 2. Shift process under different control strategies. (**a**) Direct-shift method. (**b**) Bridge-shift method.

2.3. Modeling of the Swash Plate Axial Piston Units

The swash plate axial piston units are the core speed-regulating components of the CVT, consisting of a variable-displacement pump and a fixed-displacement motor. Its pressure, torque, flow, and speed meet the following equations:

$$T_p = \frac{e\Delta P_p V_p}{2\pi} \tag{1}$$

$$T_m = \frac{\Delta P_m V_m}{2\pi} \tag{2}$$

$$Q_p = \frac{e V_p n_p}{1000} \tag{3}$$

$$Q_m = \frac{V_m n_m}{1000} \tag{4}$$

where T_p and T_m are the theoretical torques of the pump shaft and motor shaft, respectively, N·m; ΔP_p and ΔP_m are the pressure differences between the inlet and outlet of the pump and motor, respectively, MPa; Q_p and Q_m are the theoretical flows of the pump and motor, respectively, L/min; n_p and n_m are the rotation speeds of the pump shaft and motor shaft, respectively, r/s; V_p and V_m are the rated displacements of the pump and motor, respectively, cm^3/r; and e is the displacement ratio of the pump to the motor.

In actual systems, it is necessary to consider the torque loss and flow loss caused by mechanical friction and oil leakage:

$$T_{pr} = \frac{T_p}{\eta_{mp}} \tag{5}$$

$$T_{mr} = T_m \eta_{mm} \tag{6}$$

$$Q_{pr} = Q_p \eta_{vp} \tag{7}$$

$$Q_{mr} = \frac{Q_m}{\eta_{vm}} \tag{8}$$

where T_{pr} and T_{mr} are the real torques of the pump shaft and motor shaft, respectively, N·m; η_{mp} and η_{mm} are the mechanical efficiencies of the pump and motor, respectively; Q_{pr} and Q_{mr} are the real flows of the pump and motor, respectively, L/min; and η_{vp} and η_{vm} are the volume efficiencies of the pump and motor, respectively.

2.4. Modeling of the Power-Shift System

The power-shift system consists of wet clutches and a corresponding hydraulic circuit. The frictional torque that the clutch can transmit is:

$$T_c = \mu F_n n_p \times \frac{2\left(r_o^3 - r_i^3\right)}{3\left(r_o^2 - r_i^2\right)} \tanh\left(2 \times \frac{R_n}{d_v}\right) \tag{9}$$

where T_c is the frictional torque transmitted in the clutch plates, N; μ is the coulomb friction coefficient; F_n is the normal force acting on the clutch plates, N; n_p is the number of clutch contact faces; r_o and r_i are the outside radius and inside radius of the friction plates, respectively, mm; R_n is the relative velocity, r/min; and d_v is the rotary stick velocity threshold, r/min.

The normal force F_n is determined by the combination of the oil pressure, centrifugal force, and spring force:

$$F_n = P_c A_c + F_c - k_c(x_{ci} + \Delta x_c) \tag{10}$$

where P_c is the oil pressure, MPa; A_c is the effective area of the piston, mm^2; F_c is the centrifugal force, N; k_c is the stiffness of the spring, N/mm; and x_{ci} and Δx_c are the initial compression and relative displacement of the spring, respectively, mm.

The rotary hydraulic cylinder is a typical coupling element used in multi-plate wet clutches in the transmission. Its rotational speed is high, so its oil chamber is subjected to centrifugal acceleration. The structure of the clutch hydraulic cylinder is shown in Figure 3, and the centrifugal force acting on its piston is calculated as follows:

$$F_c = \frac{\pi \rho \omega^2 \left[r_p^4 - r_r^4 - 2 \times r_i^2 \left(r_p^2 - r_r^2\right)\right]}{4} \tag{11}$$

where ρ is the bulk density of the hydraulic oil, kg·m^3; ω is the angular velocity, r/min; r_p and r_l are the outside radius and inside radius of the fluid volume acting on the piston, respectively, mm; and r_r is the inside radius of the piston, mm.

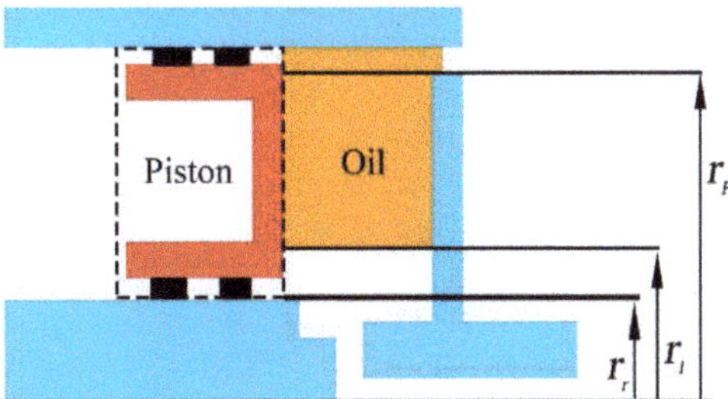

Figure 3. Schematic diagram of the clutch hydraulic cylinder.

A proportional pressure valve is used to control the working pressure of the clutch, consisting of a proportional electromagnet and a three-way spool valve, as shown in Figure 4. When the electromagnetic force increases, the valve spool moves to the right to increase the opening of the outlet port, causing the output pressure to increase. When the electromagnetic force decreases, the valve spool moves to the left, causing the output pressure to decrease. On this basis, when the output pressure increases, the valve spool moves left to allow excess oil to flow back to the tank through the oil return port, thereby reducing the output pressure. When the output pressure decreases, the piston moves right to increase the output pressure. According to its working principle, the force equation of the valve spool is expressed as:

$$A_s p_{out} + F_{jet} + k_s(x_{si} + \Delta x_s) = F_g \tag{12}$$

where A_s is the effective area of the valve spool, mm^2; p_{out} is the output pressure of the valve, MPa; F_{jet} is the jet force, N; k_s is the stiffness of the spring, N/mm; x_{si} and Δx_s are the initial compression of the spring and the displacement of the valve spool, respectively, mm; and F_g is the electromagnetic force, N.

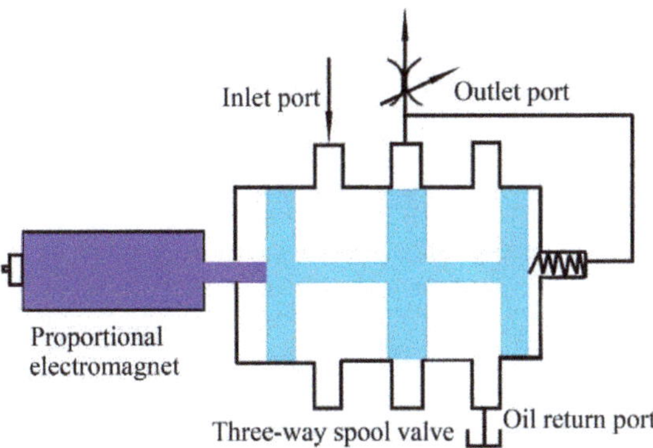

Figure 4. Schematic diagram of the proportional pressure valve.

The jet force is calculated using the following equation:

$$F_{jet} = 2C_q \pi d_s \Delta x_s |\Delta p_v| \cos \alpha_{jet} \tag{13}$$

where C_q is the flow coefficient; d_s is the equivalent diameter of the valve spool, mm; Δp_v is the pressure difference between the inlet and outlet of the valve, MPa; and α_{jet} is the jet angle, rad.

The displacement of the valve spool obtained by simultaneous Equations (12) and (13) is as follows:

$$\Delta x_s = \frac{F_g - A_s p_{out} - k_s x_{si}}{k_s + 2C_q \pi d_s |\Delta p_v| \cos \alpha_{jet}} \tag{14}$$

According to the above analysis, the output pressure of the proportional valve depends on the electromagnetic force F_g, which is controlled by a current signal. To clarify the corresponding relationship between the input current and output pressure, a signal generator is used to calibrate the valve, and the results are shown in Figure 5. It can be seen that within the current range of 4–20 mA, the input current of the proportional valve exhibits a clear linear relationship with oil pressure. Therefore, we used the calibration data to construct an electromagnetic model of the valve.

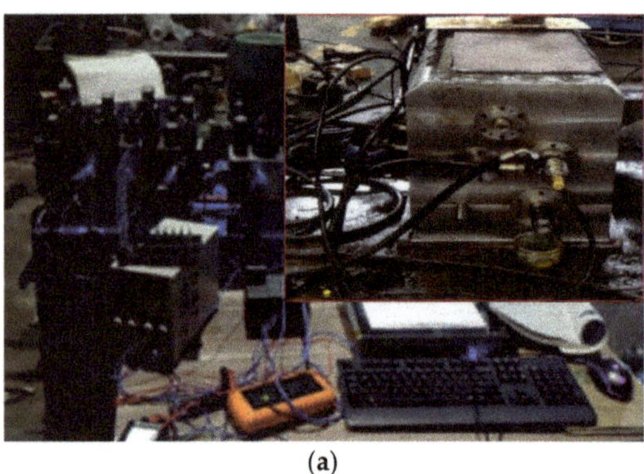

(a)

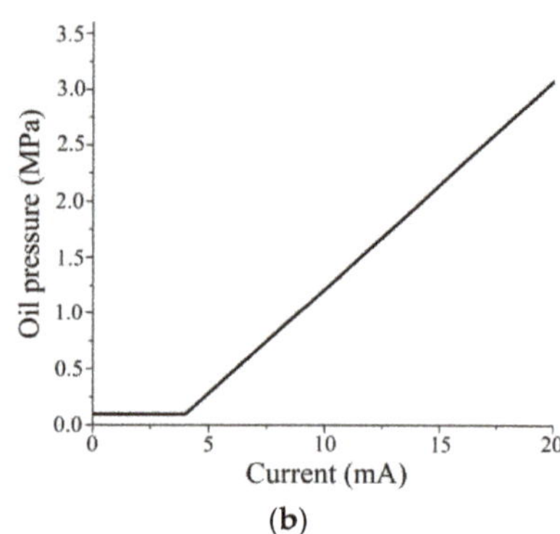

(b)

Figure 5. Calibration of the proportional pressure valve. (**a**) Hydraulic system used for calibration testing. (**b**) Calibration results of the proportional valve.

The hydraulic circuit constructed based on clutches and proportional valves is the core of the power-shift system and requires independent experimental verification of its mathematical model. We closed the outlet of the proportional valve before the experiment, and the PLC controlled its AD module to output a step signal corresponding to the rated pressure of the clutch. Note that the output signal of the AD module used was the voltage, which needed to be converted into a 4–20 mA current through a converter module to control the proportional valve. At the same time, the Labview program controlled the data acquisition card (NI USB-6009) to capture the output pressure of the proportional valve feedback from the sensor. The input signal of the simulation model was consistent with the experiment, that is, the input current was modulated from 0 to the maximum value in the experiment in a very short time to observe the pressure response of the model. The simulation and measurement results of the step response of the proportional pressure valve are shown in Figure 6. From the figure, it can be seen that the simulation results of the mathematical model constructed in this study were highly consistent with the experimental results and could meet the needs of subsequent dynamic analysis.

We connected the model of the proportional valve with the model of the wet clutch, further constructed the model of the power-shift system and conducted an experimental verification of it. The simulation and measurement results are shown in Figure 7. The figure shows that under the same input signal of the proportional valve, the clutch pressure response of the simulation model was basically consistent with the experimental results, thus proving the reliability of the constructed model.

The models of the proportional valve and wet clutch were relatively complex; for some models not covered in this article, please refer to the AMESim manual. The key parameters used in the simulation calculations are shown in Tables 2 and 3.

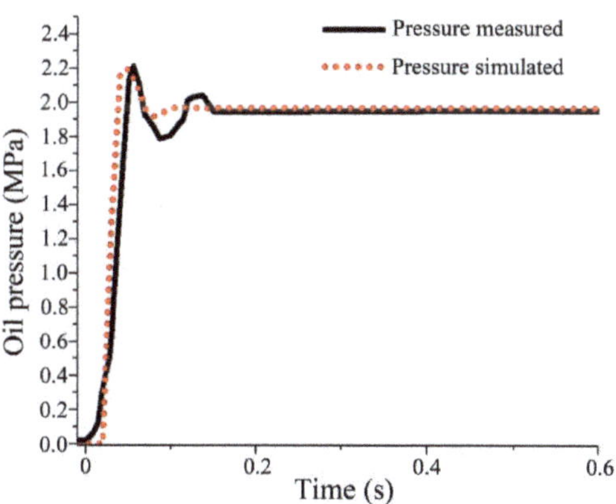

Figure 6. Simulation and measurement results of step response of proportional pressure valve.

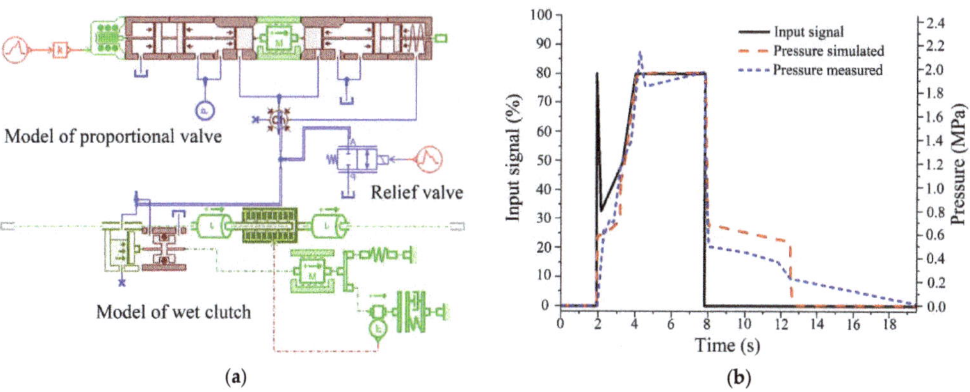

(a)

(b)

Figure 7. Experimental verification of the simulation model of the shift hydraulic system. (a) Simulation model of shift hydraulic system. (b) Simulation and measurement results of the response of clutch pressure to control signals. Note: the relief valve was designed for rapid pressure relief and was consistent with the actual hydraulic circuit.

Table 2. Configuration parameters of proportional valves.

Input Signal/mA	Output Pressure/MPa	Spring Preload/N	Spring Stiffness/(N/mm)	Flow Coefficient	Mass of Spool/kg	Hole Diameter/mm
4~20	0~2	20.0	19.9	0.6	0.52	1.2

Table 3. Configuration parameters of wet clutches.

Clutch	Area of Friction Plate/(mm²)	Area of Piston/(mm²)	Number of Friction Plates	Spring Stiffness/(N/mm)	Frictional Coefficient
C_1	7780	7210	7	19.6	0.12
C_2	6900	6090	8	10.42	0.11
C_3/C_4	4780	3810	7	6.7	0.08

2.5. Modeling of Gears and Shafts

The torque and speed of the two meshing gears satisfy the following equations:

$$n_2 = \frac{n_1}{i_{12}} \tag{15}$$

$$T_2 = i_{12}T_1 \tag{16}$$

where i_{12} is the transmission ratio of the gear pairs; n_1 and n_2 are the speeds of the two gears, r/min; and T_1 and T_2 are the torques of the two gears, N·m.

The speed and torque between the three basic components of the planetary gear, the sun gear, the ring gear, and the carrier satisfy the following equations:

$$n_s + kn_r - (1+k)n_c = 0 \tag{17}$$

$$T_s : T_r : T_c = 1 : k : (1+k) \tag{18}$$

where n_s, n_r, and n_c are the speeds of the sun gear, ring gear, and carrier, respectively, r/min; T_s, T_r, and T_c are the torques of the sun gear, ring gear, and carrier, respectively, N·m; and k is the standing ratio of the standard planetary gear.

In this study, the moment of inertia of each component is calculated by the SolidWorks 2016 software and is equivalent to the transmission shaft. Its influence on the torque of each shaft is as follows:

$$T_a = T_0 + J\frac{d\omega}{dt} \tag{19}$$

where T_a and T_0 are the actual torque and theoretical torque of the shaft, respectively, N·m; J is the moment of inertia, kg·m^2; ω is the angular velocity of the shaft, rad/s; and t is the time, s.

2.6. Modeling of Tractor

Based on the above equations, a shift dynamics model of the entire continuously variable transmission tractor was constructed using AMESim, as shown in Figure 8.

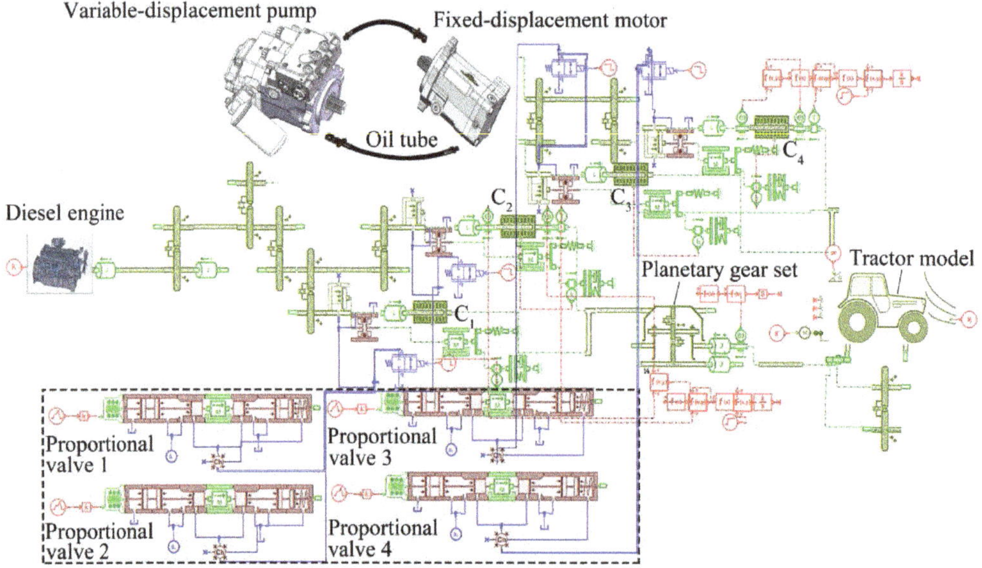

Figure 8. Shifting dynamics model of continuously variable transmission tractor.

3. Results and Discussion

3.1. Evaluation Indicators

The speed drop is defined as the difference between the output speed of the transmission before the shift and the lowest output speed during the shift:

$$\delta_1 = \omega_1 - \omega_2 \tag{20}$$

where δ_1 is the speed drop, r/min; ω_1 is the output speed before the shift, r/min; and ω_2 is the lowest output speed during the shift, r/min.

According to Duncan and Wegscheid [23], the peak acceleration of a tractor in the longitudinal direction can well reflect its driving comfort during shifting. Therefore, this study took the peak acceleration as one of the indicators for evaluating the shifting quality of the CVT, and its expression is as follows:

$$\delta_2 = \max\left(\frac{dv}{dt}\right) \tag{21}$$

where δ_2 is the peak acceleration of the tractor during the shift, m/s^2.

When the clutch is engaged, a large amount of heat will be generated due to friction, and in severe cases, it may burn out the clutch. The power loss during the aforementioned process is as follows:

$$\delta_3 = \max\left(\frac{T_c|\Delta\omega|}{9550}\right) \tag{22}$$

where δ_3 is the maximum power loss during the shift, kW; T_c is the friction torque, N·m; and $\Delta\omega$ is the difference in the angular speed of the clutch driving and driven disc, r/min.

On this basis, sliding friction work is defined as the integral of the power loss over time:

$$\delta_4 = \int_{t_1}^{t_2} \frac{T_c|\Delta\omega|}{9550} dt \tag{23}$$

where δ_4 is the sliding friction work, kJ; and t_1 and t_2 are the start and end times of the shift, s.

3.2. Direct-Shift Method

3.2.1. Determination of Shift Points

The process of direct shifting is relatively simple, with clutches C_1 and C_3 being separate while clutches C_2 and C_4 engage. During the shift process, the displacement ratio is synchronously adjusted, and its initial and final values need to meet the following relationship:

$$i_{HM1} = \frac{i_1 i_2 i_4 i_6 (1+k)}{k i_1 i_4 + e i_2 i_6} \tag{24}$$

$$i_{HM2} = \frac{i_1 i_3 i_5 i_6 (1+k)}{k i_1 i_5 + e i_3 i_6} \tag{25}$$

$$i_{HM1} = i_{HM2} \tag{26}$$

where i_{HM1} and i_{HM2} are the transmission ratios in HM$_1$ and HM$_2$, respectively; and i_x is the transmission ratio of the gear pair g_x.

From the perspective of transmission efficiency and energy consumption, the authors have demonstrated in previous research that the optimal shift point for this CVT in the range HM$_1$ is $e = 1$. Based on the above equations, the displacement ratio after the shift was calculated to be $e = -0.8034$.

3.2.2. Optimization of Shifting Quality

The pressure control signal of the proportional valve based on the direct-shift method is shown in Figure 9. If we define the pressure relief time of proportional valves 1 and 3 as t_{s1} (i.e., reference time, 10 s), then ΔT is the start time of the pressure rise for proportional valves 2 and 4 relative to t_{s1}. T_4 and T_2 are the times corresponding to the two inflection points in the pressure rise curve of proportional valve 2, respectively. T_1 and T_5 are the times corresponding to the two inflection points in the pressure rise curve of proportional valve 4, respectively. K_2 and K_1 are the percentages of the input signals corresponding to the second inflection point in the pressure rise curve of proportional valves 2 and 4, respectively. The starting time of T_1 and T_4 is the same, and the starting time and duration of the reverse change in the displacement ratio are T_s and T_d, respectively. On this basis, we adopted an orthogonal experiment with nine factors and four levels to optimize the above control parameters. The schedule of the experiment is shown in Table 4, and the results are shown in Tables 5 and 6. Considering that the peak acceleration directly affects the driving comfort, we only optimized the parameters for this indicator when designing the orthogonal experiments, but we considered other indicators together when analyzing the results.

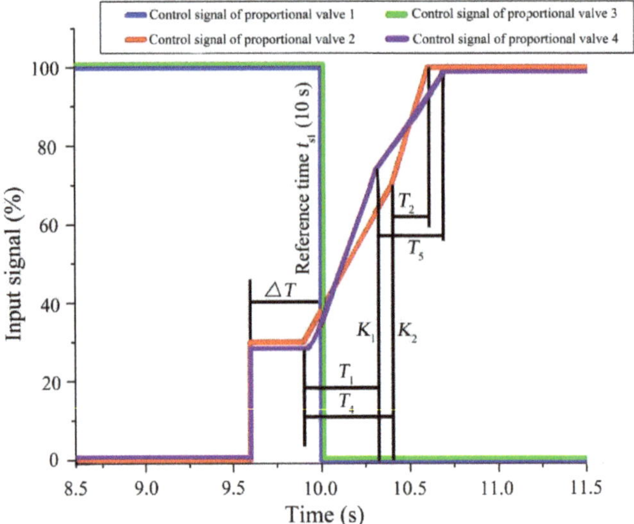

Figure 9. Pressure control signal of the clutch under direct-shift strategy. Note: considering that the rated pressure of the proportional valve and the clutch are not consistent, the input signal has been redefined here, with the maximum signal corresponding to the rated pressure of the clutch.

Table 4. Factors and levels used for direct-shift optimization.

Level	T_1/ms	T_2/ms	K_1/%	T_4/ms	T_5/ms	K_2/%	ΔT/ms	T_s/s	T_d/ms
1	200	400	40	200	400	40	350	9.9	500
2	250	500	50	250	500	50	400	10	650
3	300	600	60	300	600	60	450	10.1	800
4	350	700	70	350	700	70	500	10.2	950

Table 5. Orthogonal simulation sequence for direct-shift optimization.

Factor Number	A	B	C	D	E	F	G	H	I	Peak Acceleration
Test 1	1	1	1	1	1	1	1	1	1	2.754091
Test 2	1	2	2	2	2	2	2	2	2	2.218880
Test 3	1	3	3	3	3	3	3	3	3	0.746190
Test 4	1	4	4	4	4	4	4	4	4	1.523540
Test 5	2	1	1	2	2	3	3	4	4	1.526142
Test 6	2	2	2	1	1	4	4	3	3	1.028317
Test 7	2	3	3	4	4	1	1	2	2	2.135518
Test 8	2	4	4	3	3	2	2	1	1	2.862647
Test 9	3	1	2	3	4	1	2	3	4	1.061769
Test 10	3	2	1	4	3	2	1	4	3	0.627891
Test 11	3	3	4	1	2	3	4	1	2	3.136779
Test 12	3	4	3	2	1	4	3	2	1	0.614615
Test 13	4	1	2	4	3	3	4	2	1	2.631731
Test 14	4	2	1	3	4	4	3	1	2	3.048925
Test 15	4	3	4	2	1	1	2	4	3	1.001010
Test 16	4	4	3	1	2	2	1	3	4	0.674891
Test 17	1	1	4	1	4	2	3	2	3	1.952227
Test 18	1	2	3	2	3	1	4	1	4	2.758566
Test 19	1	3	2	3	2	4	1	4	1	0.501519
Test 20	1	4	1	4	1	3	2	3	2	0.814421
Test 21	2	1	4	2	3	4	1	3	2	0.627852
Test 22	2	2	3	1	4	3	2	4	1	0.584379
Test 23	2	3	2	4	1	2	3	1	4	2.766946
Test 24	2	4	1	3	2	1	4	2	3	1.894956
Test 25	3	1	3	3	1	2	4	4	2	0.901438
Test 26	3	2	4	4	2	1	3	3	1	1.022129
Test 27	3	3	1	1	3	4	2	2	4	1.486624
Test 28	3	4	2	2	4	3	1	1	3	2.839502
Test 29	4	1	3	4	2	4	2	1	3	2.871530
Test 30	4	2	4	3	1	3	1	2	4	1.853240
Test 31	4	3	1	2	4	2	4	3	1	0.999228
Test 32	4	4	2	1	3	1	3	4	2	0.398879

Table 6. Range analysis of orthogonal optimization for direct-shift strategy.

Factor	T_1	T_2	K_1	T_4	T_5	K_2	ΔT	T_s	T_d
1	1.569	1.791	1.644	1.502	1.467	1.628	1.502	2.880	1.496
2	1.678	1 643	1.681	1.573	1.731	1.626	1.613	1.848	1.660
3	1.461	1.597	1.411	1.609	1 518	1.767	1.510	0.872	1.620
4	1.685	1.453	1.747	1.799	1.768	1.463	1.859	0.883	1.706
Range	0.224	0.338	0.336	0.297	0.301	0.304	0.357	2.008	0.210

According to the results of the orthogonal range analysis, when switching from HM_1 to HM_2, the degree of influence of each factor on the direct-shift method was ranked as follows: the reverse starting point T_s, the time difference ΔT, time T_2, current K_1, current K_2, time T_5, time T_4, time T_1, and the reverse duration T_d. The best combination of factors was $A_3B_4C_3D_1E_1F_4G_3H_3I_1$. By substituting the optimized parameters into the simulation model, the various indicators for the direct-shift method were obtained as follows: the speed drop was 30.67 r/min (no power interruption), the peak acceleration was 0.384535 m/s^2, the power loss of clutch C_2 was 22.7383 kW, the sliding friction work of clutch C_2 was 8.0752 kJ, the power loss of clutch C_4 was 18.1166 kW, and the sliding friction work of clutch C_4 was 2.3906 kJ.

3.3. Bridge-Shift Method

3.3.1. Determination of Shift Points

The process of bridge shifting is divided into two stages. In the first stage, the transmission shifts from the low-speed range HM_1 to the transition range, where clutch C_1 separates while clutch C_2 engages. In the second stage, the transmission shifts from the transition range to the high-speed range HM_2, where clutch C_3 separates and clutch C_4 engages. During the shift process, the displacement ratio is synchronously adjusted, and its initial and final values are the same as those of the direct-shift method. However, the displacement ratio of the transition range will be used as an optimization variable, which will be discussed later.

3.3.2. Optimization of Shifting Quality

The pressure control signal of the proportional valve based on the bridge-shift method is shown in Figure 10. The factors T_1, T_2, K_1, T_4, T_5, and K_2 in the bridge-shift method are the same as those specified in the direct-shift method. We define the pressure relief times for proportional valves 1 and 3 as t_{s1} (10 s) and t_{s2} (10 s + ΔT_3), respectively; then, ΔT_1 is the start time of the pressure increase for proportional valve 2 relative to t_{s1}. ΔT_2 is the start time of the pressure increase for proportional valve 4 relative to t_{s2}. The starting time of the reverse change in displacement ratio is T_s. The duration of the two stages of the reverse change in the displacement ratio are T_{d1} and T_{d2}, respectively. The displacement ratio of the transition range is e_t. On this basis, we adopted an orthogonal experiment with thirteen factors and three levels to optimize the above control parameters. The schedule of the experiment is shown in Table 7, and the results are shown in Tables 8 and 9.

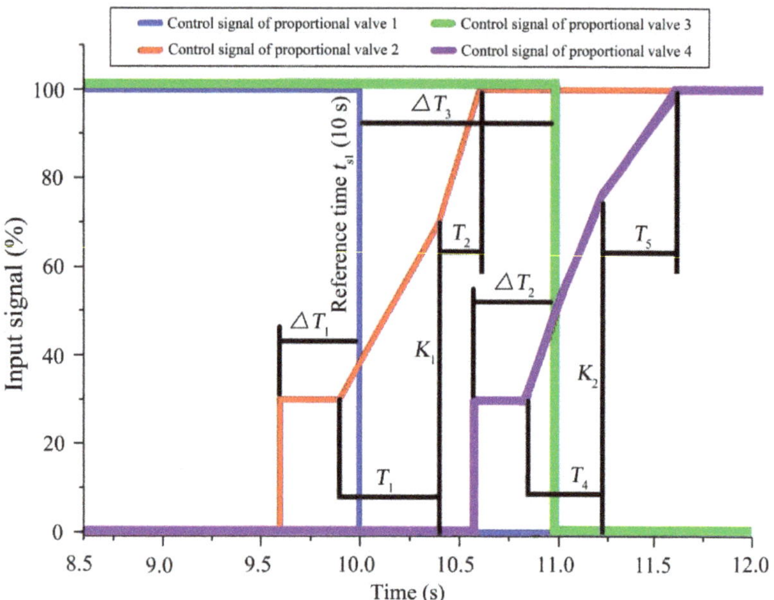

Figure 10. Pressure control signal of the clutch under bridge-shift strategy. Note: considering that the rated pressure of the proportional valve and the clutch are not consistent, the input signal has been redefined here, with the maximum signal corresponding to the rated pressure of the clutch.

Table 7. Factors and levels used for bridge-shift optimization.

Level	T_1/ms	T_2/ms	K_1/%	T_4/ms	T_5/ms	K_2/%	ΔT_1/ms	ΔT_2/ms	ΔT_3/ms	T_s/s	T_{d1}/ms	T_{d2}/ms	e_t/s
1	250	500	50	250	500	50	400	400	500	9.9	150	350	−1
2	300	600	60	300	600	60	450	450	750	10	200	500	−0.95
3	350	700	70	350	700	70	500	500	1000	10.1	250	650	−0.9

Table 8. Orthogonal simulation sequence for bridge-shift optimization.

Factor Number	A	B	C	D	E	F	G	H	I	J	K	L	M	Peak Acceleration
Test 1	1	1	1	1	1	1	1	1	1	1	1	1	1	2.693358
Test 2	1	1	1	1	2	2	2	2	2	2	2	2	2	2.655476
Test 3	1	1	1	1	3	3	3	3	3	3	3	3	3	1.527961
Test 4	1	2	2	2	1	1	1	2	2	2	3	3	3	1.152763
Test 5	1	2	2	2	2	2	2	3	3	3	1	1	1	1.169344
Test 6	1	2	2	2	3	3	3	1	1	1	2	2	2	2.620924
Test 7	1	3	3	3	1	1	1	3	3	3	2	2	2	0.986310
Test 8	1	3	3	3	2	2	2	1	1	1	3	3	3	2.600973
Test 9	1	3	3	3	3	3	3	2	2	2	1	1	1	1.178958
Test 10	2	1	2	3	1	2	3	2	2	3	1	2	3	0.840597
Test 11	2	1	2	3	2	3	1	2	3	1	2	3	1	2.284741
Test 12	2	1	2	3	3	1	2	3	1	2	3	1	2	1.072096
Test 13	2	2	3	1	1	2	3	2	3	1	3	1	2	2.606293
Test 14	2	2	3	1	2	3	1	3	1	2	1	2	3	1.374078
Test 15	2	2	3	1	3	1	2	1	1	3	2	3	1	1.139140
Test 16	2	3	1	2	1	2	2	3	1	2	2	3	1	1.244164
Test 17	2	3	1	2	2	3	3	1	2	3	3	1	2	0.545924
Test 18	2	3	1	2	3	1	1	2	3	1	1	2	3	2.701621
Test 19	3	1	3	2	1	3	2	1	3	2	1	3	2	1.371793
Test 20	3	1	3	2	2	1	3	2	1	3	2	1	3	0.451569
Test 21	3	1	3	2	3	2	1	3	2	1	3	2	1	2.626360
Test 22	3	2	1	3	1	3	2	2	1	3	3	2	1	0.417362
Test 23	3	2	1	3	2	1	3	3	2	1	1	3	2	2.697961
Test 24	3	2	1	3	3	2	1	1	3	2	2	1	3	1.339399
Test 25	3	3	2	1	1	3	2	3	2	1	2	1	3	2.665314
Test 26	3	3	2	1	2	1	3	1	3	2	3	2	1	1.008765
Test 27	3	3	2	1	3	2	1	2	1	3	1	3	2	0.411107

Table 9. Range analysis of orthogonal optimization for bridge-shift strategy.

Factor	T_1/ms	T_2/ms	K_1/ms	T_4/ms	T_5	K_2	ΔT_1	ΔT_2	ΔT_3	T_s	T_{d1}	T_{d2}	e_t
1	1.843	1.726	1.758	1.787	1.553	1.545	1.492	1.573	1.432	2.612	1.604	1.525	1.530
2	1.535	1.613	1.471	1.543	1.644	1.722	1.755	1.541	1.722	1.378	1.711	1.692	1.663
3	1.443	1.483	1.593	1.492	1.624	1.555	1.575	1.707	1.667	0.832	1.506	1.605	1.628
Range	0.400	0.243	0.287	0.295	0.091	0.177	0.263	0.166	0.290	1.780	0.205	0.167	0.133

According to the results of the orthogonal range analysis, when switching from HM_1 to HM_2, the degree of influence of each factor with the bridge-shift method was ranked as follows: the reverse starting point T_s, time T_1, time T_4, time difference ΔT_3, current K_1, time difference ΔT_1, time T_2, the reverse duration T_{d1}, the reverse duration T_{d2}, current K_2, time difference ΔT_2, displacement ratio e_t, and time T_5. The optimum level combination was $A_3B_3C_2D_3E_1F_1G_1H_2I_1J_3K_3L_1M_1$, and after substituting the parameters into the simulation model, the various indicators for the bridge-shift method were obtained as follows: the speed drop was 16.035 r/min (no power interruption), the peak acceleration was 0.385807 m/s^2, the power loss of clutch C_2 was 17.5495 kW, the sliding friction work of clutch C_2 was 6.8700 kJ, the power loss of clutch C_4 was 4.6241 kW, and the sliding friction work of clutch C_4 was 0.5775 kJ.

The shifting results under the two control strategies are shown in Figure 11. Compared to the direct-shift method, the shifting quality of the tractor based on the bridge-shift method was greatly improved: the speed drop was reduced by 47.72%, the peak acceleration was increased by 0.33% (which can be ignored), the power loss of clutch C_2 was reduced by 22.82%, the sliding friction work of clutch C_2 was reduced by 14.92%, the power loss of clutch C_4 was reduced by 74.48%, and the sliding friction work of clutch C_4 was reduced by 75.84%.

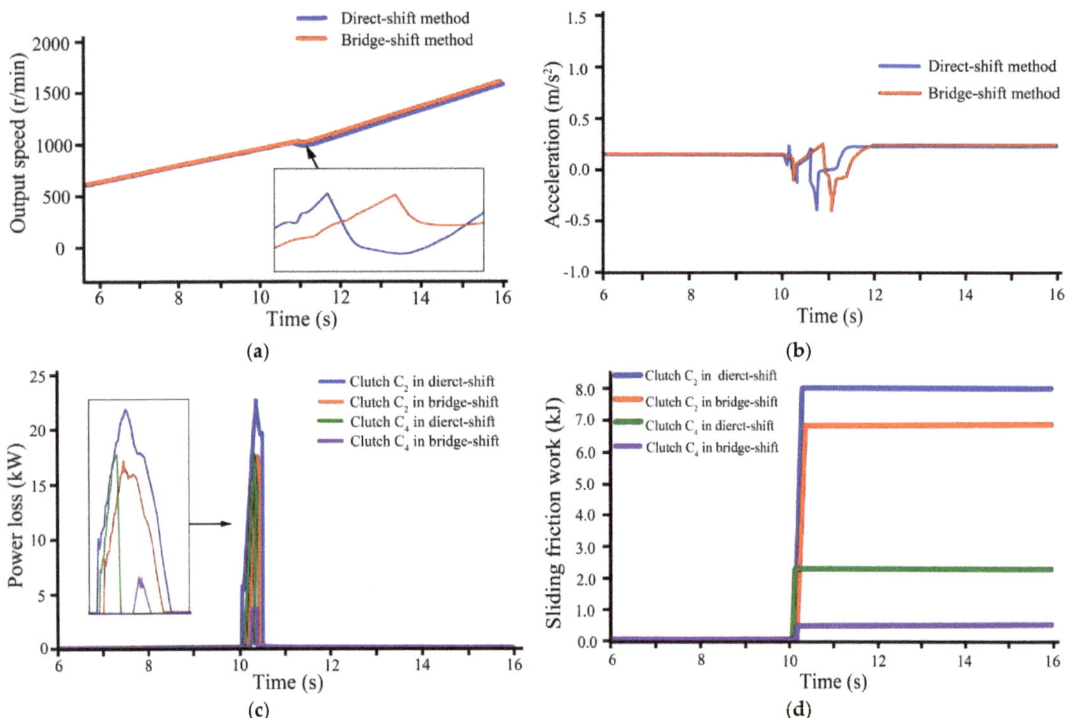

Figure 11. Comparison of shifting quality under different control strategies. (**a**) Output speed of the transmission during shift. (**b**) Acceleration of the tractor during shift. (**c**) Power loss of the clutches during shift. (**d**) Sliding friction work of the clutches during shift.

3.4. Discussion

Due to the involvement of multiple clutch actions, the generation of parasitic power is inevitable. To determine the direction of the power flow, we observed the power loss of all the clutches during shifting, as shown in Figure 12a,b. The figure shows that in the direct-shift method, parasitic power mainly flowed back along clutch C_3, while in the bridge-shift method, parasitic power mainly flowed back along clutch C_1. However, compared to the power flowing in the forward direction in clutches C_2 and C_4, the parasitic power generated under both shifting strategies was not significant, so the energy loss of clutches C_1 and C_3 was not discussed in this study.

Tractors can operate within a larger range of loads, so it was necessary to further analyze the shifting quality of the tractor under different tractive forces, as shown in Figure 13.

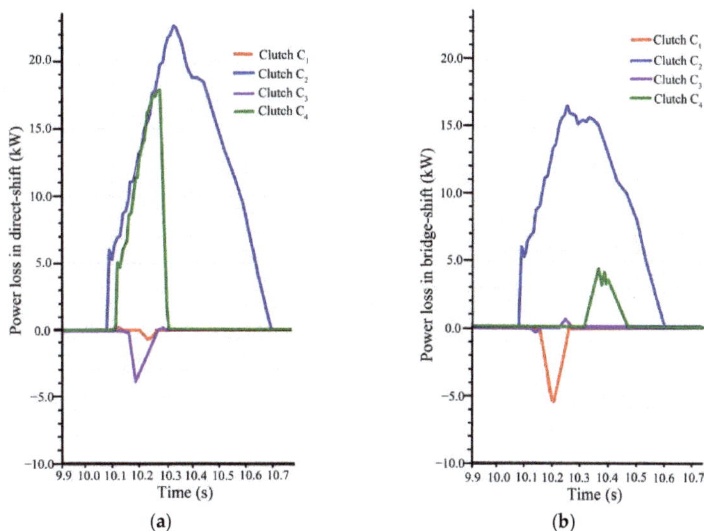

Figure 12. Power loss of clutches. (**a**) Power loss in direct-shift method. (**b**) Power loss in bridge-shift method.

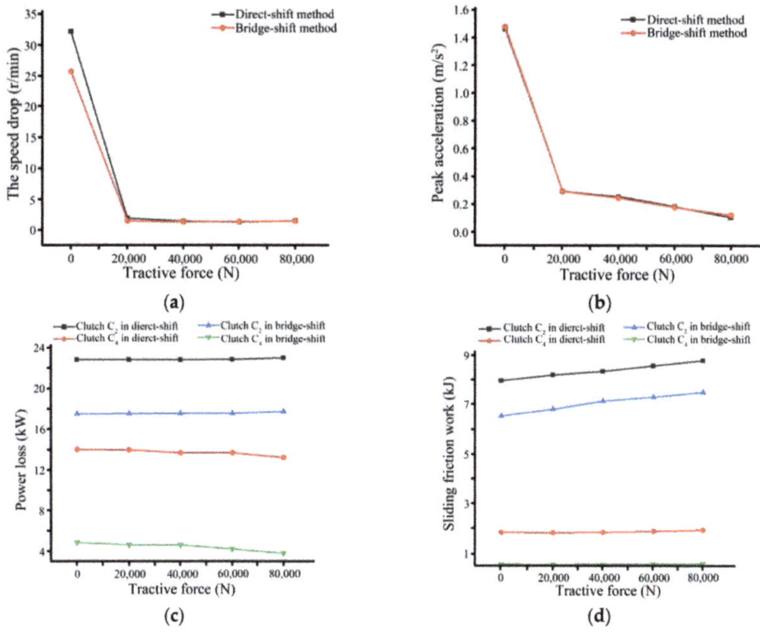

Figure 13. Shifting quality under different tractive forces. (**a**) Speed drop of the transmission during shifting. (**b**) Peak acceleration of the tractor during shifting. (**c**) Power loss of the clutches during shifting. (**d**) Sliding friction work of the clutches during shifting. Note: The clutch still loses energy under no load because the model takes into account the effects of cab air resistance and tire rolling resistance. Moreover, inertial loads can also cause energy losses.

From Figure 13a,b, it can be seen that when the tractive force was less than 20,000 N, the speed drop and peak acceleration both rapidly decreased with the increase in the load.

When the tractive force was greater than 20,000 N, the speed drop no longer changed significantly, while the peak acceleration still decreased slightly with the increase in the load. Figure 13c shows that the power loss of clutch C_4 decreased with increasing the load, while the variation law of clutch C_2 was opposite. However, the above law was very insignificant, especially since the variation in clutch C_2 with the load was very small. Figure 13d shows that the sliding friction work of clutch C_2 increased with increasing the load, while clutch C_4 showed the same law but was not significant.

In response to the above laws, we provide the following explanation: the speed impact and energy losses were mutually affected and formed a causal relationship, that is, the clutch absorbed the speed and acceleration impact of the transmission system through its sliding process. Therefore, the load reduced the speed impact, and its cost was the severe sliding of the clutch friction plates and high energy losses.

In addition, by comparing the response of the two shifting strategies to the load, it was found that bridge-shift method had a significantly lower speed drop, power loss, and sliding friction work than the direct-shift method, except for its peak acceleration, which was comparable to that of the direct-shift method. It should be further emphasized that the tractor did not experience power interruption in all the simulation results. Therefore, the bridge-shift method proposed in this study is widely applicable to various load conditions of tractors.

4. Conclusions

This study conducted a shift dynamics analysis of a hydrostatic power-split tractor transmission with a single standard planetary gear set. Two power-shift strategies were proposed and compared, and the conclusions obtained are as follows:

(1) The degree of influence of each factor with the direct-shift method is ranked as follows: the reverse starting point T_s, the time difference ΔT, time T_2, current K_1, current K_2, time T_5, time T_4, time T_1, and the reverse duration T_d. The best combination of factors is $A_3B_4C_3D_1E_1F_4G_3H_3I_1$.

(2) The degree of influence of each factor with the bridge-shift method is ranked as follows: the reverse starting point T_s, time T_1, time T_4, time difference ΔT_3, time difference ΔT_1, time T_2, the reverse duration T_{d1}, time difference ΔT_2, displacement ratio e_t, swash plate axial piston unit's reversal start point, current K_2, the reverse duration T_{d2}, and time T_5. The optimum level combination is $A_3B_3C_2D_3E_1F_1G_1H_2I_1J_3K_3L_1M_1$.

(3) Compared with the direct-shift method, the bridge-shift method reduces the speed drop by 47.72%, the power loss of clutch C_2 by 22.82%, the sliding friction work of clutch C_2 by 14.92%, the power loss of clutch C_4 by 74.48%, and the sliding friction work of clutch C_4 by 75.84%. In addition, the influence of the two control strategies on the peak acceleration can be ignored.

(4) Under different tractive forces, the quality of the bridge-shift method is better than that of the direct-shift method, and no power interruption phenomenon was observed in all the simulation calculations.

Author Contributions: Conceptualization, G.W.; formal analysis, Z.X.; investigation, J.W., Y.Y. and G.W.; resources, G.W.; data curation, Z.X.; writing—original draft preparation, Z.X., G.W., J.W. and Y.Y.; writing—review and editing, S.F. and G.W.; visualization, Z.X.; supervision, G.W.; project administration, G.W.; funding acquisition, G.W. All authors have read and agreed to the published version of the manuscript.

Funding: This research was funded by the Shandong Provincial Natural Science Foundation, grant number ZR2020QE163, and the Shandong Provincial Key Research and Development Program, grant number 2018GNC112008.

Institutional Review Board Statement: Not applicable.

Data Availability Statement: The data presented in this study are available on request from the corresponding author. The data are not publicly available due to ongoing research.

Acknowledgments: We thank Wanqiang Chen and Yue Song for their contributions to the development and testing of the first-generation prototype of the power-split CVT.

Conflicts of Interest: The authors declare no conflict of interest.

References

1. Guo, X.; Vacca, A. Advanced design and optimal sizing of hydrostatic transmission systems. *Actuators* **2021**, *10*, 243. [CrossRef]
2. Dario, R.; Macro, C. Power losses in power-split CVTs: A fast black-box approximate method. *Mech. Mach. Theory* **2018**, *128*, 528–543. [CrossRef]
3. Kim, J.Y.; Bae, D.S. Development of 3D dynamic and 1D numerical model for computing pulley ratio of chain CVT transmission. *Int. J. Auto. Technol.* **2022**, *23*, 1045–1053. [CrossRef]
4. Chen, Y.; Cheng, Z.; Qian, Y. Fuel consumption comparison between hydraulic mechanical continuously variable transmission and stepped automatic transmission based on the economic control strategy. *Machines* **2022**, *10*, 699. [CrossRef]
5. İnce, E.; Güler, M.A. On the advantages of the new power-split infinitely variable transmission over conventional mechanical transmissions based on fuel consumption analysis. *J. Clean. Prod.* **2020**, *244*, 118795. [CrossRef]
6. Zhang, M.; Wang, J.; Wang, J.; Guo, Z.; Guo, F.; Xi, Z.; Xu, J. Speed changing control strategy for improving tractor fuel economy. *Trans. Chin. Soc. Agric. Eng.* **2020**, *36*, 82–89.
7. Wang, G.; Zhao, Y.; Song, Y.; Xue, L.; Chen, X. Optimizing the fuel economy of hydrostatic power-split system in continuously variable tractor transmission. *Heliyon* **2023**, *9*, e15915. [CrossRef]
8. Renius, K.T. *Fundamentals of Tractor Design*; Springer Nature: Cham, Switzerland, 2020.
9. Renius, K.T.; Resch, R. Continuously variable tractor transmissions. In Proceedings of the 2005 Agriculture Equipment Technology Conference, Louisville, KY, USA, 14–16 February 2005.
10. Xia, Y.; Sun, D.; Qin, D.; Zhou, X. Optimisation of the power-cycle hydro-mechanical parameters in a continuously variable transmission designed for agricultural tractors. *Biosyst. Eng.* **2020**, *193*, 12–24. [CrossRef]
11. Liu, F.; Wu, W.; Hu, J.; Yuan, S. Design of multi-range hydro-mechanical transmission using modular method. *Mech. Syst. Signal Process.* **2019**, *126*, 1–20. [CrossRef]
12. Wang, J.; Xia, C.; Fan, X.; Cai, J. Research on transmission characteristics of hydromechanical continuously variable transmission of tractor. *Math. Probl. Eng.* **2020**, *2020*, 6978329. [CrossRef]
13. Li, B.; Pan, J.; Li, Y.; Ni, K.; Huang, W.; Jiang, H.; Liu, F. Optimization method of speed ratio for power-shift transmission of agricultural tractor. *Machines* **2023**, *11*, 438. [CrossRef]
14. Wang, J.; Xia, C.; Fan, X.; Cai, J. Research on the influence of tractor parameters on shift quality, based on uniform design. *Appl. Sci.* **2022**, *12*, 4895. [CrossRef]
15. Li, B.; Ni, K.; Li, Y.; Pan, J.; Huang, W.; Jiang, H.; Liu, F. Control strategy of shuttle shifting process of agricultural tractor during headland turn. *IEEE Access* **2023**, *11*, 38436–38447. [CrossRef]
16. Bao, M.; Ni, X.; Zhao, X.; Li, S. Research on the HMCVT gear shifting smoothness of the four-speed self-propelled cotton picker. *Mech. Sci.* **2020**, *11*, 267–283. [CrossRef]
17. Chen, Y.; Qian, Y.; Lu, Z.; Zhou, S.; Xiao, M.; Bartos, P.; Xiong, Y.; Jin, G.; Zhang, W. Dynamic characteristic analysis and clutch engagement test of HMCVT in the high-power tractor. *Complexity* **2021**, *2021*, 8891127. [CrossRef]
18. Iqbal, S.; Al-bender, F.; Ompusunggu, A.P.; Pluymers, B.; Desmet, W. Modeling and analysis of wet friction clutch engagement dynamics. *Mech. Syst. Signal Process.* **2015**, *60–61*, 420–436. [CrossRef]
19. Wang, G.; Xue, L.; Zhu, Y.; Zhao, Y.; Jiang, H.; Wang, J. Fault diagnosis of power-shift system in continuously variable transmission tractors based on improved Echo State Network. *Eng. Appl. Artif. Intel.* **2023**, *in press*.
20. Xiang, Y.; Li, R.; Brach, C.; Liu, X.; Geimer, M. A novel algorithm for hydrostatic-mechanical mobile machines with a dual-clutch transmission. *Energies* **2022**, *15*, 2095. [CrossRef]
21. Li, G.; Görges, D. Optimal control of the gear shifting process for shift smoothness in dual-clutch transmissions. *Mech. Syst. Signal Process.* **2018**, *103*, 23–38. [CrossRef]
22. Li, J.; Dong, H.; Han, B.; Zhang, Y.; Zhu, Z. Designing comprehensive shifting control strategy of hydro-mechanical continuously variable transmission. *Appl. Sci.* **2022**, *12*, 5716. [CrossRef]
23. Duncan, J.R.; Wegscheid, E.L. Determinants of off-road vehicle transmission 'shift quality'. *Appl Ergon.* **1985**, *16*, 173–178. [CrossRef] [PubMed]

Article

Optimization Design and Experiment for Precise Control Double Arc Groove Screw Fertilizer Discharger

Guoqiang Dun [1,*], Xingpeng Wu [2,*], Xinxin Ji [1], Wenhui Liu [2] and Ning Mao [2]

[1] Intelligent Agricultural Machinery Equipment Engineering Laboratory, Harbin Cambridge University, Harbin 150069, China; jixinxin1994@nefu.edu.cn

[2] College of Mechanical and Electrical Engineering, Northeast Forestry University, Harbin 150040, China; wyh_1996@nefu.edu.cn (W.L.); mn@nefu.edu.cn (N.M.)

* Correspondence: dunguoqiang@nefu.edu.cn (G.D.); wxp13941934871@163.com (X.W.)

Abstract: In order to solve the problem of uniform and precise fertilizer discharge, based on experimental analysis of the uneven nature of single-screw fertilizer discharge, a double arc groove screw fertilizer discharger was designed based on the principle of the half-cycle superposition of the fertilizer discharge curve. The fertilizer discharge amount and the instantaneous fertilizer discharge characteristics of the double arc groove screw fertilizer discharger were theoretically analyzed, and the factors affecting the fertilizer discharge uniformity of the double arc groove screw fertilizer discharger were obtained, taking the pitch S, arc groove radius R_p and center distance as the test factors. Using the uniformity variation coefficient and fertilization accuracy as test indexes, the experimental indicators were evaluated through a quadratic universal rotation combination design experiment with three factors and five levels. The optimal parameters were pitch S = 35 mm, arc groove radius R_p = 17 mm and center distance a = 40 mm. The fertilizer discharger was produced based on the optimal parameter combination, and a bench verification test and a comparative test were carried out. The test results show that the uniformity variation coefficient of the bench test and the relative error between the fertilization accuracy and the simulation test are 5.60% and 5.52%, respectively, and there is little difference between them, which verifies the correctness of the simulation. The comparative test results show that the uniformity variation coefficient of the optimized double arc groove screw fertilizer discharger is 7.16%, the fertilization accuracy is 3.44% and the fitting curve equation R^2 of fertilizer discharge flow is 0.998, all of which are significantly better than in the single-screw fertilizer discharger. We developed an electronic fertilizer discharge controller and conducted bench verification tests on it. The test results show that the average deviation between the measured fertilizer discharge capacity and the preset value of the double arc groove screw fertilizer discharger based on our self-developed controller is 2.78%. This fertilizer discharge device can precisely control fertilizer discharge, effectively solving the problem of uneven fertilizer discharge in single-screw fertilizer dischargers.

Keywords: single screw fertilizer discharger; double arc groove screw fertilizer discharger; precision fertilization; structural optimization; discrete element

Citation: Dun, G.; Wu, X.; Ji, X.; Liu, W.; Mao, N. Optimization Design and Experiment for Precise Control Double Arc Groove Screw Fertilizer Discharger. *Agriculture* **2023**, *13*, 1866. https://doi.org/10.3390/agriculture13101866

Academic Editor: Massimiliano Varani

Received: 27 August 2023
Revised: 18 September 2023
Accepted: 21 September 2023
Published: 24 September 2023

1. Introduction

China is one of the leading countries in fertilizer applications, but the utilization rate of domestic fertilizer is only 33% [1–3]. Precision fertilization is a method of precise and efficient fertilization according to the supply and demand relationship of crop nutrients [4]. It is of great significance to improve the fertilizer utilization rate and reduce the amount of fertilizer application [5] since the fertilizer discharger is a vital part of precision fertilization. Therefore, improving the accuracy and uniformity of the fertilizer discharged from a fertilizer discharger is significant for achieving precise fertilization [6–8].

At present, the fertilizer dischargers on the market are mainly divided into screw fertilizer dischargers and external groove wheel fertilizer dischargers. As a common fertilizer discharger, screw fertilizer dischargers have the advantages of a simple structure, adjustable fertilizer discharge amount and low price [9–11]. In recent years, relevant scholars have carried out a great deal of research on improving the precision and uniformity of fertilizer discharged from traditional single-screw fertilizer dischargers. Xue et al. [12] conducted simulation and bench experiments on the stability and uniformity of fertilizer discharge via fertilizer dischargers, and they optimized the optimal rotating speed of the fertilizer discharger. However, it is not suitable for situation where the fertilizer output is adjusted by changing the rotational speed. Kretz et al. [13] carried out simulation and bench tests on the influence of screw design parameters and installation inclination angle on the stability of the material flow rate at the screw outlet, and the optimization resulted in the flow rate at the screw outlet being more uniform. Li et al. [14] designed a vertical helical quantitative fertilizer discharger for operating environments in hilly and mountainous areas, and they optimized the structural parameters of the helical blade through theoretical calculations and experiments. With the development of computer technology, the discrete element method and its numerical simulation software EDEM have been widely used in the field of agricultural engineering. Yang et al. [15] used the Hertz–Mindlin Contact Model to simulate the fertilization process by changing the structure size of the sheave and optimized the sheave. Song et al. [16] optimized the screw conveyor mechanism based on the EDEM simulation method using a double-line helical structure at the entrance, and the pulsation peak value of the material flow at the outlet was reduced. However, the ratio between the pulsation amplitude and the average value was still significant, which is not suitable for precision fertilization scenarios. The above research is mainly aimed at single-screw precision fertilization to meet the need for uniform fertilization. However, there are few reports on the optimization of precision fertilization in double-screw fertilizer dischargers.

Our team previously studied and designed a double arc-groove screw fertilizer discharger (DAGSFD). Through optimization, it was determined that its optimal structure is a pitch of 35 mm, arc groove radius of 17.5 mm and center distance of 35 mm. But we did not explore the essence of improving the uniformity and accuracy of fertilizer discharge via the double arc groove screw fertilizer discharger. Therefore, based on the previous analysis, our team calculated and measured the fertilizer output of a single-screw fertilizer discharger (SSFD) through a combination of theoretical analysis and modeling software based on the reasons for uneven fertilizer discharge. We further optimized its factor range through EDEM simulation experiments and three-factor five-level quadratic universal rotation combination design experiments. Three-dimensional-printing technology was used to process the optimized double arc groove screw fertilizer discharger, and a bench test was carried out to verify the correctness of the simulation results. Simultaneously, we designed an electronic fertilizer discharge control system that accurately adjusts the discharge flow rate of the double arc groove screw fertilizer discharger by controlling the rotational speed. We also provide a reference for the optimization design of the double arc groove screw fertilizer discharger.

2. Materials and Methods

2.1. Design and Working Principle of Double Arc Groove Screw Fertilizer Discharger

2.1.1. Structure and Working Principle of Double Arc Groove Screw Fertilizer Discharger

The structure of the double arc groove screw fertilizer discharger is shown in Figure 1. It consists of a pair of gears, a pair of fertilizer discharge wheels, a shell and an end cover. A pair of mutually meshing gears rotate in the center to drive a pair of fertilizer discharge wheels to rotate in the center. The fertilizer enters from the fertilizer inlet due to its own gravity and the centering and mixing action of the double fertilizer discharge wheels. Under the combined action of the double fertilizer discharge wheels and the shell, the fertilizer is transported to the fertilizer outlet. The fertilizer falls into the fertilizer discharge pipe due to gravity, and the fertilizer discharge process is completed.

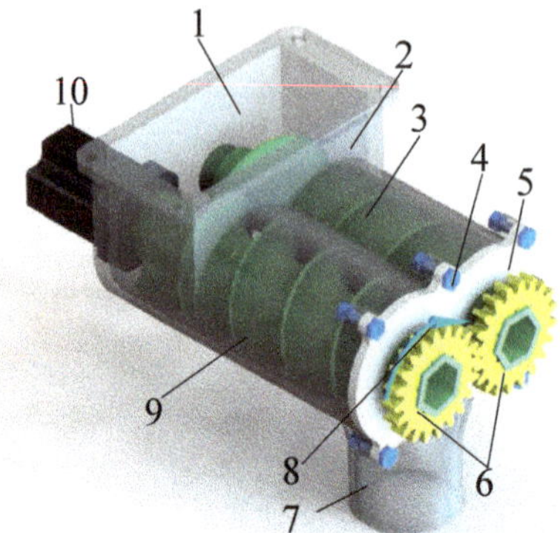

Figure 1. Schematic diagram of the double arc groove screw fertilizer discharger: 1. Fertilizer inlet. 2. Shell. 3. Left-handed fertilizer discharge wheel. 4. Bolt fertilizer discharge wheel. 5. End cover. 6. Gear. 7. Fertilizer outlet. 8. Direction of rotation. 9. Right- handed fertilizer discharge wheel. 10. Motor.

2.1.2. Research on Fertilizer Discharge Characteristics of Single-Screw Fertilizer Discharger

In order to study the transition to fertilizer discharge characteristics of the single-screw fertilizer discharger, EDEM 2018 discrete element simulation software was used to simulate and analyze the fertilizer discharge process of the single-screw fertilizer discharger (rotating speed 30 r/min [17]). We determined the parameters of the fertilizer discharge wheel according to the commonly used amount of fertilizer [18]. The large radius of the screw blade is $R = 25$ mm; the small radius of the screw blade is $r = 7.5$ mm; the pitch is $S = 35$ mm; and the thickness of the screw blade is $b = 2$ mm. The test fertilizer particles were made of Stanley compound fertilizer (average radius, 1.64 mm; density, 1.86 g/cm^3). Consulting the relevant literature [19–21], the simulation parameters are set according to Table 1.

Table 1. Global variable parameter settings.

Project	Attributes	Value
Fertilizer granules	Poisson's ratio	0.25
	Shear modulus Pa	1.0×10^7
	Density kg·m^{-3}	1861
Fertilizer discharge wheel, shell	Poisson's ratio	0.394
	Shear modulus Pa	3.18×10^8
	Density kg·m^{-3}	1240
Granules—Granules	Coefficient of restitution	0.11
	Static friction coefficient	0.3
	Rolling friction coefficient	0.1
Granules—Fertilizer discharge wheels, Housings	Coefficient of restitution	0.41
	Static friction coefficient	0.32
	Rolling friction coefficient	0.18

The change in fertilizer flow with the opening and closing of the fertilizer outlet in one rotation cycle is shown in Figure 2. By analyzing the instantaneous fertilizer flow at the fertilizer outlet during the single-cycle fertilizer discharge process, it is found that the transient fertilizer discharge amount of the fertilizer outlet changes periodically. The periodic opening and closing of the fertilizer outlet are reasons for the uneven fertilizer discharge of the screw fertilizer discharger.

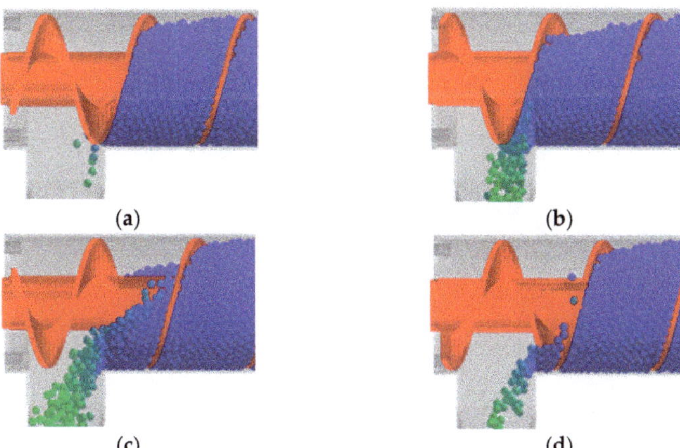

Figure 2. Opening size of fertilizer discharge outlet in different phases: (**a**) 0°, (**b**) 90°, (**c**) 180°, (**d**) 270°.

2.1.3. Design of Double Arc Groove Screw Fertilizer Discharger

The fertilizer discharge curve of the single-screw fertilizer discharger changes periodically. The fertilizer discharge curve is idealized as a sinusoidal function. According to the properties of the sinusoidal function of Formula (1), it can be known that increasing the number of fertilizer discharge wheels and adjusting the installation angle can achieve the purpose of a uniform fertilizer discharge.

Each time a fertilizer discharge curve is added, the number of corresponding transmission gears, fertilizer discharge wheels, and cavities are all increased, which reduces the reliability of the system and increases the manufacturing costs. Therefore, under comprehensive consideration, a fertilizer discharger is manufactured by superimposing two fertilizer discharge curves. However, the single-screw fertilizer discharge curve is not a strict sinusoidal curve, so it is necessary to optimize the structure of the double-screw fertilizer discharger to further improve its precise fertilization.

$$
\begin{cases}
A sint + A sin(t+\pi) = 0 \\
A sint + A sin\left(t + \frac{2}{3}\pi\right) + A sin\left(t + \frac{4}{3}\pi\right) = 0 \\
A sint + A sin\left(t + \frac{\pi}{2}\right) + A sin(t + \pi) + A sin\left(t + \frac{3}{2}\pi\right) = 0 \\
\quad \cdots\cdots
\end{cases}
\tag{1}
$$

In the formula: t—rotation time of the fertilizer discharger, s; A—coefficient of the fertilizer discharge curve, g.

2.2. Double Arc Groove Screw Fertilizer Discharger

2.2.1. Theoretical Fertilizer Discharge Amount of Double Arc Groove Screw Fertilizer Discharger

The theoretical fertilizer discharge amount of the double arc groove screw fertilizer discharger is mainly determined by the difference between the volume of the inner cavity and the volume of the fertilizer discharge wheel in a single rotation cycle (hereinafter referred to as the volume difference). The volume difference in a single rotation cycle is shown in Figure 3.

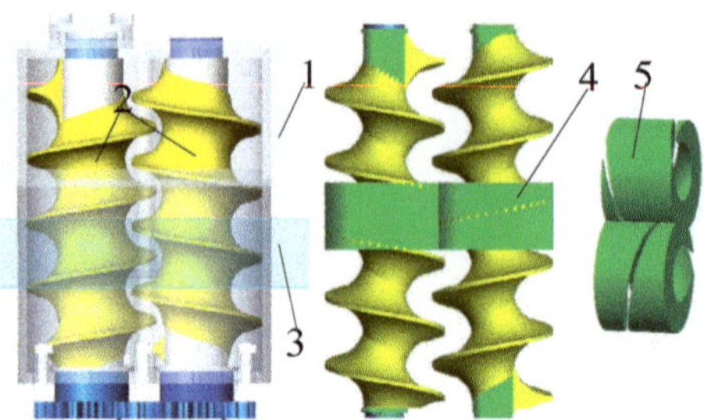

Figure 3. Schematic diagram of the difference in volume: 1. Shell. 2. Fertilizer discharge wheel. 3. Pitch. 4. Pitch length inner cavity volume. 5. Volume difference.

The traditional method for calculating volume difference is to calculate the difference between the volume of the double-screw cavity with a single pitch length and the volume of the fertilizer discharge wheel with a single pitch length. As shown in Figure 4, its volume can be calculated according to Formula (4).

$$\alpha = arccos\frac{a}{2R} \tag{2}$$

$$S_{shell} = 2\pi R^2\left(1 - \frac{\alpha}{360}\right) + \frac{R}{2}\sqrt{R^2 - \frac{a^2}{4}} - 2\pi r^2 \tag{3}$$

$$V_{shell} = S_{shell} \times S \tag{4}$$

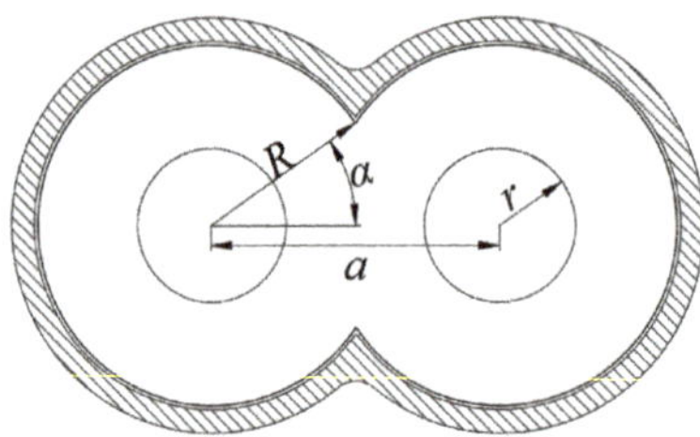

Figure 4. Schematic diagram of the inner cavity area.

In the formula: S_{shell}—circumferential cross-sectional area of shell, mm^2; R—outer radius of helical vane, mm; r—inner radius of helical vane, mm; S—pitch, mm; α—included angle of 1/2 coincidence zone, °; a —center distance, mm.

The cross-section of the screw blade of the fertilizer discharge wheel is shown in Figure 5, and its area can be divided into the sum of the volume V_1 of the black area and the volume V_2 of the shaded area.

$$V_1 = 2\int_{-\frac{S-b}{2}}^{\frac{S-b}{2}}\left(r_1 + R_p + \sqrt{R_p^2 - (x - R_p)^2}\right)^2 dx \tag{5}$$

$$V_1 = \frac{8\pi(r_1 + R_p)}{3}\left(\frac{(S-b)^2}{4} - R_p(S-b)\right)^{\frac{3}{2}} \tag{6}$$

$$V_2 = \pi\left(R_1^2 - r_1^2\right)b \tag{7}$$

$$V_{blade} = (V_1 + V_2) \tag{8}$$

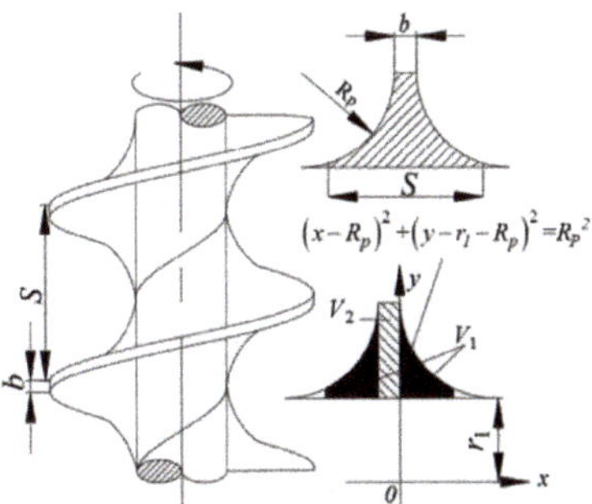

Figure 5. Screw blade section.

In the formula: V_{blade} —volume of the helical blade in one revolution, mm^3; V_1—volume of the black area in one revolution, mm^3; V_2—volume of the shadow area in one revolution, mm^3; x—horizontal arc groove curve in the xy coordinate system coordinate, mm; y—ordinate of the arc groove curve in the xy coordinate system, mm; R_p—arc groove radius, mm; b—thickness of the helical blade, mm.

Because manual calculation of theoretical fertilizer volume is cumbersome and complicated, the combined-delete command in the computer software SolidWorks was used to solve 4 (pitch length inner cavity volume) in Figure 3, the main entity; 2 (fertilizer discharge wheel) in Figure 3, the entity to be combined; and 5 in Figure 3 (volume difference) is obtained. The mass attribute (Figure 6) command is used in the evaluation to obtain the volume difference, which is simple and convenient.

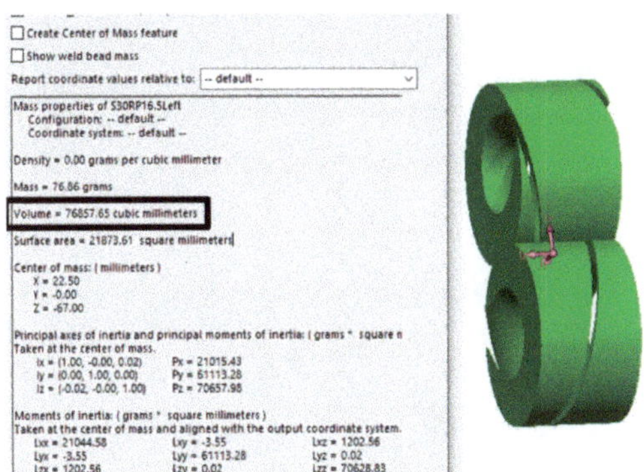

Figure 6. Volume recognition.

The theoretical fertilizer discharge amount of the double arc groove screw fertilizer discharger can be calculated according to Formula (9).

$$Q = (V_{shell} - V_{blade})\omega t \varphi \rho \tag{9}$$

In the formula: Q—fertilizer discharge amount, mm^3/s; ω—fertilizer rotating speed, r/min; ρ—fertilizer bulk density, g/cm^3; φ—fertilizer filling factor (0.7) [22].

In order to prevent the occurrence of gaps in the arc of the groove, resulting in residual fertilizer particles, and because the arc groove radius must meet the structural design constraints, as shown in Figure 5, the pitch S, the thickness of the screw blade b, and the arc groove radius R_p should satisfy Formula (10); the large and small radius s R and r of the helical blade and the arc groove radius R_p should satisfy Formula (11).

$$S \leq 2R_p + b \tag{10}$$

$$Rp \leq R - r \tag{11}$$

From Formulas (10) and (11), the arc groove radius is $16.5 \leq R_p \leq 17.5$ mm, so the experimental optimization interval of the arc groove radius R_p is set as $16.5 \leq R_p \leq 17.5$ mm.

2.2.2. Uniformity Analysis of Double Arc Groove Double-Spiral Fertilizer Discharger

Since the fertilizer granules are not fully filled in the fertilizer discharger [23], the effective fertilizer volume ΔV will fluctuate when the screw blades are in different positions, and the variation in the effective fertilizer volume ΔV determines whether the fertilizer discharge amount is uniform.

As shown in Figure 7, when there is a large and small radius R and r of the screw blade, the thickness b of the screw blade and the rotational speed ω of the fertilizer discharger are constant values. Therefore, reducing the fluctuation of effective fertilizer discharge volume ΔV can be realized by increasing the circumferential cross-sectional area S_{axial} of the fertilizer discharger and reducing the axial cross-sectional area S_{wheel} of the screw blade. The fluctuation in the effective fertilizer discharge volume ΔV can be reduced. It can be seen from Formula (12) that the circumferential cross-sectional area S_{axial} of the fertilizer discharger is only related to the center distance a, the screw pitch S, and the arc groove radius Rp.

$$S_{axial} = \frac{(S \cdot S_{shell} - V_{blade})}{S} \tag{12}$$

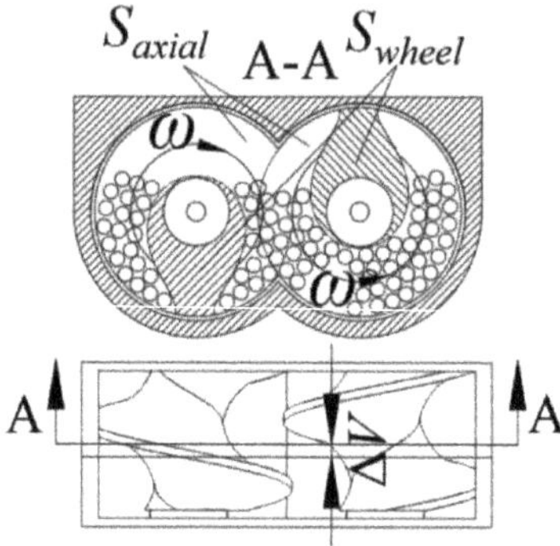

Figure 7. Circumferential section of fertilizer ejector.

In the formula: S_{axial}—the circumferential cross-sectional area of the fertilizer discharger, mm^2.

Center distance a should ensure that the screw blades coincide with each other without collision. The center distance a should satisfy Formula (13), $35 \leq \alpha \leq 40$ mm.

$$R + \frac{r}{2} \leq \alpha \leq 2R \tag{13}$$

In the studies of [11,18], it can be seen that the optimal range of the pitch S is between $(0.5{\sim}0.7)$ $2R$, but if the pitch S is too small, the rotating speed ω of the fertilizer discharger needs to be increased to increase the fertilizer discharge amount. However, the increase in the rotating speed ω of the fertilizer discharger will aggravate the wear of the fertilizer discharger and reduce the service life of the fertilizer discharger. So, the pitch S should not be too small, and $32 \leq S \leq 35$ mm should be taken into comprehensive consideration.

3. Simulation Test of Double Arc Groove Screw Fertilizer Discharger

3.1. Simulation Test

In order to determine the optimal structural parameters of the double arc groove screw fertilizer discharger, parameter optimization of the double arc groove screw fertilizer discharger is implemented through simulation experiments. As shown in Figure 8, we constructed a simulation model of the double arc groove screw fertilizer discharger and simplified its unnecessary structure. The structure is simplified, and after the simplification, the fertilizer discharger is mainly composed of five parts: the shell, fertilizer box, double-fertilizer discharge wheel, particle factory and fertilizer collection tank. The fertilizer particles used in the experiment were selected from the Stanley fertilizer particles selected in Section 2.1.2 above. The contact model between particles and the fertilizer discharger in the EDEM was set as the built-in contact model Hertz Mindlin (no slip). Two fertilizer factories are set up at the inlet of the double arc groove screw fertilizer discharger. The fertilizer particles generated in the factory are all the same. After the simulation starts, the fertilizer particles naturally fall into the fertilizer discharger along the negative Z-axis direction. Because the fertilizer has the same motion law in each fertilizer discharge cycle of the fertilizer discharger, in order to facilitate the parameter setting and the extraction of data from the simulation monitoring area, and to compare and analyze the data with the single-screw fertilizer discharger above, the rotating speed of the fertilizer discharge wheel is set to 60 r/min, the moving speed of the fertilizer discharger is 0.2 m/s, the simulation step size is 9.25×10^{-6}s and the data recording interval is 0.01 s. The total simulation process is set to 5S. We also set up a monitoring area at the fertilizer collection tank and fertilizer outlet below the fertilizer discharger to monitor the trend of the particle flow rate per second and the trend of the particle total quality changes.

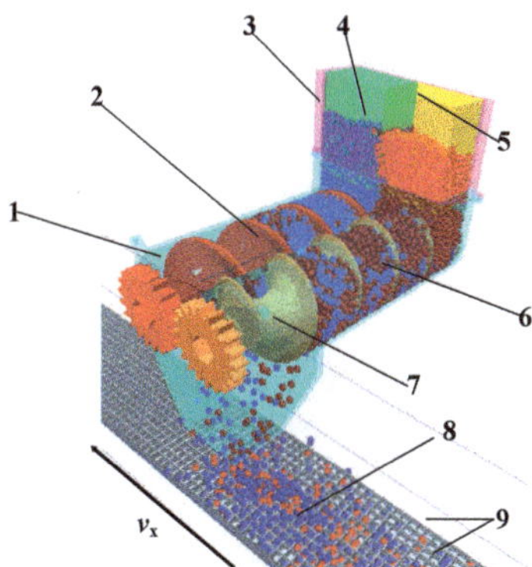

Figure 8. EDEM simulation: 1. Shell. 2. Left arc-groove-type fertilizer discharge wheel. 3. Fertilizer box. 4. Left fertilizer particle factory. 5. Right fertilizer particle factory. 6. Fertilizer particles. 7. Right arc-groove-type fertilizer discharge wheel. 8. Fertilizer collection tank 9. Monitoring area. Note: v_x is the movement direction of the fertilizer discharger.

3.2. Experimental Factors

Since the center distance, arc groove radius and screw pitch are important parameters that affect the working performance of the fertilizer discharger, they play a decisive role in the fertilization accuracy and uniformity. This studies [24,25] have adopted three factors and five levels of quadratic universal rotation combination in their design tests. The table of test factor levels is shown in Table 2. Design-expert8.0.6 software was used for data processing and statistical analysis.

Table 2. Test factor level table.

Level	Pitch S/(mm)	Center Distance a/(mm)	Arc Groove Radius R_p/(mm)
1.682	35.00	40.00	17.50
1	34.39	38.99	17.25
0	33.50	37.50	17.00
−1	32.61	36.01	16.70
−1.682	32.00	35.00	16.50

3.3. Analysis of Fertilizer Discharge Performance

To facilitate the observation of the mixing process of fertilizer particles, as shown in Figure 9, a timing chart is used to display the fertilizer discharge process of the double arc groove screw fertilizer discharger. We established a fertilizer quality monitoring area at the fertilizer outlet of the fertilizer discharger to monitor the quality of red and blue fertilizer particles. The results are shown in Figure 10. We calculated the average, standard deviation and coefficient of variation of particle mass at each monitoring point, as shown in Table 3.

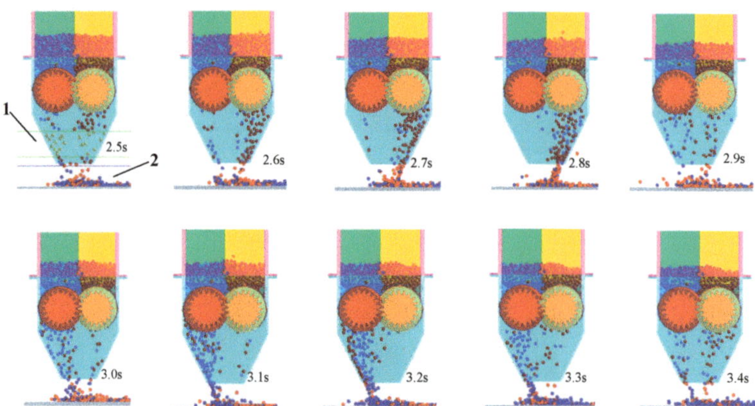

Figure 9. Time sequence diagram of working simulation status of double arc groove screw fertilizer discharger: 1. Fertilizer discharge flow monitoring area. 2. Particle quality monitoring area.

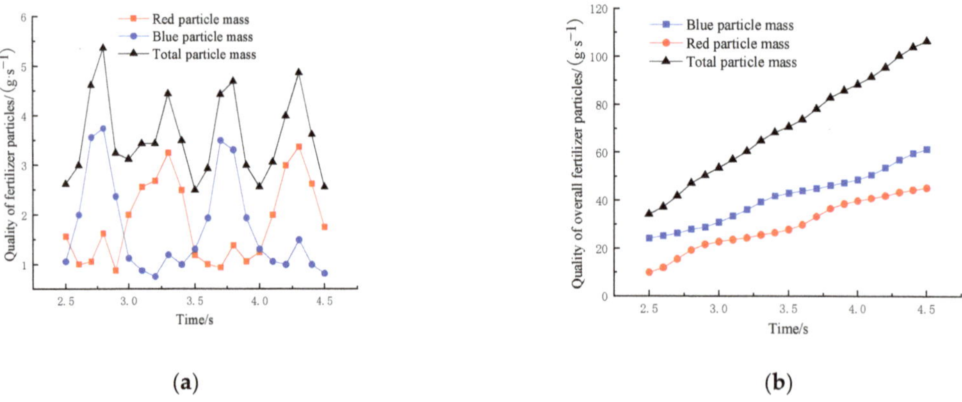

Figure 10. Particle mass change trend chart: (**a**) Trend of quality change per second; (**b**) Trend of overall quality change.

Table 3. Particle quality results.

Project	Red Particle	Blue Particle	Total Particle
average mass/g·s^{-1}	1.839	1.729	3.568
standard deviation/g·s^{-1}	0.800	0.967	0.837
coefficient of variation/%	43.52	55.93	23.46

Figure 9 shows that the blue particles generated by the left fertilizer box and the red particles generated by the right fertilizer box are alternately discharged under the action of the left and right arc groove fertilizer spirals. As shown in Figure 10a, the mass of blue and red particles fluctuates in a wavy pattern over time. When the mass of blue particles is the highest, the mass of red particles is the lowest. When the mass of blue particles rises, the mass of red particles declines, and vice versa. Figure 10b shows that there are significant fluctuations in the total mass of both red and blue particles, and the overall particle mass fluctuation decreases significantly after the combination of the two types of particles. The results from Table 3 show that the variation coefficients of red particle mass, blue particle mass and total particle mass are 43.52%, 55.93% and 23.46%, respectively, indicating that this fertilizer discharge method ensures uninterrupted fertilizer discharge and improves the uniformity and precision of fertilizer particle discharge.

3.4. Test Indicators

The grid method [26] was used to conduct statistics on the uniformity of fertilization. As shown in Figure 11, the smaller the coefficient of variation of fertilizer uniformity between grid units, the better the uniformity of fertilization [27].

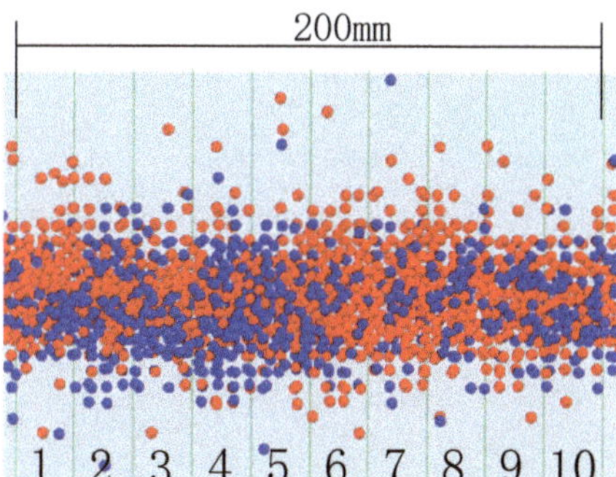

Figure 11. Schematic diagram of grid division.

In fertilizer discharge processes of different rotating speeds for any single cycle of a fertilizer discharge wheel, the fertilizer particles have the same motion characteristics. To analyze the uniformity of fertilizer discharge during a single fertilizer discharge cycle, the experimental index was determined as the coefficient of variation of fertilizer discharge uniformity, referring to NY/T1003-2006 "Technical Specification for Fertilization Machinery Quality Evaluation" [28]. According to the fertilizer discharge time of a single cycle and the total length of the grid, the forward rotating speed of the fertilizer discharger is calculated, the grid is divided into 10 parts and the quality of the fertilizer in each grid is counted. The following steps are undertaken: change the grid position, measure the quality of fertilizer particles in the grid at the same rotating speed and different fertilization time and repeat the steps three times. The uniformity coefficient of variation is obtained via Formulas (14) and (15).

$$\overline{m_u} = \frac{\sum\limits_{i=1}^{10} m_i}{n} \, (i = 1, 2, \cdots, 10) \tag{14}$$

$$\sigma_u = \sqrt{\frac{\sum\limits_{i=1}^{10}(m_i - \overline{m_u})^2}{\overline{m_u}^2(n-1)}} \times 100\% \ (i=1,2,\ldots,10) \tag{15}$$

In the formula: m_i—total mass of fertilizer particles in the i-th grid cells, g; n—number of grid cells, $n = 10$; $\overline{m_u}$—average mass of fertilizer particles in the grid cells, g; σ_u—each grid's uniformity coefficient of variation between grid cells.

The fertilization accuracy represents the difference between the theoretical fertilization amount and the actual fertilization amount. The higher the fertilization accuracy, the higher the fertilization accuracy of the fertilizer discharger [28]. The experiment was repeated three times. The fertilization accuracy is calculated using Equation (17).

$$m_{reality} = \sum\limits_{i=1}^{10} m_i (i=1,2,\cdots,10) \tag{16}$$

$$\sigma = \left| \frac{m_{theory} - m_{reality}}{m_{theory}} \right| \times 100\% \tag{17}$$

In the formula: m_{theory}—the theoretical fertilization amount in a single circle, g; $m_{reality}$—the actual simulated fertilization amount in a single circle, g; σ—the fertilization accuracy.

3.5. Test Results and Analysis

The test results are shown in Table 4, where x_1, x_2 and x_3 represent the factor code values.

Table 4. Test and results.

Serial Number	Factor Level			Test Results	
	x_1	x_2	x_3	Uniformity Coefficient of Variation y_1/%	Fertilization Accuracy y_2/%
1	−1.000	−1.000	−1.000	18.32	3.15
2	1.000	−1.000	−1.000	14.69	2.14
3	−1.000	1.000	−1.000	18.41	3.86
4	1.000	1.000	−1.000	12.46	3.63
5	−1.000	−1.000	1.000	13.95	2.28
6	1.000	−1.000	1.000	14.73	1.68
7	−1.000	1.000	1.000	12.43	2.33
8	1.000	1.000	1.000	13.01	2.58
9	−1.682	0.000	0.000	17.36	2.76
10	1.682	0.000	0.000	14.15	2.03
11	0.000	−1.682	0.000	13.37	1.85
12	0.000	1.682	0.000	12.22	3.33
13	0.000	0.000	−1.682	16.58	3.76
14	0.000	0.000	1.682	15.33	2.63
15	0.000	0.000	0.000	23.74	3.17
16	0.000	0.000	0.000	24.16	3.11
17	0.000	0.000	0.000	23.76	2.7
18	0.000	0.000	0.000	23.57	2.93
19	0.000	0.000	0.000	23.94	2.88
20	0.000	0.000	0.000	25.86	3.07

3.6. Influence of Test Factors on Test Indicators

The variance analysis of the uniform coefficient of variation of the model is shown in Table 5. The model significance test shows that $F = 61.48$, $p < 0.01$, the regression model is extremely significant and the lack of fit test results are not significant ($p > 0.05$), indicating that the regression model fits well in the test range.

Table 5. ANOVA for coefficient of variation of uniformity.

Evaluation Indicators	Source of Variance	Sum of Square	DF	Mean Square	F Value	p Value
Coefficient of variation for uniformity	Model	425.46	9	47.27	61.48	<0.0001
	x_1	13.58	1	13.58	17.66	0.0018
	x_2	3.92	1	3.92	5.09	0.0476
	x_3	10.30	1	10.30	13.40	0.0044
	x_1x_2	0.79	1	0.79	1.03	0.3335
	x_1x_3	14.96	1	14.96	19.46	0.0013
	x_2x_3	0.15	1	0.15	0.20	0.6668
	x_1^2	119.05	1	119.05	154.83	<0.0001
	x_2^2	221.53	1	221.53	288.11	<0.0001
	x_3^2	113.26	1	113.26	147.31	<0.0001
	Residual	7.69	10	0.77		
	Lack of fit	4.07	5	0.81	1.12	0.4510
	Pure error	3.62	5	0.72		
	Total variation	433.15	19			

For the uniform coefficient of variation model, x_1, x_3, x_1x_3, x_1^2, x_2^2 and x_3^2 had extremely significant effects on the equation ($p < 0.01$), and x_2 had a significant effect on the equation ($0.01 < p < 0.05$). The other factors had no effect on the equation ($p > 0.05$), and the factors that had no significant effect on the coefficients in the regression equation were excluded. The regression equation of each factor and the coefficient of variation of uniformity are the following:

$$y_1 = -1265.39 + 153.28x_1 + 132.71x_2 + 902.71x_3 + 5.16x_1x_3 - 3.61x_1^2 - 1.77x_2^2 - 31.72x_2^2 \tag{18}$$

The variance analysis of the fertilization accuracy model is shown in Table 6. In the model significance test, $F = 32.62$, $p < 0.01$, the regression model is extremely significant and the lack of fit test result is not significant ($p > 0.05$), indicating that the regression model fits well in the test range.

Table 6. ANOVA for fertilization accuracy.

Evaluation Indicators	Source of Variance	Sum of Square	DF	Mean Square	F Value	p Value
Fertilization accuracy	Model	6.99	9	0.78	32.62	<0.0001
	x_1	0.58	1	0.58	24.41	0.0006
	x_2	2.33	1	2.33	97.75	<0.0001
	x_3	2.47	1	2.47	103.78	<0.0001
	x_1x_2	0.33	1	0.33	13.94	0.0039
	x_1x_3	0.10	1	0.10	4.16	0.0688
	x_2x_3	0.20	1	0.20	8.20	0.0169
	x_1^2	0.62	1	0.62	25.86	0.0005
	x_2^2	0.27	1	0.27	11.49	0.0069
	x_3^2	0.08	1	0.08	3.50	0.0908
	Residual	0.24	10	0.02		
	Lack of fit	0.09	5	0.02	0.57	0.7252
	Pure error	0.15	5	0.03		
	Total variation	7.23	19			

For the fertilization accuracy model, x_1, x_2, x_3, x_1x_2, x_1^2 and x_2^2 had extremely significant effects on the equation ($p < 0.01$), and x_2x_3 had a significant effect on the equation ($0.01 < p < 0.05$). The other factors had no effect on the equation ($p > 0.05$), and the factors that had no significant effect on the coefficients in the regression equation were excluded.

The regression equation of each factor and the coefficient of variation of uniformity are as follows:

$$y_2 = -402.49 + 12.06x_1 + 6.07x_2 + 11.83x_3 + 0.15x_1x_2 - 0.35x_2x_3 - 0.27x_1{}^2 - 0.07x_2{}^2 \quad (19)$$

By analyzing the response (Figure 12) of the pitch and the arc groove radius to the fertilization accuracy, it can be seen that when the pitch S is at a low level, the uniformity variation coefficient y_1 first increases and then decreases with the increase in the arc groove radius R_p. When the pitch S is at a high level, the uniformity variation coefficient y_1 first increases and then decreases with the increase in the arc groove radius R_p.

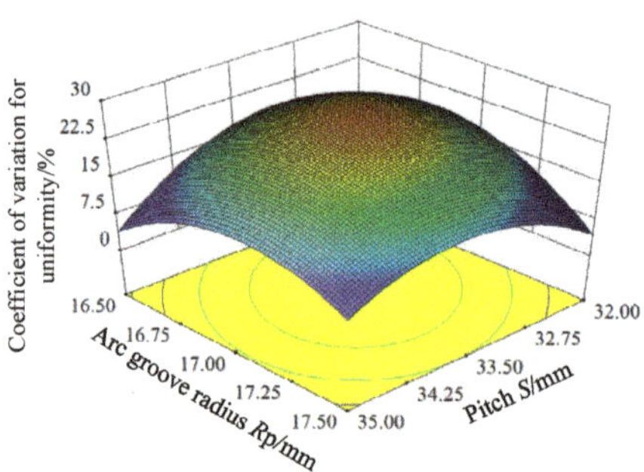

Figure 12. Response surface of arc groove radius and center distance to uniformity variation coefficient.

By analyzing the response (Figure 13) of the center distance and the pitch to the fertilization accuracy, it can be seen that when the center distance a is at a low level, the fertilization accuracy y_2 decreases with the increase in the pitch S, and when the center distance a is at a high level, the fertilization accuracy y_2 increases with the increase in the pitch S, which first increases and then decreases.

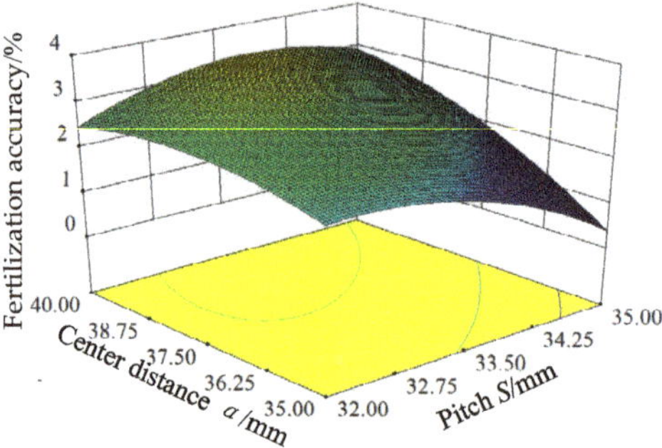

Figure 13. Response surface of center distance and screw pitch on fertilization accuracy.

When the pitch S is at a high level, the fertilization accuracy y_2 increases with the center distance a; when the pitch S is at a low level, the fertilization accuracy y_2 increases with the center distance, first as an increase and then a decrease.

By analyzing the response (Figure 14) of the center distance and the arc groove radius on the fertilization accuracy, it can be seen that when the center distance a is at a low level, the fertilization accuracy y_2 slightly increases with the increase in the arc groove radius R_p; when a is at a high level, the fertilization accuracy y_2 increases with the increase in the arc groove radius R_p.

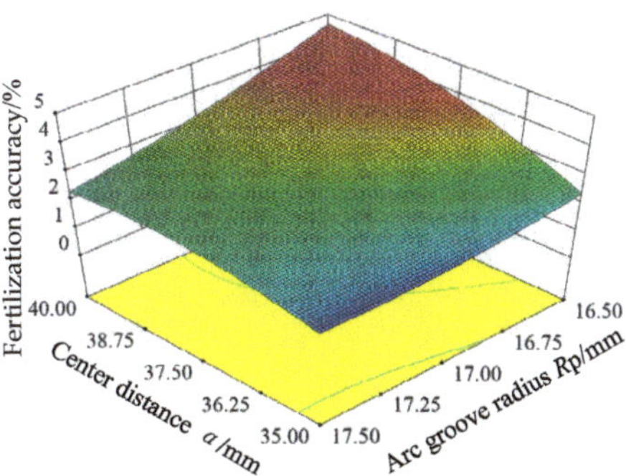

Figure 14. Response surface of arc groove radius and center distance on fertilization accuracy.

When the arc groove radius R_p is at a high level, the fertilization accuracy y_2 first increases and then decreases slightly with the increase in the center distance a; when the arc groove radius R_p is at a low level, the fertilization accuracy y_2 increases as the center distance a increases.

When the center distance a is large, the fertilizer discharger is equivalent to two independent fertilizer discharge systems. Therefore, the screw blade has little effect on fertilizer mixing and high fertilization accuracy. When the center distance is small, the overlapping screw blades are equivalent to the screw blades with half the pitch reduced, so the fertilization accuracy is high.

When the arc groove radius R_p is at a low level, the uniformity variation coefficient y_1 first increases and then decreases slightly with the increase in the pitch S; when the arc groove radius R_p is at a high level, the uniformity variation coefficient y_1 increases as the pitch S increases.

3.7. Parameter Optimization

Since the material of the fertilizer discharger is mostly plastic, under the same fertilizer volume, the larger the center distance is, the larger the fertilizer volume is, so the smaller the required rotation speed and the smaller the wear of the fertilizer discharger, meaning that the center distance is set to 40 mm. In order to obtain a parameter combination of the best combination of fertilizer spreaders when the center distance is the largest, a multi-objective optimization method is used in Design-Expert. The objective functions y_1, y_2 are Equations (18) and (19), and the optimization equation is shown in Equation (20):

$$\begin{cases} 32 \leq x_1 \leq 35 \\ x_2 = 40 \\ 16.5 \leq x_3 \leq 17.5 \\ y_1 = f(x_1, x_2, x_3) \\ y_1 \leq 10\% \\ y_2 \leq 3\% \end{cases} \tag{20}$$

Based on the above optimization (Equation (20), the yellow shadow optimization interval is obtained, as shown in Figure 15. The optimization interval obtained is that when the pitch is within a range of 32–35 mm and the arc groove radius is within a range of 16.83–17.5 mm, the fertilization accuracy is less than 3%, and the coefficient of variation of uniformity is less than 10%. The larger the pitch and the larger the arc groove radius, the smaller the volume of the fertilizer discharge wheel and the lower the manufacturing costs. Based on the optimization results, the selected parameter combination is that the center distance a is 40 mm, the pitch S is 35 mm and the arc groove radius R_p is 17 mm.

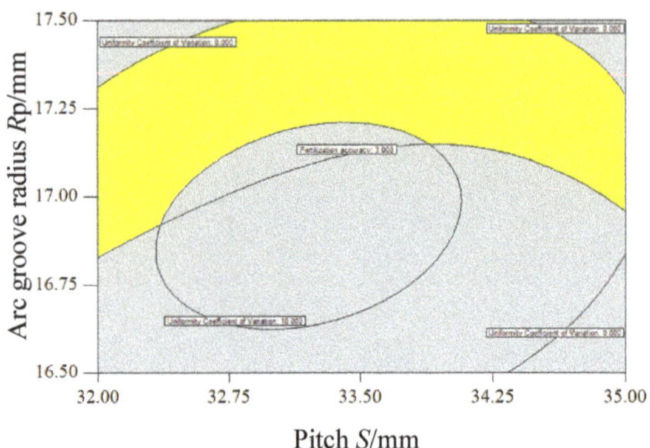

Figure 15. Parameter optimization analysis.

4. Bench Verification and Comparison Test

4.1. Bench Test

The experiment was carried out in July 2021 at the 200 Experimental Center of the Banking School of Northeast Forestry University in Heilongjiang Province. In this experiment, the fertilizer granules used Stanley compound fertilizer (average radius 1.64 mm, density 1.86 g/cm^3), and the parameters were a center distance a of 40 mm, a pitch S of 35 mm and an arc groove radius R_p of 17 mm. In a range of 30~105 r/min (excluding the case of low rotating speed), the precise fertilization performance of different types of fertilizer discharger was tested under the condition of a gradient of 15 r/min. After the fertilizer discharge was stable, the motor was started to control the conveyor belt to move at a speed of 0.2 m/s. The length of the fertilizer collecting box was 20 mm. The fertilizer discharger collects 10 copies in a single discharge cycle. The fertilizer was weighed using a precision electronic scale. We replaced the single-screw fertilizer for the comparative test. The same method as the simulation test was used for data statistics. In the bench test, a single grid was replaced by a fertilizer collecting box with a width of 20 mm. Each group of tests was repeated five times to obtain the average value. The test is shown in Figure 16.

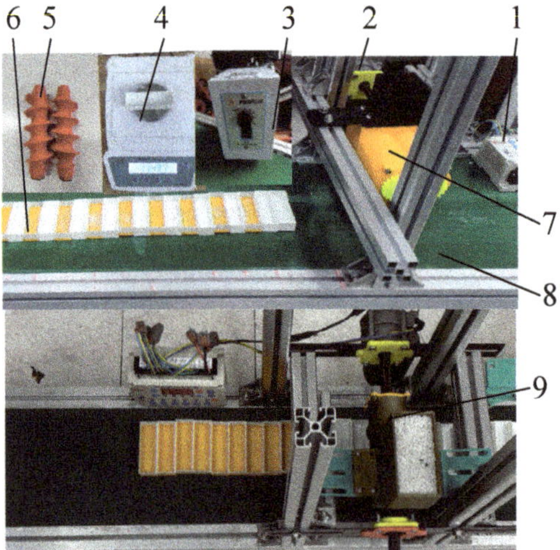

Figure 16. Bench test. 1. Motor controller. 2. Drive motor. 3. Conveyor belt controller. 4. Precision electronic scale. 5. Optimized arc groove screw fertilizer discharge wheel. 6. Fertilizer collection box. 7. Fertilizer discharger. 8. Conveyor belt. 9. Single screw fertilizer discharger.

4.2. Test Results

According to the test results (Figures 17–19), it can be seen that the optimized double arc groove screw fertilizer discharger has a fertilization accuracy of 3.15% at 60 r/min, which is a relative error of 5% compared to the optimization validation requirements. And within a rotating speed range of 30~105 r/min, the fertilization accuracy is significantly better than that of the single-screw fertilizer discharger at different speeds, and the coefficient of variation of fertilization uniformity is less than 10%, which meets the requirements of fertilization uniformity and precision listed in NY/T1003-2006 "Technical Specification for Fertilization Machinery Quality Evaluation" [28]. The fitting functional equation of the fertilizer discharge flow rate of the optimized double arc groove screw fertilizer discharger is $y = 1.084x - 5.076$, and the coefficient of determination R^2 is 0.998. The fitting equation for the fertilizer discharge flow rate of the single-screw fertilizer discharger is $y = 0.797x - 7.275$, and R^2 is 0.982. By analyzing Figures 15–17, it can be seen that the fertilizer discharge performance of the optimized double arc groove screw fertilizer discharger is better than that of the single-screw fertilizer discharger.

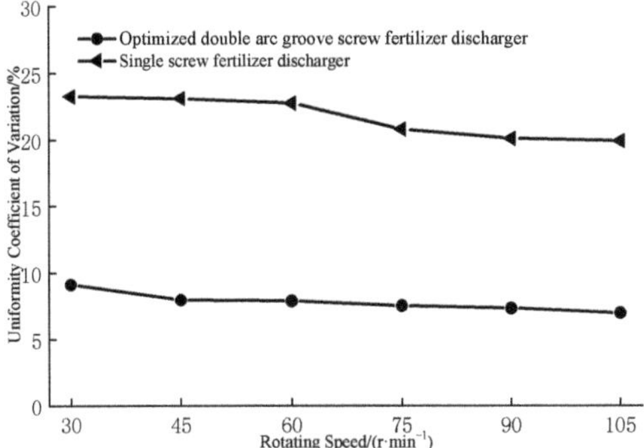

Figure 17. Uniformity coefficient of variation comparison test.

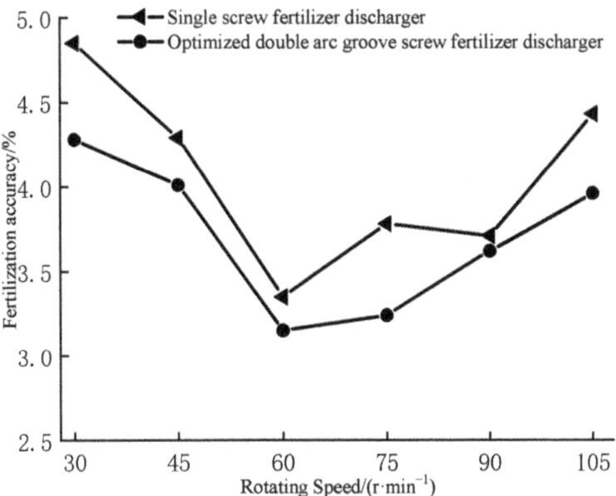

Figure 18. Fertilization accuracy comparison test.

Table 7 shows that the uniformity variation coefficient of the bench test, the uniformity variation coefficient of the simulation test and the relative error of the fertilization accuracy are 5.60% and 5.52%, respectively, indicating that the simulation optimization result is correct. Moreover, the main cause of the error is that spherical fertilizer granules are used in the simulation test, but the sphericity of the compound fertilizer in the bench test is less than 100%, and there is occasionally agglomerated fertilizer. There is an error in the distribution characteristics of fertilizer particles in the bench test compared to the simulation test, but the error is small. This proves that factors such as agglomerated fertilizer and sphericity can have a negligible impact on the optimized double arc groove screw fertilizer discharger's fertilizer discharge performance.

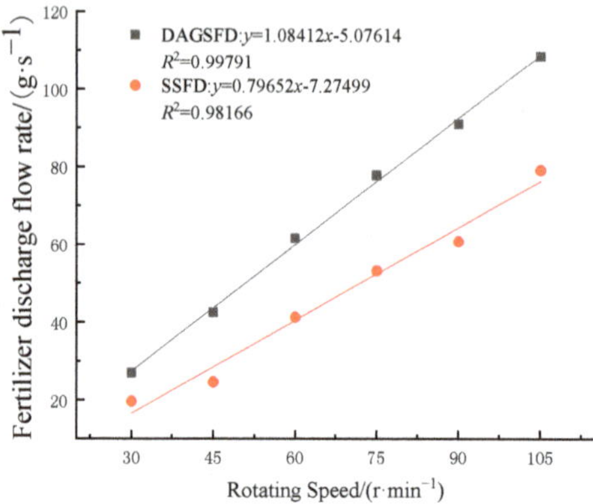

Figure 19. Fertilization discharge flow rate comparison test.

Table 7. Bench verification and comparison test results.

Fertilizer Type	Test Type	Test Index	
		Uniformity Coefficient of Variation/%	Fertilization Accuracy/%
Optimized DAGSFD	Simulation test	6.78%	3.26%
	Bench test	7.16%	3.44%
SSFD	Bench test	21.66%	4.28%

From the data in Table 7, it can be seen that the fertilization accuracy of the optimized double arc groove screw fertilizer discharger is 3.44%, and the fertilization accuracy is higher. The coefficient of variation for uniformity is reduced by 14.50% on average, and the optimized double arc groove screw fertilizer discharger has good fertilizer discharge uniformity, which effectively solves the problem of uneven fertilizer discharge of a single-screw fertilizer discharger.

To achieve precise control of the fertilizer discharge flow rate, based on equation $y = 1.084x - 5.076$, an electrically controlled fertilizer discharge system was developed and designed to adjust the speed according to the required fertilizer amount accurately. As shown in Figure 20, the electrically controlled fertilizer discharge system was tested and verified in the bench test. We randomly selected rotational speeds of 35, 49, 57, 75, 82 and 108 r/min from 30 to 105 r/min. We then carried out three-fold repetition tests. After stable fertilizer discharge, we recorded the data, measured the mass of fertilizer particles in the fertilizer collection box within 5 s, calculated the fertilizer discharge flow rate and compared it with the preset values of the electrically controlled fertilizer discharge system. The results are shown in Figure 21. Analysis shows that the average error between the discharge flow rate and the preset value of the optimized double arc groove screw fertilizer discharger controlled by the electric-controlled fertilizer discharge system is 2.78%, achieving the goal of automatically adjusting the rotational speed through the electronic control system to achieve precise fertilization.

Figure 20. Bench test of electronic fertilizer discharge system. 1. Fertilizer collection box. 2. Optimized arc groove screw fertilizer discharge wheel. 3. Fertilizer box. 4. Motor controller. 5. GPS receiver. 6. Electronic fertilizer discharge system.

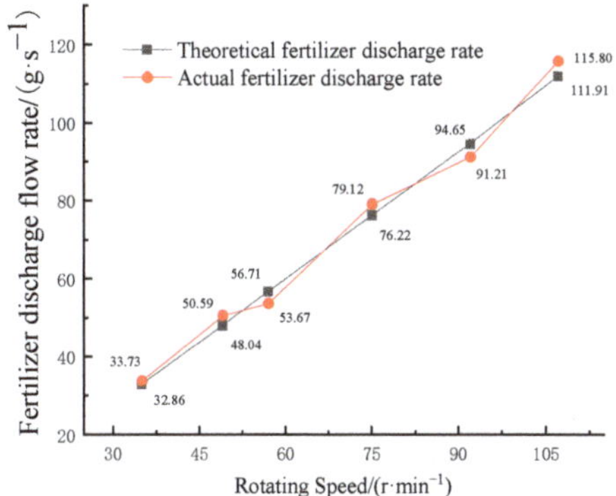

Figure 21. Comparison curve of fertilizer discharge flow rate in electric control fertilizer discharge test.

5. Conclusions

(1). Through the bench test of the single-screw fertilizer discharger, the uneven fertilizer discharge phenomenon of the single-screw fertilizer discharger was analyzed, and an improvement plan for the double arc groove screw fertilizer discharger was put forward. The instantaneous fertilizer discharge characteristics of the fertilizer discharger were theoretically analyzed, and the factors affecting the fertilizer discharge uniformity of the double arc groove screw fertilizer discharger were obtained, respectively, as the pitch S, the center distance a and the arc groove radius R_p.

(2). Taking the pitch S, the center distance a and the arc groove radius R_p as the test factors, and the uniformity coefficient of variation as the test index, the three factors and five levels of quadratic universal rotation combination design test were carried out. According to the established uniformity coefficient of variation, fertilization accuracy regression model and the use of Design-expert8.0.6 software to obtain the relationship between the influence of test factors on test indicators, the effect of the screw pitch and arc groove radius on the uniformity coefficient of variation is extremely significant ($p < 0.01$). The effect of center distance on the uniformity variation coefficient was significant ($0.01 < p < 0.05$). The effect of screw pitch, center distance and arc groove radius on the fertilization accuracy was extremely significant ($p < 0.01$), and in the center distance 40 mm, $y_1 \leq 10\%$, $y_2 \leq 3\%$. On the basis of the minimum manufacturing cost, the optimal parameters of the fertilizer discharger were optimized as a pitch S of 35 mm and an arc groove radius R_p of 17 mm.

(3). In order to verify the accuracy of the optimized analysis results, a bench verification test was carried out with Stanley compound fertilizer as the test material. The test results show that the coefficient of variation of uniformity, fertilization accuracy of bench test and the relative error of simulation test are 5.60% and 5.52%, respectively. The optimized fertilization accuracy of the double arc groove screw fertilizer discharger is 3.44%, and the fertilization accuracy is high. The coefficient of variation of the uniformity of the double arc groove screw fertilizer discharger is 14.50% lower than that of the single-screw fertilizer discharger. Moreover, the fitting equation of the fertilizer discharge flow rate of the double arc groove screw fertilizer discharger is significantly better than that of the single-screw fertilizer discharger. Therefore, an electric control fertilizer discharge system was designed based on the optimal parameter fertilizer discharge flow rate fitting equation. The bench test results show that the electric control system can accurately adjust the double arc groove screw fertilizer discharger fertilizer discharge flow rate by adjusting the rotational speed. The double arc groove screw fertilizer discharger controlled by an electronic control fertilizer discharge system has good uniformity in fertilizer discharge and a controllable fertilizer discharge flow, which effectively solves the problem of the uniformity of the fertilizer discharged by the single-screw fertilizer discharger.

Author Contributions: G.D. and X.W. designed the research. X.W., X.J., N.M. and W.L. participated in the measurements and data analysis. X.W. wrote the first draft of the manuscript. G.D. and X.W. revised and edited the final version of the manuscript. G.D. is responsible for funding acquisition. All authors have read and agreed to the published version of the manuscript.

Funding: This research was funded by the Heilongjiang Province Natural Science Foundation of China, grant number LH2023E025, and the Harbin Cambridge University Key Scientific Research Application Research Project, grant number JQZKY2022021.

Institutional Review Board Statement: Not applicable.

Data Availability Statement: Not applicable.

Conflicts of Interest: The authors declare no conflict of interest.

References

1. Gao, J.J.; Peng, C.; Shi, Q.H. Study on the High Chemical Fertilizers Consumption and Fertilization Behavior of Small Rural Househo in China: Discovery from 1995~2016 National Fixed Point Survey Data. *Manag. World* **2019**, *35*, 120–132.
2. Zhang, W.L.; Wu, S.X.; Ji, H.J.; Kolbe, H. Estimation of Agricultural Non-Point Source Pollution in China and the Alleviating Strategies III. A Review of Policies and Practices for Agricultural Non-Point Source Pollution Control in China. *Scientia Agric. Sin.* **2004**, *2004*, 1026–1033.
3. Ma, C.B. Development trend and Prospect of agricultural fertilizer application in China. *China Agric. Technol. Ext.* **2016**, *32*, 6–10.
4. Luo, X.; Liao, J.; Zang, Y.; Zhou, Z. Improving agricultural mechanization level to promote agricultural sustainable development. *Trans. Chin. Soc. Agric. Eng.* **2016**, *32*, 1–11.
5. Lan, Z.L.; Muhammad, N.K.; Tanveer, A.S. Effects of 25-year located different fertilization measures on soil hydraulic properties of lou soil in Guanzhong area. *Trans. Chin. Soc. Agric. Eng.* **2018**, *34*, 100–106.
6. Shi, Y.Y.; Chen, M.; Wang, X.C.; Mo, O.; Zhang, Y.; Ding, W. Analysis and experiment of fertilizing performance for precision fertilizer applicator in rice and wheat fields. *Trans. Chin. Soc. Agric. Mach.* **2017**, *48*, 97–103.
7. Anton, F.; Hubert, Z.; Georg, B. Mass flowmeter for screw conveyors based on capactitive secsing. In Proceedings of the 2007 IEEE Instrumentation & Measurement Technology Conference IMTC 2007, Warsaw, Poland, 1–3 May 2007; 2007.
8. Luo, S.; Zhang, X.; Xu, J.; Ma, K. Structural Optimization and Performance Simulation of Spiral Discontinuous Feeding Device. *Trans. Chin. Soc. Agric. Eng.* **2013**, *29*, 250–257.
9. Minglani, D.; Sharma, A.; Pandey, H.; Dayal, R.; Joshi, J.B.; Subramaniam, S. A review of granular flow in screw feeders and conveyors. *Powder Technol.* **2020**, *366*, 369–381. [CrossRef]
10. Van Liedekerke, P.; Tijskens, E.; Ramon, H. Discrete element simulations of the influence of fertilizer physical properties on the spread pattern from spinning disc spreaders. *Biosyst. Eng.* **2009**, *102*, 392–405. [CrossRef]
11. Yu, S.H. Optimization Design of Performance Parameters of Vertical Screw Conveyor. *Mach. Design Manuf.* **2015**, *11*, 215–218.
12. Xue, Z.; Zhao, L.; Wang, F.H.; Wang, S.; Wang, G.; Pan, R. Performance simulation test of the spiral fertilizer distributor based on discrete element method. *J. Hunan Agric. Univ.* **2019**, *45*, 548–553.

13. Kretz, D.; Callau-Monje, S.; Hitschler, M.; Hien, A.; Raedle, M.; Hesser, J. Discrete element method (DEM) simulation and validation of ascrew feeder system. *Powder Technol.* **2015**, *287*, 131–138. [CrossRef]
14. Li, X.X.; Zhao, J.; Ren, Z.Y.; Chen, C.; Wang, Y.; He, P.Y. The design of small vertical screw precision sliver. *Trans. Chin. Soc. Agric. Eng.* **2018**, *40*, 75–79.
15. Yang, Z.; Zhu, Q.C.; Sun, J.F.; Chen, Z.C.; Zhang, Z.W. Study on the performance of fluted roller fertilizer distributor based on EDEM and 3D printing. *J. Agric. Mech. Res.* **2018**, *40*, 175–180.
16. Song, H. Optimization Design and Simulation of Quantitative Screw Conveyor. Master's Thesis, Qingdao University of Science and Technology, Qingdao, China, 2016.
17. *GB/T35487-2017*; General Administration of Quality Supervision, Inspection and Quarantine of the People's Republic of China. Variable Fertilization Planter Control System; China Standard Press: Beijing, China, 2017.
18. Chen, X.F.; Luo, X.W.; Wang, Z.M.; Zhang, M.; Hu, L.; Yang, W.; Zeng, S.; Zang, Y.; Wei, H.; Zheng, L. Design and experiment of fertilizer distribution apparatus with double-level screws. *Trans. Chin. Soc. Agric. Eng.* **2015**, *31*, 10–16.
19. Hu, K.Y.; Dai, L.L.; Pi, Y.A. Theories and Calculation of the Auger-Type Conveyer. *J. Nanchang Univ. Eng. Technol.* **2000**, *22*, 29–33.
20. Lü, J.Q.; Wang, Z.M.; Sun, X.S.; Li, Z.H.; Guo, Z.P. Design and experimental study of feed screw potato planter propulsion. *J. Agric. Mech. Res.* **2015**, *6*, 194–196.
21. Yu, Y.W.; HenrikSaxe, N. Experimental and DEM study of segregation of ternary size particles in a blast furnace top bunker model. *Chem. Eng. Sci.* **2010**, *65*, 5237–5250. [CrossRef]
22. XU, H. *Mechanical Engineer's Manual*; Machinery Industry Press: Beijing, China, 2000.
23. Moysey, P.A.; Thompson, M.R. Modelling the Solids Inflow and Solids Conveying of Single-screw Extruders Using the Discrete Element Method. *Powder Technol.* **2005**, *153*, 95–107. [CrossRef]
24. Ren, L.Q. *Experiment Optimization Design and Analysis*, 2nd ed.; Higher Education Press: Beijing, China, 2003.
25. Pan, L.J. *Experimental Design and Data Processing*; Southeast University Press: Nanjing, China, 2008.
26. Yang, W.W.; Fang, L.Y.; Luo, X.W.; Li, H.; Ye, Y.Q.; Liang, Z.H. Experimental study on the effect of the parameters of the spiral fertilizer feeder on the performance of the fertilizer. *Trans. Chin. Soc. Agric. Eng.* **2020**, *36*, 1–8.
27. Sugirbay, A.; Zhao, J.; Nukeshev, S.; Chen, J. Determination of pin-roller parameters and evaluation of the uniformity of granular fertilizer application metering devices in precision farming. *Comput. Electron. Agric.* **2020**, *179*, 105835. [CrossRef]
28. Zhang, D.H. Optimization of Screw Conveyor. Master's Thesis, Dalian University of Technology, Dalian, China, 2006.

 *agriculture*

Article

Design and Experimental Testing of a Centrifugal Wheat Strip Seeding Device

Xingcheng An [1], Xiupei Cheng [1], Xianliang Wang [1], Yue Han [2], Hui Li [3], Lingyu Liu [1], Minghao Liu [1], Meng Liu [1] and Xiangcai Zhang [1,*]

[1] School of Agricultural Engineering and Food Science, Shandong University of Technology, Zibo 255000, China; 18253640804@163.com (X.A.); cheng2021@sdut.edu.cn (X.C.); wxl1990@sdut.edu.cn (X.W.); 19862513616@163.com (L.L.); liumh99@163.com (M.L.); mengliu166@163.com (M.L.)

[2] Zibo Investment Promotion Development Center, Zibo 255086, China; zbstzcjfzzx@zb.shandong.cn

[3] Shandong Academy of Agricultural Machinery Sciences, Jinan 250010, China; lihuictrc@163.com

* Correspondence: zxcai0216@sdut.edu.cn

Abstract: Wheat sowing has the characteristics of wide and short sowing periods, and there are situations in which the suitable sowing period is missed. In order to meet the needs of high-speed sowing, a centrifugal wheat strip seeding device was designed, the principle of which is that rotating parts were mainly composed of centrifugal concave plate and guide strip rotating in the shell to provide the mechanical force and drive the airflow and then realize high-speed seeding. The influence of the rotational speed of the seed discharging plate, the seed feed rate, and the dip angle of the guide strip on the distribution of the flow field and trajectory of seeds in the device was analyzed. The aerodynamic characteristics of seeds and the distribution of the gas-phase flow field inside the seed displacer under airflow were analyzed by CFD–DEM coupled simulation. The effects of three operating parameters on the coefficient of variation of sowing uniformity (CVSU) and the row-to-row seeding amount coefficient of variation (RSCV) were clarified, and the simulation results were verified by bench experiments after secondary optimization. When the centrifugal concave plate rotational speed, seed feed rate, and guide strip angle were 408 rpm, 4938 grains/s, and 69°, the results showed that CVSU and RSCV were 1.12% and 2.39%, respectively, which was in line with the standards for grain strip seeders stipulated. The designed seed discharge device can sow 3.4 ha per hour. This study provides a reference for research of centrifugal airflow-assisted high-speed seeding devices for wheat.

Keywords: wheat seeding; high-speed seeding; centrifugal; CFD–DEM coupling

Citation: An, X.; Cheng, X.; Wang, X.; Han, Y.; Li, H.; Liu, L.; Liu, M.; Liu, M.; Zhang, X. Design and Experimental Testing of a Centrifugal Wheat Strip Seeding Device. *Agriculture* **2023**, *13*, 1883. https://doi.org/10.3390/agriculture13101883

Academic Editor: Massimiliano Varani

Received: 6 September 2023
Revised: 17 September 2023
Accepted: 19 September 2023
Published: 26 September 2023

1. Introduction

The sowing period is the first part of the wheat production process, the optimization of which improves sowing speed, ensures sowing quality, and plays a very important role in increasing and stabilizing the yield of a crop. China has a large wheat planting area; the planting area of winter wheat in 2022 was about 23,518,500 hectares [1]. Sowing wheat during the appropriate sowing period can make full use of heat resources, cultivate strong seedlings, ensure the formation of robust large tillers and developed root systems, manufacture and accumulate more nutrients, and enhance resistance to adversity to improve the rate of spikes and cultivate strong stalks and large spikes to lay a sound foundation. However, the suitable sowing period for wheat is short [2], and there are cases when wheat sowing is not carried out within the suitable sowing period due to the efficiency of seed dischargers, affecting the wheat yield. At present, the wheat seeders are mainly mechanical and pneumatic. Mechanical seeders are mainly external groove wheel seeders, the advantage of which is a stable and adjustable sowing volume, and the disadvantage of which is the existence of pulsation, which can easily produce uneven sowing, crowded seedling growth, lack of seedlings, and the broken ridge phenomenon. In addition, there are also problems such as difficult seed filling, uneven seed filling, and a high leakage rate arising from high-speed seeding. The adopted pneumatic seeder is a high-speed and precision pneumatic seed discharger, and the seed filling method is mainly through the equipped fan, which rotates to generate negative pressure airflow: the seed is sucked into the seed disk hole when the seed moves to the negative pressure domain [3–5]. However, the increase in the total negative pressure required for seed filling will cause an increase in the rotational speed of the fan system in the wide operation mode, resulting in increased vibration and a substantial increase in energy consumption. At the same time, the pneumatic seed dispenser has poor stability of instantaneous speed of seed drop in the seed dispensing process, which leads to poor grain spacing [6–8].

The quality of seed sowing is related to the performance of the seed discharger. The seed filling process determines the quality and speed of seed discharge. Lei et al. [9] designed a Venturi-based airflow collector-exhaust seed discharger to improve the seed filling rate by driving the seeds into the centrifugal mixer through a fan blast. Liao et al. [10] designed a double-cylinder seed discharger to accomplish high-speed seed discharging of six rows by negative pressure filling and positive pressure discharging. This method can meet the agronomic requirements of rapeseed and wheat seeds.

During the working process of a seed discharger, the seed-to-seed and seed-to-guide strip interactions are a more complicated seed flow problem than a discontinuous medium problem. Deng et al. [11] simulated wheat seed movement characteristics under different guide strips of a centrifugal seed discharger using EDEM; Liu et al. [12] designed a cone-guided horizontal plate wheat seeder and conducted a one-way experiment on the effect of no guide strip, a straight guide strip, and an involute guide strip on the seeding effect using EDEM; Xu et al. [13] analyzed the performance of four structures of slotted-wheel seed dischargers for discharging rice seeds using EDEM simulation. At the same time, the rotation of the centrifugal concave plate and the guide strip causes airflow and the role of the airflow in the complex area of the aerodynamic characteristics of the seeds is routinely difficult to calculate. Cheng et al. [14] optimized the precision seeding performance of the seed discharger by analyzing the flow field conditions at different seed suction ports using the FLUENT 2020R2 software. Xing et al. [15] designed and optimized the negative pressure duct of the rice seed discharger and optimized the duct structure by analyzing the suction hole configuration on the duct using the FLUENT 2020R2 software.

In this study, a centrifugal wheat strip seeding device was studied to address the above problems, including the design of a centrifugal force-based combination of seed plate and dip guide wheel. The flow field characteristics of the seed discharge device were simulated following the following steps: (1) selecting reasonable parameter intervals by using ANSYS-FLUENT. (2) The motion characteristics of seeds and the consistency of seed discharge under different guide strips, rotational speeds, and seed feed rates of a centrifugal concave plate were simulated using a CFD–DEM coupled analog simulation. (3) Experiments were conducted on a seed rowing experimental bench and the simulation results were verified with data collected by a multichannel real-time weighing device, and finally design parameters were optimized to improve the performance of the system for seed discharging to meet the production requirements of high-speed seed discharge.

2. Materials and Methods

2.1. Structure and Principle of Centrifugal Wheat Strip Seeding Device

The structure of the centrifugal wheat strip seeding device is shown in Figure 1. This device consists of seed feed tubes, centrifugal concave plates, dip guide strips, seed discharge notches, a shell, shell outlets, and inoculation capsules. Among them, the dip guides make an angle of θ with the centrifugal concave plate. There is an airflow channel outside seed feed tube, and the centrifugal concave plate rotates to create a negative pressure inside seed discharger. When the seeding device works, the seed flows from the seed box into the seed feed tube and is then accelerated under the effect of gravity and negative pressure to fall onto the centrifugal concave plate in the cone angle for seed distribution. Seeds in the cone angle are evenly distributed and accelerated away from the wheat seed feed gap between seed tube and cone angle by the combined effect of the support force of guide strips and concave plate and airflow drag force. At the end of the centrifugal concave plate set seed notch, with the fluid phase of the forces, the ends of guide strips on the seed support force seeds through the shell outlet into the inoculation capsule. The seed decelerates and relieves pressure in the inoculation capsule and leaves the seed dispenser after the direction changes toward the ground, completing a uniform 12-row row. The CFD–DEM coupled [16–19] simulation simulates the movement of seeds during the seed discharge work, as shown in Figure 2.

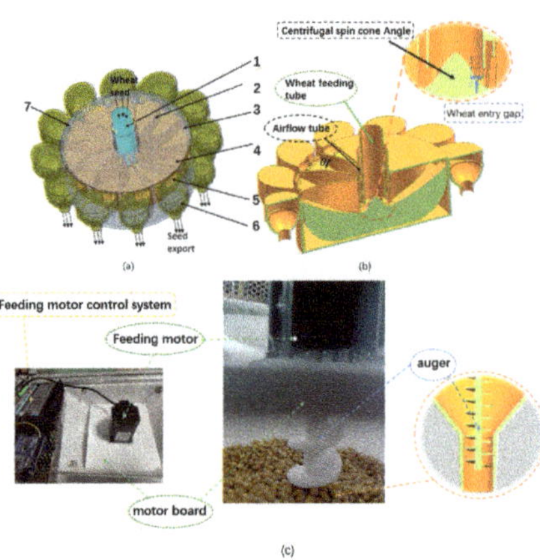

Figure 1. Structure of the centrifugal wheat strip seeding device and seed feeding device: (**a**) is a perspective view, where 1—seed feed tube; 2—dip guide strip; 3—seed outlet notch; 4—centrifugal concave plate; 5—shell outlet; 6—inoculation capsule; 7—shell; (**b**) is a sectional view; (**c**) is a seed feeding device.

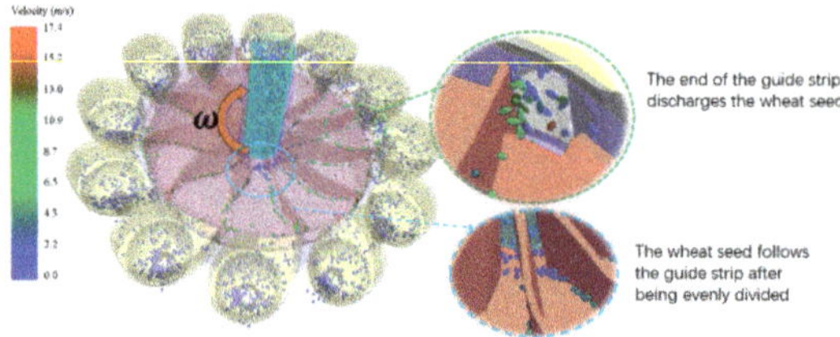

Figure 2. Schematic diagram of the seed filling and discharging process.

2.2. Mathematical Modeling Analysis of CFD–DEM Coupling

Air conveying can be categorized into dilute-phase flow and dense-phase flow based on the gas–solid mass ratio (ψ), and the ψ of the centrifugal wheat strip seeding device is less than 10%, which belongs to dilute-phase flow. The coupled CFD–DEM simulation method [20–22] was used to investigate the motion characteristics and distribution process of wheat seeds.

The gas motion in the CFD part was analyzed in the Eulerian framework of ANSYS-FLUENT and the effect of gas phase was analyzed by adding a void fraction α_g to the conservation equation. The gas-phase fluid dynamics was expressed in terms of a continuity equation and the momentum conservation equation, as shown in Equations (1) and (2) below:

$$\frac{\partial}{\partial t}\left(\rho_g \alpha_g\right) + \nabla\left(\rho_g \alpha_g \mathbf{u}_g\right) = 0 \tag{1}$$

$$\frac{\partial}{\partial t}\left(\rho_g \alpha_g u_g\right) + \nabla\left(\alpha_g \rho_g u_g u_g\right) = -\alpha_g \nabla p_g + \nabla\left(\alpha_g \tau_g\right) + \alpha_g \rho_g g - F_P \tag{2}$$

where ρ_g, u_g, p_g, t, g, α_g are gas-phase density, gas-velocity vector, gas-phase pressure, instantaneous time, gravitational acceleration, and void fraction, respectively. F_P is the gas–solid two-phase flow resistance, which is the exchange of momentum between the solid and gas phases. τ_g represents the fluid stress tensor, calculated as:

$$\tau_g = \mu_g \nabla u_g + \mu_g \left(\nabla \mu_g\right)^T \tag{3}$$

where μ_g is the dynamical viscosity, $\left(\nabla \mu_g\right)$ and $\left(\nabla \mu_g\right)^T$ are the gradient matrix of the velocity field and its transpose matrix, respectively.

In the DEM model, real-time information about the position and velocity of particles is explicitly tracked by Newton's second law of motion in a Lagrangian manner. Considering particle–particle and particle–excluder collisions, the collision force term ($F_{C,ij}$) is added for solving the momentum exchange. The translational and rotational equations of motion for a single particle i are given by:

$$m_i \frac{du_i}{dt} = m_i g + F_{d,i} + F_{Mag,i} + F_{Saff,i} + \sum_j^m F_{c,ij} \tag{4}$$

$$I_i \frac{d\omega_i}{dt} = T_i = \sum_{j=1}^{n_i} (T_t + T_r + T_n) \tag{5}$$

where m_i, u_i, and ω_i are the particle mass, translational velocity, and rotational velocity of particle i. $F_{d,i}$ is the drag force on particles in a gas–solid two-phase flow. $F_{Mag,i}$ is the Magnus lift generated by the rotation of the particles. $F_{Saff,i}$ is the Saffman lift force on particles in a gas–solid two-phase flow. $\Sigma F_{c,ij}$ is the collision force between particles and between particles and the seed discharger. n_i denotes the number of particles interacting with particle i, I_i and T_i represent the moment of inertia and total torque of particle I, respectively, T_t and T_r are rolling friction moments, generated by the moment of particle j acting on particle i and the tangential force. T_n is the moment, which is generated by the normal force when it is not directed towards the center of mass of the particle.

The gas-particle interaction is mainly calculated numerically through the momentum exchange between the gas and solid phases, where the trailing force plays an important role in the momentum exchange between the two phases. In this paper, the interaction between gas and particles is simulated by the Di Felice drag model, Equation (6):

$$F_{d,i} = \frac{1}{8}\rho_g C_d \pi d_i^2 \alpha_g{}^{2-\beta}\left(u_g - u_i\right)\left|u_g - u_i\right| \tag{6}$$

where d_i is the diameter of the wheat seed particle. $\left|u_g - u_i\right|$ is the slip velocity between the gas and the particles C_d is the drag coefficient, β is the model coefficient described by the following equation:

$$C_d = \begin{cases} \frac{24}{R_{ep}}(\text{stokes})\left(R_{ep} \leq 1\right) \\ \frac{24}{R_{ep}}\left(1 + 0.1875 R_{ep}\right)\left(1 < R_{ep} \leq 5\right) \\ \frac{24}{R_{ep}}\left(1 + 0.15 R_{ep}^{0.687}\right)\left(5 < R_{ep} \leq 800\right) \\ \frac{24}{R_{ep}}\left(1 + 0.167 R_{ep}^{0.667}\right)\left(800 < R_{ep} \leq 1000\right) \\ 0.4392\left(1000 < R_{ep} \leq 30000\right) \end{cases} \tag{7}$$

$$R_{ep} = \frac{\rho_g d_i \alpha_g \left|u_g - u_i\right|}{\mu_g} \tag{8}$$

where μ_g is the airflow viscosity and R_{ep} is the Reynolds number, which is taken as 1600 for this experiment [16]. The remaining force in Equation (4) can be expressed as:

$$F_{Mag,i} = \frac{\pi}{8} C_l \rho_g d_i^2 |u_i - u_g|^2 \frac{\omega_i - \nabla \times (u_i - u_g)}{|u_i - u_g||\omega_i - \nabla \times u_g|} \tag{9}$$

$$F_{saff,i} = 1.615 d_i^2 \sqrt{\rho_g \mu_g}(u_g - u_i) \sqrt{\left|\frac{du_g}{dy}\right|} \tag{10}$$

$$\Sigma F_{c,ij} = \Sigma F_{n,ij}^d + \Sigma F_{t,ij}^d \tag{11}$$

where C_l is an empirical coefficient, related to R_{e_i}. F_n^d is the damping force of the particles against the wall and the wheat seed; F_t^d is the tangential damping force.

2.3. Mechanical Analyses of Wheat Seeds during Operation of Seed Discharging Equipment

2.3.1. Theoretical Analysis of Wheat Seed Particle Filling Process

Seed flow from the wheat feeding tube falls on the cone angle of the centrifugal concave plate and is evenly diverted to the gap of the guide strip to achieve seed filling. Seeds at the top of the guide strip are subjected to gravity, gas-phase drag force, Magnus lift force, collision force between the seed particles, the support force of the guide strip, etc., among which the Magnus force has much less effect on the particles than the other forces, so it is ignored [23–25]. In addition, the gas-phase drag force is related to the velocity and density of fluid affected by the air pressure, and the drag force is related to the air pressure. The barometric pressure formula is as follows:

$$P_r = s\rho_g \omega arccos\theta(l_2 V_2 - l_1 V_1) \tag{12}$$

where P_r is the theoretical pressure at the inlet. s is the correction coefficient, related to the cross-sectional area between adjacent guide strips. ω is the angular velocity of the centrifugal concave plate, rpm. θ is the tilt angle of the guide strip. l_1 and l_2 are the distances of the first section of the guide strip and the end of the guide strip from the center of the axis, m, and V_1 and V_2 are the radial airflow velocities at the inlet and the outlet, m/s, respectively.

The velocity of seed particles is lower than the airflow velocity during seed filling, and the direction of trailing force F_d is the reverse of the relative motion of seed particles and airflow, N; as shown in Figure 3a, the forces in each direction on the seed particles are:

$$X: F_{N1} \sin\theta + F_c \cos\varphi = F_{saff} \sin\beta \tag{13}$$

$$Y: F_d + F_\alpha + F_{saff} \cos\beta > F_{N2} \sin\gamma + F_f + F_c \sin\varphi \tag{14}$$

$$Z: mg + F_{N1} \cos\theta = F_{N2} \cos\gamma \tag{15}$$

where F_{N1} is the support force of the guide strip on the seed, N; F_{N2} is the support force of the centrifugal concave plate on the seed, N; F_f is the friction force of seed, N; F_a is the centrifugal force of seed, N; m is the mass of seed, Kg; θ is the installation angle of the guide strip (°): φ is the angle of collision force, F_c, with the positive X direction in the XY plane (°); β is the angle of Saffman lift force with positive Y direction (°). γ is the angle between concave disk support force F_{N2} and Z axis positive direction in the YZ plane (°).

From Equations (13)–(15), the seed moves along the guide strip with variable acceleration from the end of the guide strip to the seed exit notch area.

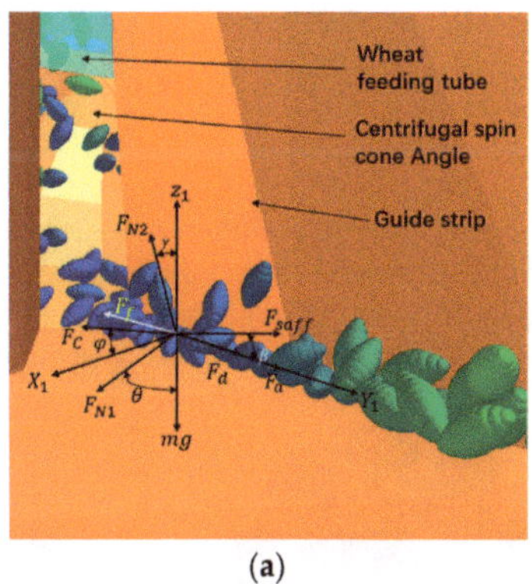

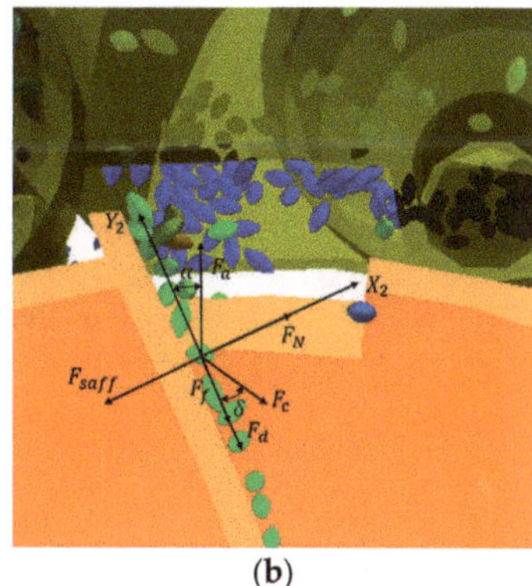

(a) (b)

Figure 3. Schematic diagram of the seed filling and discharge process: (**a**) seed filling process; (**b**) seed discharge from the centrifugal concave plate.

2.3.2. Theoretical Analyses of the Wheat Seed Particle Discharge Process

Wheat seed particles rise along the guide strip on the centrifugal concave plate due to gas-phase drag force, centrifugal force, and Saffman lift. After reaching the notch area of the seed discharge device, the gas flow rate decreases because of the pressure relief and deceleration effect of the inoculation capsule. At this time, the gas-phase resistance is in the opposite direction to the movement of the seed particles. The end of the guide is bent in the opposite direction of rotation to provide outward support to the seed, as shown in Figure 3b, and the forces in all directions on the seed particles are:

$$X : F_{saff} = F_c \sin \delta + F_N + F_a \sin \alpha \tag{16}$$

$$Y : F_a \cos \alpha \geq F_f + F_d + F_c \cos \delta \tag{17}$$

where F_N is the support force of the guide strip on the seed particles, N; δ is the angle between the collision force F_c and the forward direction of the seed in the XY plane, α is the phase angle between the centrifugal force and the forward direction of the seed (°); F_f is contact drag (the component of the combined force of gravity mg, the supporting force of the centrifugal concave plate on the seed F_{N3}, and the friction force in the Y direction in the YZ plane).

The influence of each parameter on the seed combined force during the process of wheat seed filling and discharging, in which the guide strip inclination θ affects both gas-phase drag force and support force and friction force of the guide strip on the seeds and the rotational speed ω affects gas-phase drag force and centrifugal force, was analyzed. The amount of seed per unit cross-section of the wheat seed delivery tube (i.e., seed delivery volume) affects the thickness of the seed layer, the collision force between the seed particles, and the support of seed by the guide strip. The results concluded that the rotational speed of the centrifugal concave plate, the seed feed rate, and the inclination of the guide strip had the greatest influence on the seed filling effect of the seed dispenser. Among them, the gas-phase drag force is affected by the airflow and varies in real time, so the influence of rotational speed, guide strip dip angle, and seed feed rate on the working performance of the seed discharger is investigated by coupled DEM-CFD calculation and optimization.

2.4. Materials and Methods for CFD–DEM Coupled Simulation

2.4.1. Modeling of Centrifugal Wheat Strip Seeding Device and Seed Modeling

The 3D model of the seed dispenser was drawn using the 3D drawing software UG NX 10.0. The simplified version was imported into the EDEM (v2020.2) in STP format and the rotation of the centrifugal concave plate and guide bar assembly was set by Addin Motion. A particle plant for seed generation was provided in the wheat feeding tube. The seed dispenser was assembled using 3D printing in ABS material, and the Hertz–Mindlin (no slip) standard rolling friction contact model was chosen as the contact model. The three-dimensional physical parameters of wheat seeds (Jiemai 22) were measured and modeled using the bonded particle modeling (BPM) method (Figure 4a). The static and dynamic friction coefficients between seed particles and seed discharger were verified experimentally by the stacking angle of wheat, and the modulus of elasticity was determined by the straight shearer, which had a small error and ensured the high accuracy of the simulation model. The parameters are shown in Table 1. The geometry bin group was placed at the outlet of the inoculation capsule to collect the real-time seed output of each capsule.

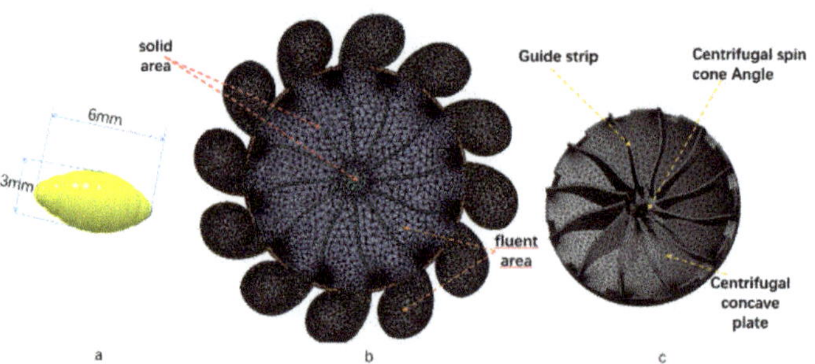

Figure 4. CFD model grid and DEM model of wheat seeds: (**a**) wheat seed modeling in EDEM; (**b**) FLUENT delineation of the model grid; (**c**) dynamic mesh through UDF-driven centrifugal concave plate and dip angle guide strip.

Table 1. Simulation parameters of ANSYS-FLUENT and EDEM.

Project	Research Target	Parameter	Numerical Value
EDEM solid phase	Wheat seed	Three-axis dimensions (mm)	$5.99 \times 3 \times 3$
		Thousand-grain weight (g)	43.6
		Poisson's ratio	0.42
		Shear modulus (PA)	5.1×10^7
		Equivalent diameter (mm)	3.83
		Densities/(Kg·m^{-3})	1.35×10^3
	Seed–seed	Crash recovery factor	0.42
		Coefficient of static friction	0.35
		Coefficient of kinetic friction	0.05
	Seed–seed dispenser	Crash recovery factor	0.75
		Coefficient of static friction	0.4
		Coefficient of kinetic friction	0.05
	Seed dispenser made of ABS	Poisson's ratio	0.384
		Shear modulus/PA	2.2×10^6
		Densities (Kg·m^{-3})	1050
ANSYS-FLUENT gas phase	Atmosphere	Gravitational acceleration (m/s^2)	9.81
		Densities (Kg·m^{-3})	1.225
		Stickiness (kg·m^{-1}·s)	1.789×10^{-5}

The UG model was imported into ANSYS Workbench through an x_t file, and the tetrahedral meshing was carried out, in which the mesh encryption at the inlet and outlet is 3 mm, the mesh encryption at the interface between the rotating body and FLUENT is 6 mm, and the rest of the area is 9 mm, as shown in Figure 4b. The number of grids is 22,593. The minimum mesh size is 3 mm and the maximum mesh size is 9 mm, both of which are three times larger than the diameter of particles. To compile the UDF, DEFINE_CG_MOTION is used to control the centrifugal concave plate and the dip guide strip (Figure 4c) to rotate around the Y axis at a set speed, defining it as the solid phase and the rest of it as the fluid phase. The fluid phase material is defined as air. The inlet of the model was set as the pressure inlet and the pressure was set to standard atmospheric pressure (101.325 kPa), the outlet was set as the pressure outlet. A k-epsilon viscous model, standard wall functions, and no-slip boundary conditions were used with a time step of 0.0002 and a step size of 20,000.

From Equations (6) and (10), it is deduced that the drag force and Saffman force in the gas phase are affected by the flow velocity, pressure of flow field, etc. Flow field simulations were performed using ANSYS-FLUENT 2020R2 software to analyze the effect of flow field on each of the flow field forces in the gas phase [26–28]. Different rotational speeds guide strip dip angles were set (Table 2), with the same inclination of different rotational speeds as a group, for a total of 5 groups of 105 tests, through the anemometer measurement experiment to verify the simulation results.

Table 2. Simulation experiments for flow field analysis of FLUENT.

Parameter	Numerical Value
Rotational speed (rpm)	200, 220, 240, 260, 280, 300, 320, 340, 360, 380, 400, 420, 440, 460, 480, 500, 520, 540, 560, 580, 600
Guide strip dip angle $(\theta)/(°)$	50, 60, 70, 80, 90

EDEM simulation experiments are carried out to study the amount of congestion of the seed feed rate at different speeds [29]. In conjunction with the speed ranges in Table 2, 21 sets of 105 experiments were carried out with different seed feeds at the same speed, with EDEM material properties as in Table 1, with a time step of 0.000001 s and a total time of 4 s.

2.4.2. CFD–DEM Coupled Seeding Simulation Experiment Based on Dense Discrete Phase Model (DDPM)

EDEM turns on the coupling switch and FLUENT selects a transient model and reads the DDPM coupling interface for CFD–DEM coupling [30]. To achieve a coupled simulation, the principle of the DDPM coupling interface is to exchange information between two mathematical models, CFD and DEM. In order to ensure calculation accuracy, after the successful connection between FLUENT and EDEM, the input mesh > reorder > rd > rd is used until the output of rd = 1 in the coding window of FLUENT, the calculation method of the coupled model is phase coupled SIMPLE, the time step is 0.00004 s, and the number of steps is 100,000. The EDEM has a time step of 0.000001 s and saves data every 0.01 s.

According to the flow distribution of FLUENT pre-experiment, combined with experimental analysis of EDEM's rotational speed and seed feed rate (Table 3), the flow field is stable at 200–600 rpm, and the seed feed does not block the seed inlet at the rate of 6000 grains/s or less. The independent variables of the Design-Expert 13 orthogonal experiment were selected to analyze the effects of three factors, namely, rotational speed (x_1), guide strip dip angle (x_2), and seed feed rate (x_3), on CVSU and RSCV through the results of three-factor, three-level orthogonal experiments, and the factors and levels of orthogonal experiments are shown in Table 4.

Table 3. Simulation experiments for seed feed analysis of EDEM.

Parameter	Numerical Value
Rotational speed (rpm)	200, 220, 240, 260, 280, 300, 320, 340, 360, 380, 400, 420, 440, 460, 480, 500, 520, 540, 560, 580, 600
Seed feed rate (grain/s)	3000, 4000, 5000, 6000, 7000

Table 4. Factors and levels of orthogonal experiments.

Level	Rotational Speed A (rpm)	Guide Strip Dip Angle B (°)	Seed Feed Rate C (Grain/s)
1	200	60	4000
2	400	70	5000
3	600	80	6000

The real-time seed output of each outlet was counted by geometry bin group at the outlet of the inoculation capsule, and the seed output of individual outlets from 1–4 s was divided into 3 equal parts by time. The seed output of each seed outlet in each period Q_{ij} ($i = 1, 2, 3, j = 1\sim12, j \in N^*$) was counted. The expressions for evaluation indicators of $CVSU$ and $RSCV$ are:

$$CVSU = \sum_{j=1}^{12} \frac{\sqrt{\frac{1}{i-1} \cdot \sum_{i=1}^{3} \left(Q_{ij} - \overline{Q_{ij}}\right)^2}}{\overline{Q_{ij}}} \times 100\% \tag{18}$$

$$RSCu = \sum_{i=1}^{3} \frac{\sqrt{\frac{1}{j-1} \sum_{j=1}^{12} \left(Q_{ij} - \overline{Q_{ij}}\right)^2}}{\overline{Q_{ij}}} \times 100\% \tag{19}$$

where $\overline{Q}_{ij}$ is the average amount of seed discharged at a particular output or at a particular time ($i = 1, 2, 3, j = 1\sim12, j \in N^*$).

3. Results and Discussion

3.1. Analysis of the Results of the Coupling Experiment

The results and evaluation metrics of the orthogonal experiment are shown in Table 5.

Table 5. Results of orthogonal experimental design and evaluation indicators.

Lab Number	Rotational Speed A (rpm)	Guide Strip Dip Angle B (°)	Seed Feed Rate C (Grain/s)	CVSU (%)	RSCV (%)
1	200	60	5000	4.78	3.25
2	600	60	5000	2.31	3.65
3	200	80	5000	10.18	3.92
4	600	80	5000	7.82	3.23
5	200	70	4000	8.00	3.98
6	600	70	4000	6.24	3.68
7	200	70	6000	6.44	3.96
8	600	70	6000	3.21	4.17
9	400	60	4000	2.66	3.58
10	400	80	4000	9.62	3.73
11	400	60	6000	1.31	3.61
12	400	80	6000	6.89	3.9
13	400	70	5000	1.03	2.17
14	400	70	5000	1.37	2.32
15	400	70	5000	1.04	2.42
16	400	70	5000	1.31	2.28
17	400	70	5000	1.26	2.23

3.2. Analysis of the Effect of Factors on CVSU

The variance results (Table 6) showed that the effects of rotational speed, guide strip dip angle, and seed feed rate were all highly significant for CVSU. The effects of three factors on the CVSU all tend to decrease and then increase with the increase in factor values, and the order of priority was B > A > C. The three factors had the same trend of influence on CVSU. For example, with a guide strip dip angle of 60° and a seed feed rate of 5000 grains/s, the rotational speed was increased from 200 rpm to 600 rpm, and the

CVSU decreased from 4.78% to 1.82% and then rose from 1.22% to 2.31%, which on the one hand means that the time for the end of the guide strip to sweep over the two adjacent seed outlets becomes shorter with the increase in rotational speed and on the other hand means that the increase in rotational speed also accelerates the air-phase force, which was dominated by a drag force, as shown by Equation (13) and CFD–DEM coupling results, which leads to change in flow field and the shell seed outlet, as indicated by the coupling results as well.

Table 6. Analysis of variance of three factors of orthogonal experiments.

Norm	Source	F-Value	p-Value	Significant
CVSU (%)	A—Rotational speed	180.72	<0.0001	***
	B—Guide strip dip angle	1030.54	<0.0001	***
	C—Seed feed rate	140.87	<0.0001	***
	AB	0.0454	0.8374	
	AC	8.10	0.0248	*
	BC	7.14	0.0319	*
	Lack of fit	4.95	0.0782	Not significant
	Primary and secondary factors B > A > C			
RSCV (%)	A—Rotational speed	5.91	0.0454	*
	B—Guide strip dip angle	8.36	0.0233	*
	C—Seed feed rate	13.29	0.0082	**
	AB	49.72	0.0002	***
	AC	23.03	0.002	**
	BC	2.02	0.198	
	Lack of fit	0.5272	0.687	Not significant
	Primary and secondary factors C > B > A			

Note: "*" denotes significant ($p < 0.05$); "**" denotes very significant ($p < 0.01$); "***" denotes extremely significant ($p < 0.001$).

From Equation (13) and the coupling results of CFD–DEM, it can be seen that the increase in rotational speed causes changes in the flow field, the pressure difference at the inlet increased as the rotational speed increased, and an unstable high-pressure region gradually appeared at the guide strip and the shell seed outlet, which affects the flow field velocity. The difference in flow velocity between the shell seed outlet and the guide strip is shown in Figure 5d–f, and the variation of flow field pressure is shown in Figure 5a–c. Seeds were dispersed away from the seed inlet in a more orderly manner by airflow forces, which was due to uniform airflow velocity and smooth pressure favoring smooth seed discharge [31]. And then the CVSU rises with the increase in rotational speed, one reason being that the increase in rotational speed increases the seed pickup frequency of the guide strip, which makes the difference in seed pickup between the guide strips increase. Another factor was that (Figure 6d) under the influence of higher rotational speed, a small number of seeds left the guide strip after impact and stayed between the outside of the guide strip and the shell and then entered into the inoculating capsule under the effect of airflow drag force, which affected the orderly discharge of seeds.

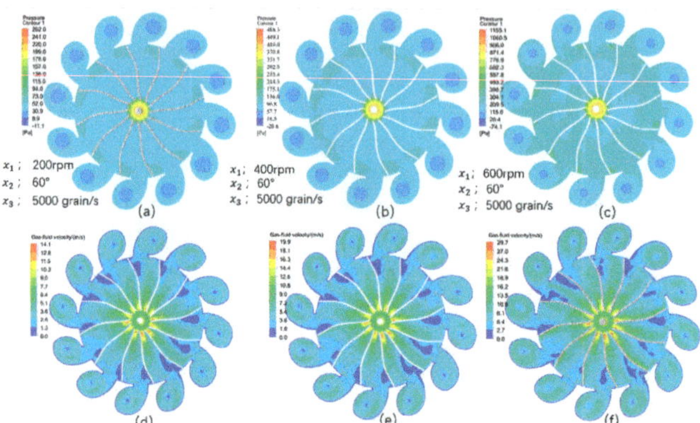

Figure 5. Flow velocity and pressure plots of the coupled CFD–DEM flow field: (**a,d**) for the 200 rpm condition; (**b,e**) for the 400 rpm condition; (**c,f**) for the 600 rpm condition.

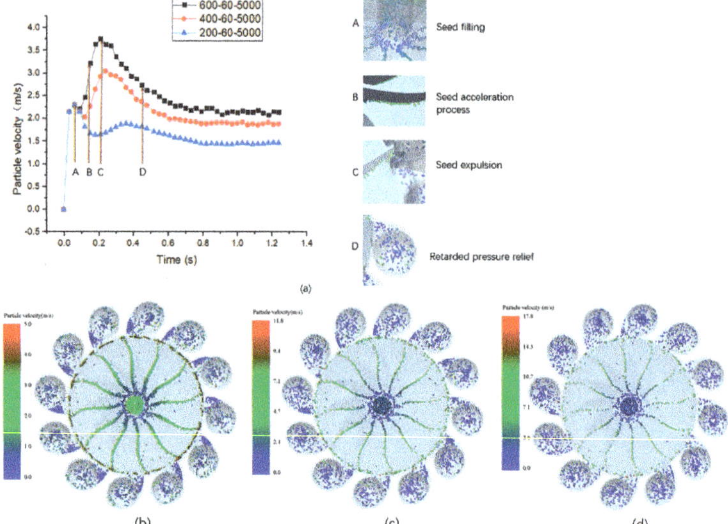

Figure 6. CFD–DEM coupled seed distribution and seed velocity: (**a**) shows the variation of seed velocity under each condition, where A, B, C, and D are the four stages of the seed discharging process, respectively; (**b–d**) are the simulation results of seed distribution in the seed discharger under the rotational speeds of 200 rpm, 400 rpm, and 600 rpm, respectively.

In addition, the effect of seed feed rate on CVSU is illustrated by an example. At a rotational speed of 400 rpm and a guide strip dip angle of 60°, the seed feed rate was increased from 4000 grains/s to 6000 grains/s, and the CVSU was reduced from 2.66% to 1.22% and then increased to 1.31%. Referring to Figure 7, it can be seen that the thickness of the seed layer diverted at the centrifugal spin cone angle of the concave plate increased with the increase in the seed feed rate, which improves the scattered seed diversion effect. However, as the seed feed rate continues to increase, the thickness of seed layer on the centrifugal spin cone angle increases, and, at the same time, the flow field changes as the amount of seed becomes larger (Figure 8), which leads to a deterioration in the seed dispersion effect. A guide strip dip angle is mainly a centrifugal concave plate tilted to the plate on the guide strip and the plate surface of different dip angles will affect the gas flow rate. A tilted guide strip is conducive to compression of air and acceleration of the speed of airflow. Through analysis, it can be seen that when the guide strip dip angle was too small, the role of seed in the airflow was strong, the airflow was low, and the seeds dispersed. When the guide strip dip angle was too large and the airflow was slightly less concentrated, the role of seed force was smaller.

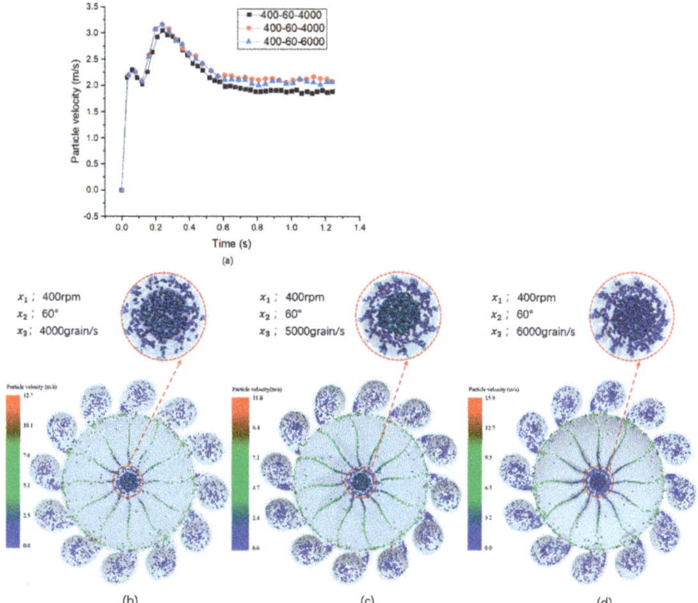

Figure 7. CFD–DEM coupling of seed distribution and seed velocity: (**a**) shows the variation of seed velocity under each condition; (**b–d**) are the simulation results of seed distribution in the seed discharger under the seed feed rate of 4000 grains/s, 5000 grains/s, and 6000 grains/s, respectively.

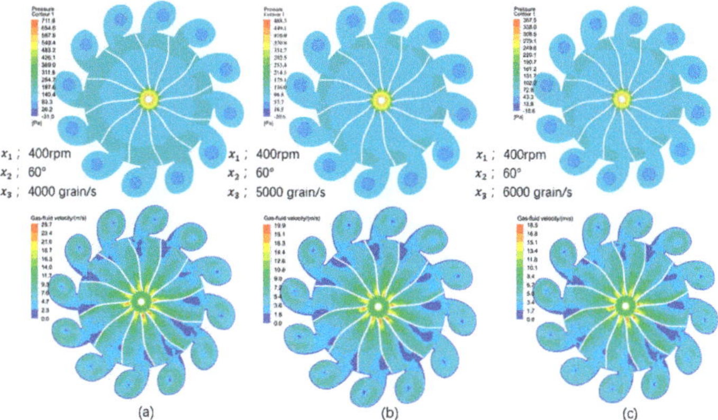

Figure 8. CFD–DEM coupled velocity and pressure maps of flow field: (**a**) velocity and pressure cloud of the flow field for a feed rate of 4000 grains/s; (**b**) velocity and pressure cloud of flow field for a feed rate of 5000 grains/s; (**c**) velocity and pressure cloud of flow field for a feed rate of 4000 grains/s.

3.3. Analysis of the Row-to-Row Seeding Amount Coefficient of Variation (RSCV) across Rows of Displacement

The experimental (Table 5) and variance results (Table 6) showed that the effect of rotational speed and guide strip dip angle on RSCV was significant, and the effect of seed feed rate on the RSCV was extremely significant, with the degree of influence being C > B > A in that order. As shown in Figure 9d–f, the trend of three factors for RSCV was the same as that for CVSU (Figure 9a–c), but in a slightly different way. RSCV decreases and then increases as the three factors increase. The increase in rotational speed, dip angle, and seed feed rate caused orderly increases in airflow velocity, seed layer thickness, and seed taking uniformity, and the wheat seeds taken between 12 guides at the same time moved to the inoculation capsule on the centrifugal concave plate by the airflow-assisted force, and at this time, the RCV had a decreasing trend. Then, as the rotational speed, dip angle, and seed feed rate increased continuously, there were some problems as the effect of scattering seed diversion became weaker and the drag force of airflow became weaker, resulting in the phenomenon that the seeds left the guide strip and were discharged by drag force alone, and the RSCV rose gradually.

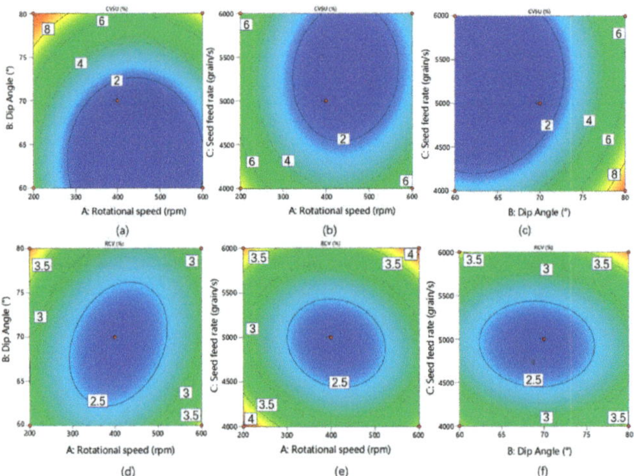

Figure 9. Two-dimensional contours of speed effects, guide strip dip angle, and seed feed rate for CVSU and RSCV: (**a**–**c**) are for CVSU; (**d**–**f**) are for RSCV.

3.4. Quadratic Optimization

3.4.1. Model Tuning and Parameter Optimization

The two-dimensional contour plots were plotted by Design-Expert to co-ordinate and analyze the pattern and degree of influence for rotational speed, guide strip dip angle, and seed feed rate on CVSU and RSCV (e.g., Figure 9). It can be seen from (a), (b), (d), and (e) that the increase in RPM favors the decrease in both metrics, and the CVSU reaches the optimal solution region in preference to the RSCV, but both metrics increase as the RPM continues to increase. As the guide strip dip angle increased, the CVSU was in an upward trend, and the RSCV showed a decreasing and then increasing trend as shown in Figure 9a,c,d,f.

According to the influence analysis of the three factors (speed, guide angle, seed feed rate) in Figure 9 on the two indicators (CVSU and RSCV), combined with the rationality analysis of the optimal value given by Design-Expert (Table 7), the rotational speed was 408 rpm, the dip angle was 69°, and the seed feed was 4938 grains/s, which was in the optimal solution area. The 1000-grain weight of wheat seed is 43.6 g, the sowing amount per hectare is 228 KG, and the optimized seeding device can sow 3.4 hectares per hour, which was determined to be the optimal solution. The parameters were modified and the optimized CFD–DEM coupled simulation experiments were carried out. The results showed that the flow velocity was increased, the flow field was stable, and the internal pressure was more uniform, as shown in Figure 10a,b.

Table 7. Optimal values given by the optimization function analysis of Design-Expert 13 software.

Name	Goal	Lower Limit	Upper Limit	Lower Weight	Upper Weight	Importance
A: Rotational speed	In range	200	600	1	1	3
B: Guide strip dip angle	In range	60	80	1	1	3
C: Seed feed rate	In range	4000	6000	1	1	3
CVSU	Minimize	1.03	10.18	1	1	3
RSCV	Minimize	2.17	4.17	1	1	3
Solutions						
1 solution						

Rotational speed	Guide strip dip angle	Seed feed rate	CVSU	RSCV	Desirability	
408.788	69.274	4938.038	1.025	2.276	0.973	Selected

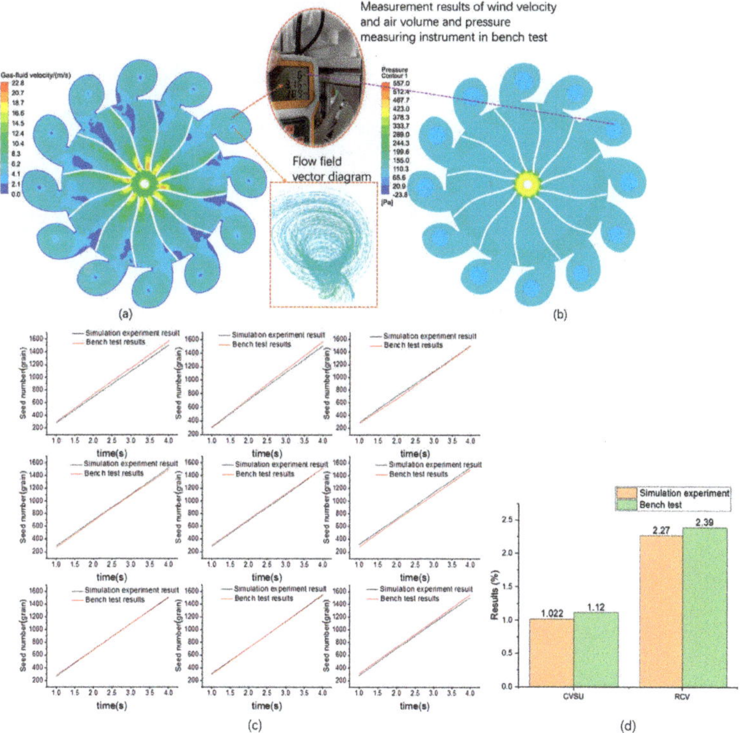

Figure 10. Comparison of flow rate graphs, pressure graphs, experimental results, and bench experiment results of simulation under optimal parameter conditions: (**a**) flow rate graphs; (**b**) pressure graphs; (**c**) comparison of simulation experiments and bench experiments in terms of seed yield; (**d**) comparison of simulation experiments and bench experiment results for CVSU, RSCV.

3.4.2. Bench Experiment Results

To verify the accuracy of optimized simulation results, a centrifugal concave plate with 69° tilted guide strips was printed by a 3D printer for bench-top experiments. The bench experiment is shown in Figure 11. The bench experiment was unable to show the internal flow field, so the simulation results were verified by measuring the flow velocity and pressure at the center of the inoculation capsule utilizing a hand-held wind speed, volume, and pressure measuring instrument. Because of the shielding of tilting guide strips, it was not possible to collect seed distribution shapes at the same time by the high-speed camera, so the choice was made to collect the real-time seed yield results from 12 ports through multiple real-time weighing devices to compare with the simulation results. Nine groups with large differences were selected from the CFD–DEM coupled simulation results of optimal values and the bench experiment results, as shown in Figure 10c. Comparing the seed yield of the simulation experiment, the evaluation index is shown in Figure 10d. The experimental results show that the flow field effect of the bench experiment was consistent with that of simulation experiment, and the difference between the number of seeds collected by the multichannel real-time weighing device and the simulation seed discharge effect was small. Due to the large radius of the centrifugal concave plate (265 mm), the slight shaking of the bench at high speeds affects the perpendicularity between the motor and centrifugal concave plate, which ultimately affects the seed discharging effect.

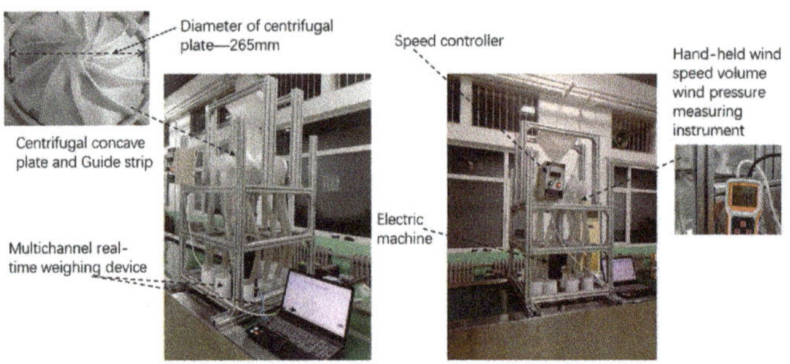

Figure 11. Diagram of bench-top experimental equipment.

4. Conclusions

In this study, a centrifugal wheat strip seeding device was designed to meet the demand for high sowing operation speed in sustainable agriculture. The magnitudes of effects for guide strip dip angle, centrifugal concave plate speed, and seed feed rate on the internal flow field, seed movement characteristics, and sowing performance were investigated using CFD–DEM coupled simulation based on the DDPM interface. The conclusions were as follows:

(1) ANSYS-FLUENT was used to simulate the flow field characteristics of the seeding device, the effects of different parameters on the distribution of the flow field, flow velocity, and pressure at the seed inlet, inside the device, and on the inoculation capsule were analyzed, and the rotational speed, the guide strip dip angle, and the seed feed rate were determined to be the main parameters. As a result of the coupled CFD–DEM simulation, it is observed that the flow field velocity increases with the speed of rotation and guide strip dip angle and decreases with the seed feed rate. As the flow rate increases, local turbulence occurs in the flow field, which affects seed discharge.

(2) The orthogonal experimental design was chosen to obtain the primary and secondary factors affecting the seeding performance indexes of the centrifugal wheat strip seeding device, and the reliability of the influence of rotational speed, guide strip dip angle, and seed feed rate on the seeding indexes CVSU and RSCV was verified. The optimal solution for quadratic optimization was determined: 408 rpm, 4938 grains/s, and 69° guide strip angle.

(3) Bench experiments were conducted to verify the simulation results by collecting real-time seed output from each port through the multichannel real-time weighing device. The optimized simulation results were CVSU of 1.022% and RSCV of 2.27% and the bench experiment results were CVSU of 1.12% and RSCV of 2.39%.

Author Contributions: Conceptualization, X.A. and X.Z.; methodology, X.W.; machine design, Y.H. and X.A.; software, H.L. and L.L.; validation, X.C. and M.L. (Minghao Liu); writing—original draft preparation, X.A. and Y.H.; writing—review and editing, X.A. and M.L. (Meng Liu); supervision, X.C.; funding acquisition, X.A. and X.Z. All authors have read and agreed to the published version of the manuscript.

Funding: This work was supported financially by the National Natural Science Foundation of China (Grant No. 51805300 and Grant No. 32101631) and the Youth Innovation Team Project of Shandong Colleges and Universities.

Institutional Review Board Statement: Not applicable.

Informed Consent Statement: Not applicable.

Data Availability Statement: All data are presented in this article in the form of figures and tables.

Acknowledgments: The authors would like to thank their teacher and supervisor for the advice and help during the experiments. We also appreciate the editor and anonymous reviewers for their valuable suggestions for improving this paper.

Conflicts of Interest: The authors declare no conflict of interest.

References

1. NBSPRC. Announcement on 2022 Grain Production Data. *Mod. Flour Milling Ind.* **2023**, *37*, 18.
2. Teng, S.H.; Li, X.X.; Liu, Q.J. Effects of Climate Change on Main Growth Stages of Winter Wheat in Linyi and Determination of Suitable Sowing Time. *J. Agric.* **2023**, *13*, 18–24.
3. Shu, C.X.; Wei, Y.P.; Liao, Y.T.; Lei, X.L.; Li, Z.D.; Wang, D.; Liao, Q.X. Influence of air blower parameters of pneumatic seed-metering system for rapeseed on negative pressure characteristics and air blower selection. *Trans. CSAE* **2016**, *32*, 26–33.
4. Chen, H.T.; Li, T.H.; Wang, H.F.; Wang, Y.; Wang, X. Design and parameter optimization of pneumatic cylinder ridge three-row close-planting seed-metering device for soybean. *Trans. CSAE* **2018**, *34*, 16–24.
5. Yong, S.; Liao, Y.T.; Liao, Q.X. Experimental study on pneumatic seed-metering system of 2BFQ-6 precision planter for rapeseed. *Trans. CSAE* **2012**, *28*, 57–62.
6. Liao, Y.T.; Huang, H.D.; Li, X.; Yu, J.J.; Yan, Q.X.; Liao, Q.X. Effects of Seed Pre-soaking on Sowing Performance by Pneumatic Precision Metering Device for Papeseed. *Trans. Chin. Soc. Agric. Mach.* **2013**, *44*, 72–76.
7. Shi, S.; Zhang, D.X.; Yang, L. Simulation and verification of seed-filling performance of pneumatic-combined holes maize precision seed-metering device based on EDEM. *Trans. CSAE* **2015**, *31*, 62–69.
8. Yan, B.X.; Zhang, D.X.; Cui, T.; He, X.T.; Ding, Y.Q.; Yang, L. Design of pneumatic maize precision seed-metering device with synchronous rotating seed plate and vacuum chamber. *Trans. CSAE* **2017**, *33*, 15–23.
9. Lei, X.L.; Liao, Y.T.; Zhang, Q.S.; Wang, L.; Liao, Q.X. Numerical simulation of seed motion characteristics of distribution head for rapeseed and wheat. *Comput. Electron. Agric.* **2018**, *150*, 98–109. [CrossRef]
10. Liao, Y.T.; Wang, L.; Liao, Q.X. Design and test of an inside-filling pneumatic precision centralized seed-metering device for rapeseed. *Int. J. Agric. Biol. Eng.* **2017**, *10*, 56–62.
11. Deng, L.J.; Qiao, T.F. Design and Experiment of Wheat Precision Metering Device Based on EDEM. *J. Agric. Mech. Res.* **2021**, *43*, 158–163.
12. Liu, C.L.; Xin, D.; Zhang, F.Y.; Ma, T.; Zhang, H.Y.; Li, Y.N. Design and Test of Cone Diversion Type Horizontal Plate Wheat Precision Seed-metering Device. *Trans. Chin. Soc. Agric. Mach.* **2018**, *49*, 56–65.
13. Xu, H.; Tao, D.C.; Tao, Y.H.; Xiao, B.W. Simulation and Experimental Research on Rice Seed Metering Device Based on EDEM. *J. Agric. Sci. Technol.* **2018**, *20*, 64–70.
14. Cheng, X.P.; Lu, C.Y.; Meng, Z.J.; Yu, J.Y. Design and parameter optimization on wheat precision seed meter with a combination of pneumatic and type hole. *Trans. CSAE* **2018**, *34*, 1–9.
15. Xing, H.; Zang, Y.; Wang, Z.M.; Luo, X.W.; Pei, J. Design and parameter optimization of rice pneumatic seeding metering device with adjustable seeding rate. *Trans. CSAE* **2019**, *35*, 20–28.
16. Manjula, E.V.P.J.; Ariyaratne, W.K.H.; Ratnayake, C.; Melaaen, M.C. A review of CFD modeling studies on pneumatic conveying and challenges in modeling offshore drill cutting transport. *Powder Technol.* **2017**, *305*, 782–793. [CrossRef]
17. Liu, M.L.; Wen, Y.Y.; Liu, R.Z.; Liu, B.; Shao, Y.L. Investigation of fluidization behavior of high density particle in spouted bed using CFD–DEM coupling method. *Powder Technol.* **2015**, *280*, 72–82. [CrossRef]
18. Oke, O.; Wacgem, B.V.; Mazzei, L. Lateral solid mixing gas-fluidized beds: CFD and DEM studies. *Chem. Eng. Res. Des.* **2016**, *114*, 148–461. [CrossRef]
19. Chaumeil, F.; Crapper, M. Using the DEM-CFD method to predict Brownian particle deposition in a constricted tube. *Particuology* **2014**, *15*, 94–106. [CrossRef]

20. Almohammed, N.; Alobaid, F.; Breuer, M.; Epple, B. A comparative study on the influence of the gas flow rate on the hydrodynamics of a gas–solid spouted fluidized bed using Euler–Euler and Euler–Lagrange/DEM models. *Powder Technol.* **2014**, *264*, 343–364. [CrossRef]
21. Gong, F.C.; Huang, H.; Babadagli, T.; Li, H.Z. A resolved CFD–DEM coupling method to simulate proppant transport in narrow rough fractures. *Powder Technol.* **2023**, *428*, 118778. [CrossRef]
22. Ma, H.Q.; Zhou, L.Y.; Liu, Z.H.; Chen, M.Y.; Xia, X.H.; Zhao, Y.Z. A review of recent development for the CFD–DEM investigations of non-spherical particles. *Powder Technol.* **2022**, *412*, 117972. [CrossRef]
23. He, L.P.; Liu, Z.X.; Zhao, Y.Z. Study on a semi-resolved CFD–DEM method for rod-like particles in a gas-solid fluidized bed. *Particuology* **2024**, *87*, 20–36. [CrossRef]
24. Lai, Z.S.; Zhao, J.D.; Zhao, S.W.; Huang, L.C. Signed distance field enhanced fully resolved CFD–DEM for simulation of granular flows involving multiphase fluids and irregularly shaped particles. *Comput. Methods Appl. Mech. Eng.* **2023**, *414*, 116195. [CrossRef]
25. Ma, H.Q.; Liu, Z.H.; Zhou, L.Y.; Du, J.H.; Zhao, Y.Z. Numerical investigation of the particle flow behaviors in a fluidized-bed drum by CFD–DEM. *Powder Technol.* **2023**, *429*, 118891. [CrossRef]
26. Ye, M.K.; Chen, H.C.; Koop, A. Verification and validation of CFD simulations of the NTNU BT1 wind turbine. *J. Wind Eng. Ind. Aerodyn.* **2023**, *234*, 105336. [CrossRef]
27. Campana, L.; Bossy, M.; Henry, C. Lagrangian stochastic model for the orientation of inertialess spheroidal particles in turbulent flows: An efficient numerical method for CFD approach. *Comput. Fluids* **2023**, *257*, 105870. [CrossRef]
28. Bumrungthaichaichan, E. A note of caution on numerical scheme selection: Evidence from cyclone separator CFD simulations with appropriate near-wall grid sizes. *Powder Technol.* **2023**, *427*, 118713. [CrossRef]
29. Bivainis, V.; Jotautienė, E.; Lekavičienė, K.; Mieldažys, R.; Juodišius, G. Theoretical and Experimental Verification of Organic Granular Fertilizer Spreading. *Agriculture* **2023**, *13*, 1135. [CrossRef]
30. Li, K.; Li, S.; Ni, X.; Lu, B.; Zhao, B. Analysis and Experimental of Seeding Process of Pneumatic Split Seeder for Cotton. *Agriculture* **2023**, *13*, 1050. [CrossRef]
31. Hu, H.J.; Zhou, Z.L.; Wu, W.C.; Yang, W.H.; Li, T. Distribution characteristics and parameter optimization of an air-assisted centralized seed-metering device for rapeseed using a CFD–DEM coupled simulation. *Biosyst. Eng.* **2021**, *208*, 246–259. [CrossRef]

Article

Analysis and Experimental Investigation of Steering Kinematics of Driven Steering Crawler Harvester Chassis

Yanxin Wang [1], Chengqian Jin [1,2,*], Tengxiang Yang [2], Tingen Wang [1] and Youliang Ni [2]

1 School of Agricultural Engineering and Food Science, Shandong University of Technology, Zibo 255000, China
2 Nanjing Institute of Agricultural Mechanization, Ministry of Agriculture and Rural Areas, Nanjing 210014, China
* Correspondence: jinchengqian@caas.cn

Abstract: In the context of automatic driving, the analysis of the steering motion characteristics is critical for enhancing the efficiency of crawler harvesters. To address issues such as the low transmission efficiency and the large steering radius encountered by traditional crawler harvesters featuring hydrostatic drives, a driven steering crawler harvester chassis was designed. This involved analysis of the chassis transmission system structure and its steering characteristics under several conditions, including differential steering, differential direction reversal, and unilateral braking steering. The steering parameters were determined based on real-time kinematic positioning–global navigation satellite system (RTK-GNSS) measurements, and they were compared with theoretical predictions based on the crawler harvester steering kinematics. The slip rates and modified models of the crawler chassis for various steering modes were then obtained. The results indicated that the increase in the ratio between the running input and steering input speeds led to larger track steering radii and smaller average rotational angular velocities. Remarkably, the slopes of the linear fits of the tracked chassis steering parameters varied significantly under differential direction reversal and differential steering modes. Compared with the actual results, the correlation coefficient of the tracked chassis steering parameters fitting model is close to 1. The steering parameter model was deemed suitable for actual operational requirements. The results provide a valuable reference for designing navigation and steering models of crawler harvesters operating on different road surfaces.

Keywords: crawler harvester; steering mode; kinematics; steering parameters; slip rate

Citation: Wang, Y.; Jin, C.; Yang, T.; Wang, T.; Ni, Y. Analysis and Experimental Investigation of Steering Kinematics of Driven Steering Crawler Harvester Chassis. *Agriculture* **2024**, *13*, 65. https://doi.org/10.3390/agriculture14010065

Academic Editor: Massimiliano Varani

Received: 30 November 2023
Revised: 17 December 2023
Accepted: 20 December 2023
Published: 29 December 2023

1. Introduction

The crawler harvester plays an unparalleled role in modern-day harvesting owing to its attributes such as small ground pressure, a tight turning radius, and a characteristic adaptability to a varied terrain [1,2]. With regard to autonomous navigation of the crawler chassis, the steering system's performance directly affects the harvester's navigation efficiency and quality [3–5]. The preeminent crawler harvester chassis traditionally employs the hydrostatic transmission system (HST), which faces such limitations as an unvarying steering mode, augmented turning radius, and limited transmission efficacy [6,7]. Furthermore, the steering process of a crawler vehicle is both dynamic and intricate, given that crawler harvesters exhibit track sag, slip, and sliding during operation [8,9]. Therefore, the differences between the practical steering radius and steering angular velocity from the theoretical values severely curtails the operational accuracy and control reliability of the crawler harvester in an autonomous scenario [10,11].

In Europe and the United States, hydraulic drive chasses have been extensively employed in self-propelled combines, which can effectively circumvent the limitations of a mechanical drive [12–15]. There are three main driving forms: a single hydraulic pump with a single hydraulic motor and a gearbox, a single hydraulic pump with double hydraulic motors, and double hydraulic pumps with double hydraulic motors [16,17]. Present-day domestic harvesters generally utilize the HST and conventional mechanical transmission systems to achieve chassis drive operations. Liu et al. [18] designed a planet-tracked agricultural power chassis with a suspension composed of parallel springs and shock absorbers that can be freely retractable in a vertical direction. Notably, Sun et al. [19] integrated the drive wheel gear transmission and cutting platform gear transmission portions into a shared box, forming a densely packed, effective closed system to enhance the transmission efficiency, built upon traditional HST methods. Establishing a vehicle steering model forms the foundation for designing tracked vehicle navigation control systems. However, early tracked vehicle steering models failed to consider track sag [20,21]. Thus, as research progressed, scholars globally began factoring in track sag, slip, roll, centroid deviations, and other factors related to the coordination between the ground and the track, ultimately establishing steady-state and transient steering models for tracked vehicles [22–26]. Fu et al. [27] studied the effects of the height, thickness, inclination Angle and spacing of track teeth on the adhesion between track soil and black soil, and optimized the design parameters of track teeth. Liu et al. [28] established a soft ground steering model for tracked vehicles and studied the influence of track and ground on the driving force. Yokoyama et al. [29] analyzed the influence of track plate spacing on the settlement, traction effect and shear deformation of tracked vehicles in dry sand soil, and designed a flexible track in the form of a grid. Nakajima et al. [30] predicted the total tractive force of a rubber track system and developed and refined an internal 2D DEM program for analyzing the total tractive force generated by a single rigid grip tooth.

After analyzing, comparing, and examining the structural principles of the prevailing steering mechanisms, a drive steering crawler harvester chassis was designed in this study. The structure and principles of the drive steering crawler chassis transmission system were determined, kinematic analysis was performed, and steering kinematic models accounting for the track and the ground were established for three steering modes, namely differential steering, differential direction reversal, and unilateral braking steering. The interplay between the steering parameters and the slip rate on both sides of the tracks was analyzed, and a model relating the slip rate, steering radius, and steering angle was established through actual real-time kinematic positioning–global navigation satellite system (RTK-GNSS)-recorded data. Finally, the offset and correction factors induced by the track chassis' skid were derived, serving as a foundation for designing the navigation and steering model of tracked vehicles.

2. Materials and Methods

2.1. Basic Structure of Driven Steering Crawler Harvester Chassis

Figure 1 shows the detailed structure of the driven steering crawler's chassis. Compared to traditional crawler chassis design, the driven steering crawler features a dual-channel HST drive and a two-channel walking–steering gear transmission, establishing the chassis transmission system. The chassis relied on diesel engines to perform operations, and belt transmissions drove the transmission. The power was transferred to the left and right driving wheels of the chassis via double HST hydraulic and internal transmission gear transmission. The movement of the driving wheels propelled the track to coil around the track frame. The tensioning wheel provided tension to the track and precisely guided the track's rotation direction alongside the supporting wheel while taking the driving direction provided by the ground into account.

Figure 2 displays the all-encompassing structure of the chassis transmission system of the driven steering crawler harvester. The combination of a fixed-shaft gear pair and a planetary gear group mechanism designed with two power flows, walking and steering, enabled the operation and switching of various steering modes, including differential-reverse, differential-speed, and unilateral-braking modes. The engine's power output was transmitted to the gearbox through belt wheel 5, and input shaft 6 was divided into two power flows, namely the walking power source running in a straight line and the steering power source, causing a speed difference between the left and right tracks. The

HST hydraulic stepless transmission device enabled independent speed control for walking and steering. Power sources on both sides input power to planet gear 12 via the internal drive shaft's gear transmission, steering power input to left and right internal gear 15, and walking power input to central gear 18. Through the planetary rack, the left and right solar gears transmit the final output power, which ultimately achieves different speeds of the left and right tracks, facilitating tracked vehicle driving and steering. The transmission system integrates the high efficiency of a mechanical transmission and the continuously variable speed of a hydraulic transmission, enabling the system to realize continuous steering at any steering radius. Consequently, the transmission performance exceeds that of the traditional transmission system.

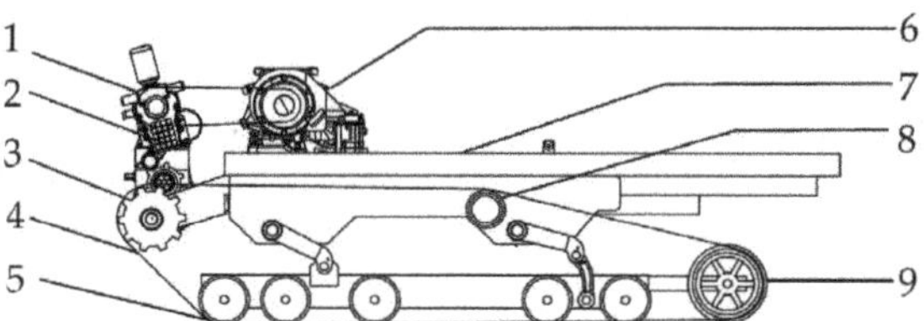

Figure 1. Schematic diagram of chassis structure. 1. Transmission, 2. Double HST assembly, 3. Driving wheel, 4. Track, 5. Return roller, 6. Engine, 7. Frame, 8. Belt wheel, 9. Tension wheel.

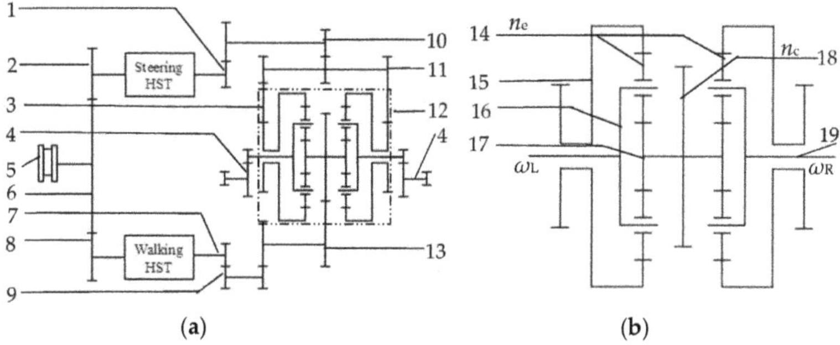

Figure 2. Schematic diagram of the transmission system. (**a**) Schematic diagram of transmission mechanism; (**b**) Schematic diagram of planetary transmission. 1. Steering drive shaft 2, 2. Steering drive shaft 1, 3. Reversing shaft, 4. Reduction shaft, 5. Belt pulley, 6. Input shaft, 7. Walking drive shaft 2, 8. Walking drive shaft 1, 9. Walking drive shaft 3, 10. Steering drive shaft 3, 11. Steering drive shaft 4, 12. Planet gear, 13. Travel drive shaft 4, 14. Pinion, 15. Internal gear, 16. Planet rack, 17. Sun gear, 18. Center gear, 19. Half shaft. In the figure, n represents the speed of the shaft, r/s; ω is the angular velocity of this axis, rad/s.

2.2. Analysis of Speed of Drive and Steering Transmission System

During straight and turning movements of tracked vehicles, the walking power source imparts various linear speeds to different gears, while the steering power source provides a speed differential between the left and right tracks. The subsequent analysis pertains to the kinematic features of these two sources in the fixed-shaft gear pair and planetary gear set.

The rotational speed of the fixed-shaft gear pair corresponds to the number of primary and secondary gear teeth and the transmission characteristics of the gears. The rotational speed of the fixed-shaft gear pair is represented by Equation (1):

$$n_{m+1} = \frac{n_m \cdot z_m}{z_{m+1}} = n_m \cdot i_{m,m+1},\tag{1}$$

where n_m denotes driving gear speed of the fixed-shaft gear pair (r/s), n_{m+1} denotes the fixed-shaft gear pair passive gear speed (r/s), z_m denotes the number of driving gear teeth of the fixed-shaft gear pair, z_{m+1} denotes the number of main and passive gear teeth of the fixed-shaft gear pair, and i_m denotes ratio of the main and passive gears.

According to the transmission characteristics of the planetary mechanism, the output speed of the planet is related to the radii of the inner gear and the solar gear, and the rotational speed of the left and right output shafts is as follows:

$$\omega_o = \frac{2\pi n_e \cdot \mathrm{r}_i + 2\pi n_c \cdot \mathrm{r}}{\mathrm{r}_i + \mathrm{r}}, \tag{2}$$

where ω_o denotes the angular velocity of the output axis (rad/s), r_i denotes the radius of the inner gear assembly (m), r denotes the radius of the solar gear (m), n_e denotes the left and right planetary gear speed (r/s), and n_c denotes the planet central gear speed (r/s).

Similarly, the rotational speed of the output shaft of the left and right hemispheres of the planet via the reduction shaft and the sprocket shaft can be calculated, i.e., the rotational speed of the driving wheel:

$$\omega_d = \frac{\omega_0 \cdot z_r}{z_s}, \tag{3}$$

where z_r denotes the number of gear teeth of the reduction shaft, z_s denotes the number of gear teeth of the sprocket shaft, and ω_d denotes drive wheel speed (r/s).

2.3. Analysis of Steering Characteristics of Driven Steering Crawler Harvester

The chassis transmission system of the driven steering crawler harvester enables the modification of the left and right driving wheel speeds by utilizing the fixed-shaft gear pair and planetary gear group mechanisms. Consequently, the crawler harvester accomplishes linear walking and steering in three modes: differential direction reversal, differential steering, and unilateral braking steering, as shown in Figure 3. The subsections below comprehensively analyze and delineate various linear and steering movements.

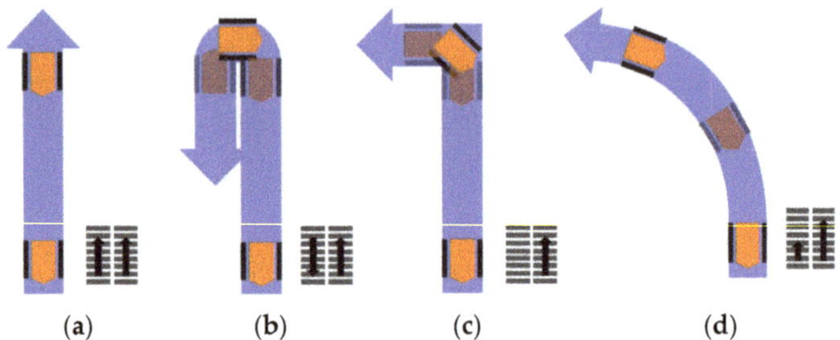

(a)   (b)   (c)   (d)

Figure 3. Schematic diagram of linear and steering motions of track chassis. (**a**) Straight-line driving; (**b**) Differential direction reversal; (**c**) Unilateral braking steering; (**d**) Differential steering.

During straight-line motion of the tracked vehicle, the fixed-shaft gear pair and planetary gear group mechanism create rotational speeds of equal magnitude and direction for the left and right driving wheels. Equations (1)–(3) show that $\omega_R = \omega_L$. At this moment, the central gear's speed is $n_c \neq 0$, and the planetary gear's speed is $n_e = 0$. Only the traveling drive from the central gear emits power, and the steering power source lacks any power input.

When the tracked vehicle reverses in differential mode, the fixed-shaft gear pair and planetary gear group mechanism generate opposing speeds for the left and right driving wheels. This condition indicates that $\omega_L \omega < 0$. Moreover, the output shaft of the left and right wheels rotate in opposite directions, causing the outer track to rotate forward and the inner track to rotate backward. In Figure 4, which provides the steering radius and input speed diagram ($R = f\,(n_c \mid n_e)$), the blue line represents a reversal when the ratio of the central gear's rotational speed to the planetary gear's speed is less than the ratio of the radius of the internal gear to the radius of the sun gear ($\frac{n_c}{n_e} < \left|\pm\frac{r_i}{r}\right|$). This condition results in a turning radius $R < B/2$.

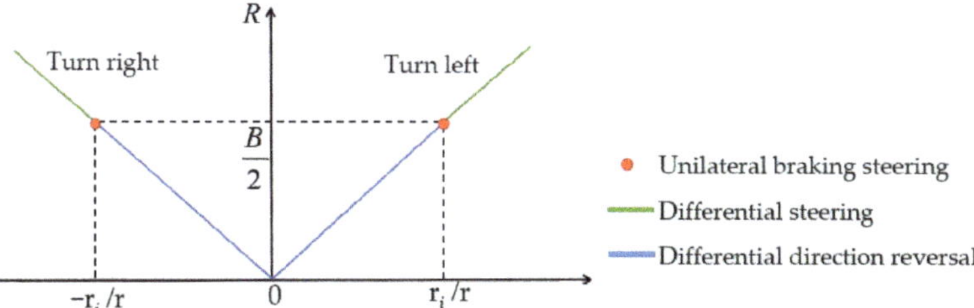

Figure 4. Relationship between steering radius and input speed.

In particular, when $\omega_R = \omega_L$, signifying an equal but opposite direction for the rotational speed of the left and right half shafts, the entire vehicle pivots in place. At this point, the traveling power source lacks any power input to the planet's central gear, leading to a rotational speed (n_c) of 0. Only the steering drive from the internal gear assembly provides a power output, $\frac{n_c}{n_e} = 0$, resulting in a steering radius $R = 0$.

During differential steering of a tracked vehicle, the fixed-shaft gear pair and planetary gear group mechanism generate motion in the same direction but with different speeds of the left and right driving wheels. This condition is indicated when $\omega_L\omega_R > 0$, leading to the same directional rotation of the inner and outer tracks, with the latter experiencing a faster speed than the former, as illustrated by the green line segment in Figure 4. At this stage, the ratio of the central gear speed to the planetary gear speed surpasses the ratio of the radius of the inner gear to the radius of the sun gear ($\frac{n_c}{n_e} > \left|\pm\frac{r_i}{r}\right|$), resulting in a steering radius of $R > B/2$. The walking drive from the central gear and steering drive from the internal gear assembly simultaneously emit power output. The steering movement of the entire vehicle represents a synthesis of the two movements.

During unilateral braking steering of a tracked vehicle, the fixed-shaft gear pair and planetary gear group mechanism generate a singular side speed for the left and right driving wheels, while the other side lacks speed, with $\omega_L = 0$ or $\omega_R = 0$. Consequently, tracked vehicles steer around tracks on the no-speed side, as depicted by the red dot in Figure 4. At this point, the ratio of the rotational speed of the central gear to that of the planetary gear equals the ratio of the radius of the internal gear to the radius of the solar gear ($\frac{n_c}{n_e} = \left|\pm\frac{r_i}{r}\right|$), resulting in a steering radius of $R = B/2$.

2.4. Kinematics Analysis of Tracked Vehicle Steering

The motion of a tracked vehicle on horizontal ground can be regarded as a combination of the relative movement between the vehicle and ground track alongside the sliding or rolling movement of the ground track. It is presumed, at this point, that the axes of the driving wheel, guide wheel, support wheel, and sprocket remain stationary. The track winds around these wheels at a certain speed propelled by the driving wheel. Obtaining this parameter is challenging and requires indirect calculation through other monitored quantities. The velocity can be expressed as follows:

$$\begin{cases} v_R = z_K l_t n_R = \frac{z_K l_t \omega_R}{60} \\ v_L = z_K l_t n_L = \frac{z_K l_t \omega_L}{60} \end{cases}, \tag{4}$$

where v_R and v_L denote the average speeds of left and right track winding motions (m/s), respectively, z_K denotes the effective number of meshed gear teeth of the driving wheel, l_t denotes the chain rail pitch (m), ω_L and ω_R denote the left and right driving wheel angular speeds (rad/s), respectively, and n_L and n_R denote the left and right driving wheel speeds (r/s), respectively.

On level terrain, a tracked vehicle attains stable steering around the steering center O, as shown in Figure 5. It is observable that precarious steering of the tracked vehicle on both track sides yields differences in the turning speeds ($|v_R - v_L|$), leading to a smaller tracked vehicle steering radius R for larger steering angular velocities of α. The mathematical formulas for calculating steering parameters R and α are as follows:

$$R = 0.5B\frac{v_R + v_L}{v_R - v_L},\tag{5}$$

$$\alpha = \frac{v_L}{R - 0.5B} = \frac{v_R}{R + 0.5B},\tag{6}$$

$$\omega = \frac{\alpha}{t},\tag{7}$$

where B denotes the track gauge (m), R denotes the theoretical turning radius of the tracked vehicle (m), α denotes the vehicle theoretical steering angle (rad), and ω denotes the average theoretical steering angular velocity (rad/s).

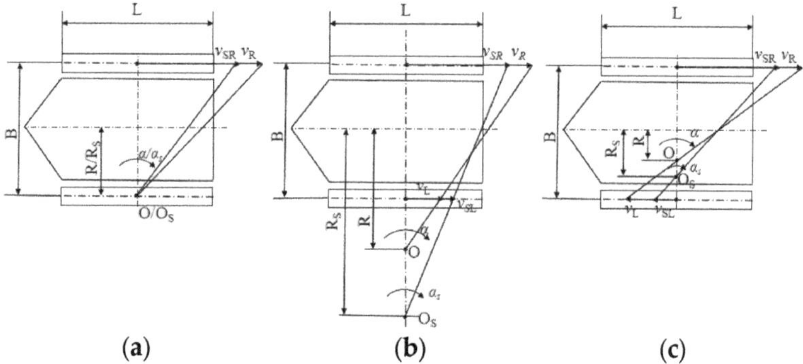

Figure 5. Schematic diagram of tracked vehicle steering movement. (**a**) Unilateral braking steering (**b**) Differential steering (**c**) Differential direction reversal.

During vehicle operation, despite not surpassing the traction force capacity of the track-ground adhesion, there remained a slight level of sliding between the track and the ground due to track shear or soil deformation caused by track compression. Figure 5 displays the scenario of track slippage, wherein the vehicle speed v denotes the actual speed, defined as the relative driving speed of the track landing zone versus the vehicle body. It combines the sliding speed of the track with the relative speed between the frame and the ground track. In cases where the steering radius R fell between $B/2$, the distance between the centers of the tracks, and the free steering radius, sliding of the outer track occurred, with its actual speed lower than the theoretical speed, while the inner track experienced skidding, with its actual speed greater than the theoretical speed. When the steering radius was less than $B/2$, a track sliding phenomenon occurred on both sides. This led to deviations of the actual instantaneous steering center from the theoretical one, causing the actual steering radius R_S to surpass the theoretical steering radius R. The actual steering angular velocity never exceeded the theoretical steering angular velocity. The sliding and skidding of the track are represented by the slip coefficients σ_1 and σ_2, respectively.

$$\begin{cases} \sigma_1 = \dfrac{S_S - S}{S_S} = 1 - \dfrac{n \cdot l_t \cdot z_k / 2\pi}{(R_S - 0.5B)\omega_S} \\ \sigma_2 = \dfrac{S - S_S}{S} = 1 - \dfrac{(R_S + 0.5B)\omega_S}{n \cdot l_t \cdot z_k / 2\pi} \end{cases},\tag{8}$$

where S_S denotes the actual traveling distance of the track (m), S denotes the track theoretical traveling distance (m), l denotes the pitch of the track plate (m), R_S denotes the theoretical turning radius of the tracked vehicle (m), and ω_s denotes the vehicle theoretical steering angular velocity (rad/s).

3. Results and Discussion

3.1. Test Conditions and Process

The tests conducted involved the linear movement of the chassis transmission system and steering performances for the three steering modes on both concrete pavement and a harvested field using a driven steering crawler harvester. The trial was conducted on 15 November 2022 at the Baima Agricultural Base in Nanjing. The RTK-GNSS antenna was installed on either side of the harvester's top in order to collect GNSS position information. Additionally, a speed sensor was installed within the transmission to gauge the speeds of the planetary gear and the central gear. Table 1 displays the various parameters associated with the entire machine, including the antenna position parameters.

Table 1. Crawler harvester and antenna position parameters.

Parameter	Numerical Value
Machine size (length × width × height)/(mm)	5640 × 2600 × 2800
Weight/(kg)	4030
Track grounding length/center distance/(mm)	2300
Track gauge/(mm)	1200
Mean ground voltage (kPa)	19.1
Track pitch × number × width	90 mm × 58 segment × 550 mm
Number of driving gear teeth	8
Spacing between antenna and central axis/(mm)	740
Spacing between antenna and track axis/(mm)	420

The actual steering radius and average steering angular velocity of a tracked vehicle can be calculated based on the position and azimuth information collected by the RTK-GNSS. The vehicle's steering trajectory was obtained by calculating the differences between the global latitude and longitude coordinates, and the actual average steering angular velocity was obtained through an integral calculation. The actual turning radius RGNSS at the RTK-GNSS receiver is presented in Figure 6.

$$R_{\text{GNSS}} = \frac{1}{m} \sum_{i=1}^{m} \sqrt{(x_i - u_1)^2 + (y_i - u_2)^2}, \tag{9}$$

where m denotes the quantity of data collected by the RTK-GNSS (each). $[u_1, u_2]$ are the least-squares-fitted center coordinates of the trajectory:

$$\begin{bmatrix} u_1 \\ u_2 \end{bmatrix} = \begin{bmatrix} \sum\limits_{i=1}^{m-1} b_i^2 \cdot \sum\limits_{i=1}^{m-1} a_i c_i - \sum\limits_{i=1}^{m-1} b_i c_i \cdot \sum\limits_{i=1}^{m-1} a_i b_i \\ \sum\limits_{i=1}^{m-1} a_i^2 \cdot \sum\limits_{i=1}^{m-1} b_i c_i - \sum\limits_{i=1}^{m-1} a_i c_i \cdot \sum\limits_{i=1}^{m-1} a_i b_i \end{bmatrix} \cdot \begin{bmatrix} \dfrac{1}{\sum\limits_{i=1}^{m-1} a_i^2 \cdot \sum\limits_{i=1}^{m-1} b_i^2 - \left(\sum\limits_{i=1}^{m-1} a_i b_i\right)^2} \end{bmatrix}, \tag{10}$$

$$\begin{cases} a_i = 2(x_{i+1} - x_i) \\ b_i = 2(y_{i+1} - y_i) \\ c_i = x_{i+1}^2 + y_{i+1}^2 - x_i^2 - y_i^2 \end{cases}, \tag{11}$$

where x and y denote the position information collected by the RTK-GNSS in the universal transverse Mercator grid system (UTM) coordinate system (m).

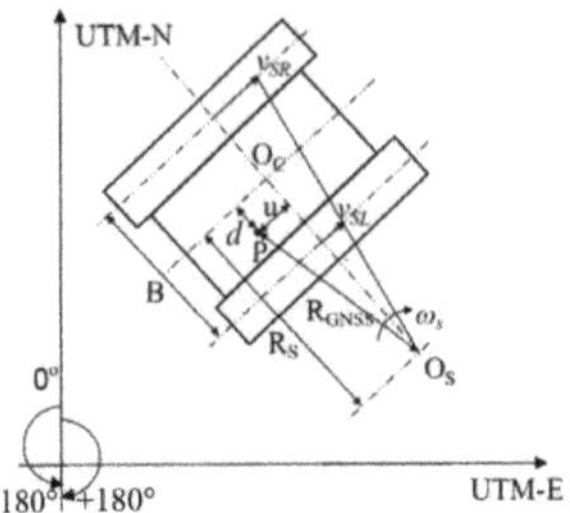

Figure 6. Steering diagram of crawler harvester. Note: UTM denotes a universal transverse Mercator grid system, where UTM-E is the projected distance from the central meridian of the longitude region and UTM-N is the projected distance from the equator. Point P is the installation position of the global navigation satellite system (GNSS). Point Os is the center of the turning moment. d is the distance between point P and the central axis. u is the distance between the antenna and the caterpillar axis. R_s is the turning radius at the central axis.

The actual turning radius R_S at the center of gravity of the harvester is calculated as follows:

$$R_S = \sqrt{R_{\mathrm{GNSS}}^2 - u^2} + d, \tag{12}$$

where u denotes the distance between the antenna and the caterpillar axis (m), and d denotes the distance between the antenna and the central axis (m).

3.2. Test Data

The harvester had three gears: high speed, middle speed, and low speed. The walking speed is determined by the position of the variable-speed operating lever. During the test, the harvester selected the moderate-speed gear when driving on concrete pavement and the low-speed gear when driving in a field. The turning radius of the combine was controlled by the steering wheel. During the test, the steering wheel was set to a fixed angle and driven at a constant speed. At the same time, the GNSS position information, heading angle, speed of the walking shaft, and speed of the steering shaft were recorded. The theoretical turning radius, turning angular velocity, actual turning radius, actual turning angular velocity, and slip/slip rate were calculated according to Equations (1)–(12). The motion parameters and test results are shown in Tables 2 and 3.

Table 2. Test results of steering motion parameters and corrected parameters of crawler harvester on concrete pavement.

No.	$\bar{n}_c$	$\bar{n}_e$	$\bar{v}_R$	$\bar{v}_L$	$\bar{n}_c/\bar{n}_e$	$\bar{R}$	$\bar{R}_S$	$\bar{\omega}$	$\bar{\omega}_s$	$\bar{\sigma}_1$	$\bar{\sigma}_2$	Steering Type
1	157.065	7.584	2.087	1.892	20.71	15.036	19.643	0.57	0.523	0.193	0.97	a
2	161.92	13.574	2.223	1.875	11.929	7.559	9.262	1.016	0.904	0.362	0.962	a
3	122.065	14.76	1.738	1.359	8.27	5.016	5.836	1.112	1.079	0.364	0.962	a
4	80.632	16.557	1.234	0.809	4.87	3.091	4.741	1.176	1.102	0.241	0.972	a
5	101.377	42.352	1.83	0.743	2.394	1.429	2.201	3.178	3.079	0.25	0.976	a
6	111.033	49.126	2.045	0.772	2.26	1.328	1.875	3.723	3.602	0.306	0.973	a
7	92.258	68.187	2.046	0.296	1.353	0.803	1.368	5.116	4.69	0.282	0.987	a
8	61.939	68.293	1.662	−0.09	0.907	0.538	1.2	5.124	4.886	0.158	0.995	a
9	66.749	74.471	1.802	−0.109	0.896	0.532	1.17	5.588	5.359	0.162	0.994	a
10	61.872	69.309	1.674	−0.104	0.893	0.529	0.99	5.36	5.19	0.215	0.992	a
11	40.452	51.422	1.171	−0.146	0.787	0.468	0.818	3.85	3.983	0.232	0.973	b
12	45.893	85.762	1.676	−0.516	0.535	0.318	0.554	6.408	6.2	0.379	0.721	b
13	17.754	92.978	1.418	−0.968	0.191	0.113	0.244	6.98	6.757	0.36	0.604	b
14	7.283	63.55	0.908	−0.721	0.115	0.068	0.343	4.758	4.402	0.272	0.743	b
15	2.839	108.739	1.424	−1.352	0.026	0.015	−0.248	8.119	8.955	0.151	0.629	b

Table 3. Test results of steering motion parameters and correction parameters of field crawler harvester.

No.	$\bar{n}_c$	$\bar{n}_e$	$\bar{v}_R$	$\bar{v}_L$	$\bar{n}_c/\bar{n}_e$	$\bar{R}$	$\bar{R}_S$	$\bar{\omega}$	$\bar{\omega}_s$	$\bar{\sigma}_1$	$\bar{\sigma}_2$	Steering Type
1	163.68	20.645	2.343	1.812	7.928	4.704	5.904	1.549	1.486	0.343	0.963	a
2	150.544	26.819	2.254	1.566	5.613	3.487	4.052	2.012	1.982	0.349	0.964	a
3	143.934	33.654	2.258	1.395	4.279	2.634	3.651	2.525	2.499	0.251	0.971	a
4	118.982	38.798	2.008	1.012	3.067	1.883	2.644	2.911	2.878	0.26	0.973	a
5	115.349	42.982	2.015	0.912	2.684	1.59	2.139	3.225	3.117	0.326	0.97	a
6	107.416	53.816	2.053	0.672	1.996	1.192	1.74	4.038	3.911	0.291	0.976	a
7	77.24	65.102	1.815	0.148	1.186	0.704	1.48	4.884	4.685	0.145	0.994	a
8	69.457	73.186	1.82	−0.058	0.949	0.564	1.097	5.491	5.265	0.219	0.996	a
9	63.411	74.089	1.755	−0.146	0.856	0.508	1.011	5.559	5.398	0.211	0.99	b
10	60.572	74.668	1.727	−0.189	0.811	0.481	0.961	5.603	4.701	0.324	0.982	b
11	34.408	54.348	1.134	−0.261	0.633	0.377	0.563	4.078	3.692	0.397	0.917	b
12	28.561	80.591	1.396	−0.672	0.354	0.21	−0.159	6.047	6.155	0.691	0.357	b
13	11.897	49.918	0.791	−0.49	0.238	0.147	−0.245	3.745	3.293	0.44	0.62	b
14	14.27	86.73	1.294	−0.931	0.165	0.097	−0.486	6.508	6.684	0.906	0.87	b

Note: $\bar{n}_c$ is the average speed of the central gear (r/s), $\bar{n}_e$ is the average speed of the planetary gear (r/s), $\bar{v}_R$ is the average left-winding speed (m/s), $\bar{v}_L$ is the right winding average speed (m/s), $\bar{R}$ is the average theoretical turning radius (m), $\bar{R}_s$ is the average actual turning radius (m), $\bar{\omega}$ is the theoretical average rotational angular velocity (rad/s), $\bar{\omega}_s$ is the actual average rotational angular velocity (rad/s), $\bar{\sigma}_1$ is the average slip rate of the high-speed side track, and $\bar{\sigma}_2$ is the average slip rate of the side track at low speeds, a is the differential steering, b is the differential direction reversal.

3.3. Test Analysis

3.3.1. Relationship between Input Speed and Steering Parameters

The actual turning radius and the actual turning angular velocity of the crawler harvester on concrete pavement and in a field for various input speeds are shown in Figures 7 and 8, respectively. As the transmission system of the harvester could realize differential direction reversal, differential steering, and unilateral braking steering, it could realize any turning radius. When the harvester was in the same gear and at the same speed, the larger the ratio between the running input speed and the steering input speed was, the larger the track steering radius and the smaller the average rotation angular velocity were. When the ratio was close to 1, the steering radius was close to half of the track gauge. On the concrete pavement, the slope of the linear fit of the theoretical turning radius was 0.707, whereas the slopes of the linear fits of the actual data were 0.913 during differential steering and 0.8 during differential direction reversal. On the field, the slope of the linear fit of the theoretical steering radius was 0.622, whereas the slopes of the linear fits of the actual data were 0.692 during differential steering and 2.129 during differential direction reversal.

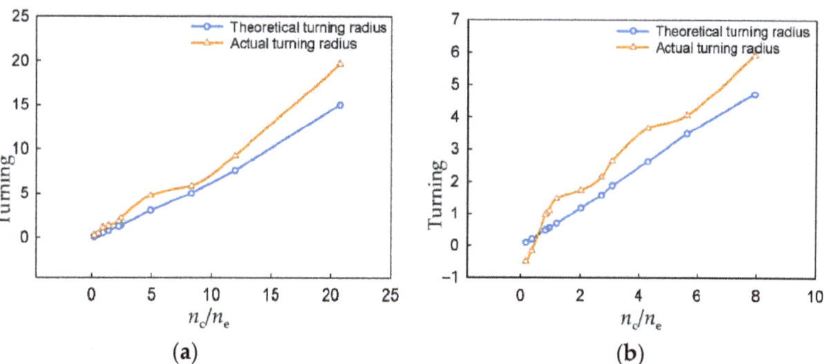

Figure 7. Variations of steering radius with input speed. (a) Concrete pavement (b) Field.

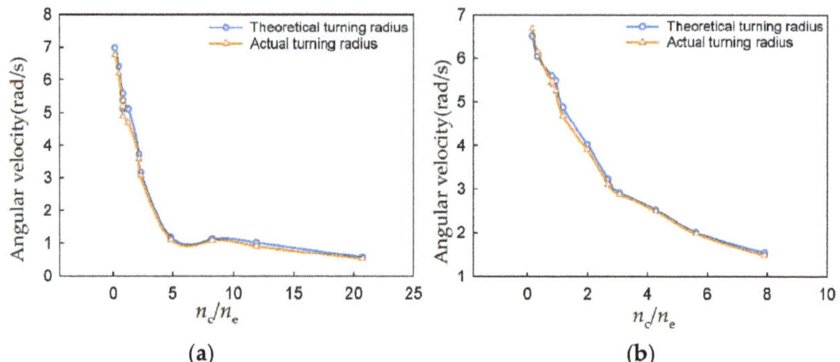

Figure 8. Average rotation angular velocity variations with input speed. (**a**) Concrete pavement (**b**) Field.

When the crawler harvester steered on the concrete pavement and the field, the relationship between the low-speed side track slip rate σ_1 and the high-speed side track slip rate σ_2 with the ratio of the input speed and the fitted curve is shown in Figure 9. For the crawler harvester during differential steering ($n_c/n_e > 1$), with the decrease in the input speed ratio, the slip rate of the low-speed side track increased and approached 1. At this time, the walking input speed n_c was close to the steering input speed n_e, and the winding speed of the low-speed side decreased and approached 0. The displacement between the track and ground came from the implicated movement of the vehicle on the ground, and the track was in a state of sliding, while the slip rate of the high-speed side track was between 0.2 and 0.4. When the crawler harvester underwent differential direction reversal ($n_c/n_e < 1$), the low-speed side track began to reverse, the traction decreased, and the motion of the track changed to skidding.

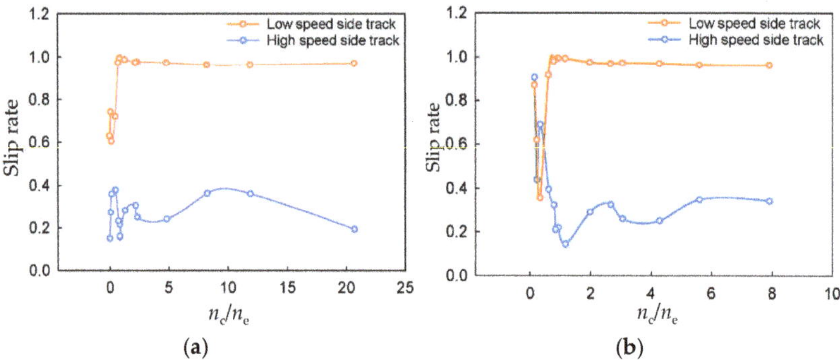

Figure 9. Sliding rate variations with input speed. (**a**) Concrete pavement (**b**) Field.

3.3.2. Correction Calculation of Steering Parameters

As can be seen from the previous analysis, the actual steering radius was larger than the theoretical steering radius due to the slip of the track, that is, the steering radius of the tracked vehicle increased. At the same time, the slip rate of the differential direction reversal and differential steering track were different for the concrete pavement and the field due to the different shear stresses, so the variation trends of the radius in the different steering modes were also different. Figures 10 and 11 show the actual turning paths of the harvester and the fitted paths of the driving wheel of the harvester at different steering angles.

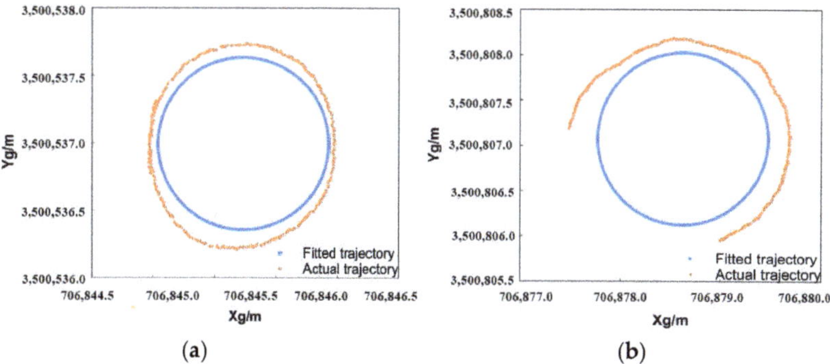

Figure 10. Differential steering path trajectory. (**a**) Concrete pavement (**b**) Field.

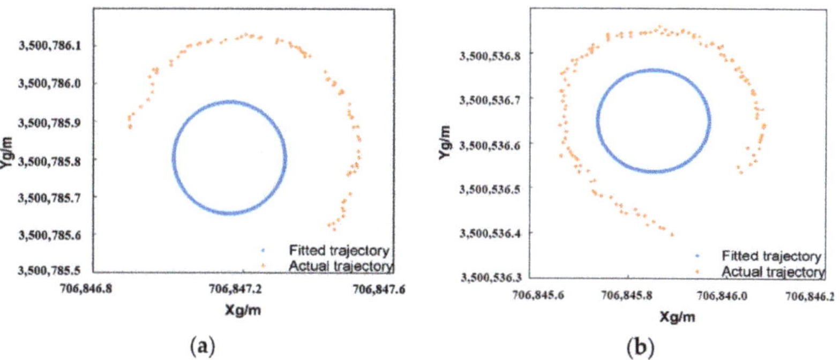

Figure 11. Trajectory of differential reversal path. (**a**) Concrete pavement (**b**) Field.

In an automatic driving scenario of a crawler harvester, the steering mode of the harvester determines the accuracy of operation and the control reliability. Reasonable analysis of the motion characteristics of different steering modes is the key to improving the working efficiency of the harvester. Due to the influence of track slippage, the actual steering motion parameters of the crawler harvester differed greatly from the theoretical values. Therefore, correction parameters of the steering radius and average steering angular velocity need to be calculated to correct the theoretical values. Figures 12 and 13 show the fitted curves of the correction parameters of the steering radius and the average steering angular velocity.

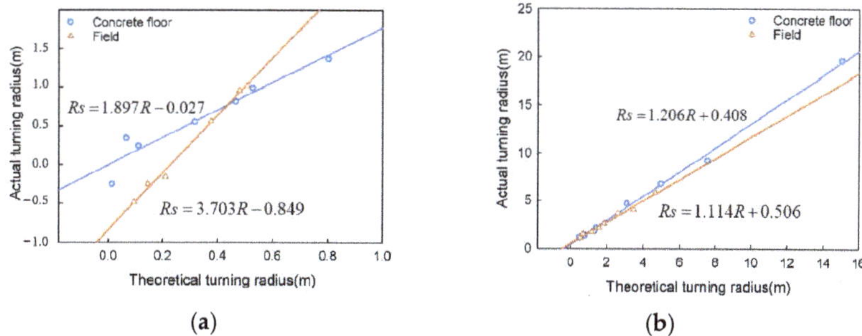

Figure 12. Correction parameters of steering radius. (**a**) Error reversal radius correction parameter (**b**) Differential steering radius correction parameter.

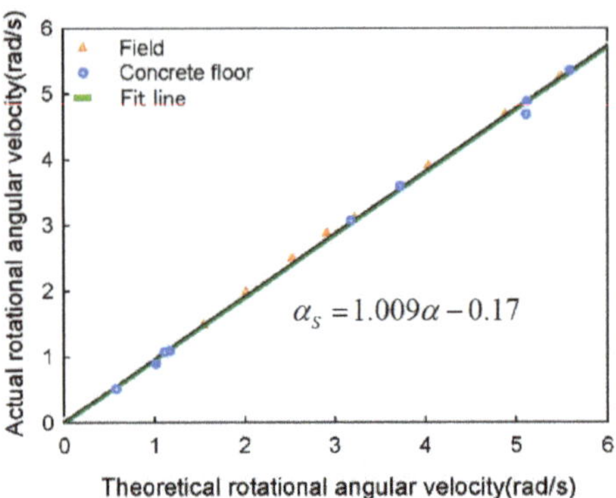

Figure 13. Average steering angular velocity correction parameters.

A linear function was used to fit the correction parameters of the steering radius and average steering angular velocity. It can be seen from the Figure 12 that the correction parameters of the steering radius differed greatly for the different road surfaces and steering modes, so they were calculated separately under the different conditions. However, the average steering angular velocities had little difference, so the fitting calculation was carried out for all the data combined. On the concrete pavement, the fitted formula of the differential turning radius was $R_s = 1.206R + 0.408$, with $R^2 = 0.997$. The fitted formula of the differential reversing turning radius was $R_s = 1.807R - 0.027$, with $R^2 = 0.871$. On the field, the fitted formula of the differential turning radius was $R_s = 1.114R + 0.506$, with $R^2 = 0.986$. The fitted formula of the differential reversing turning radius was $R_s = 3.703R - 0.849$, with $R^2 = 0.994$. The fitted formula of average turning angular velocity was $\alpha_s = 1.009\alpha - 0.17$, and the R^2 value was 0.981.

4. Conclusions

In this paper, the transmission system structure of a driven steering crawler harvester chassis was analyzed, the speed relationship and steering characteristics under conditions of differential steering, differential direction reversal, and unilateral braking steering were analyzed. Based on a comparative study of the steering parameters obtained by RTK-GNSS measurements and the calculation results from a theoretical steering kinematics model of a crawler harvester, the mathematical expressions relating the walking power source and the steering power source to the winding speed, steering radius, and steering angle on both sides of the track were determined. Finally, the slip rate and the modified model of the track chassis slippage under different steering modes on concrete pavement and a field were obtained.

The experimental results showed that the larger the ratio of the running input speed to the steering input speed was, the larger the track steering radius and the smaller the average rotation angular velocity were under differential direction reversal and differential steering modes. There was a linear relationship between the actual steering parameters and the theoretical values. Compared with the actual results, the correlation coefficient of the tracked chassis steering parameters fitting model is close to 1. In the normal running state of the harvester, the theoretical steering parameters can be calculated and corrected in real time based on the input speed of the moving steering of the gearbox, which provides a reference for the study of the steering characteristics of tracked vehicles for different road surfaces and steering modes and the verification of the theoretical steering model.

Author Contributions: Conceptualization, Data curation, Methodology, Writing—Original Draft, Y.W.; Funding acquisition, Formal analysis, Writing—review and editing, Writing—review and editing, C.J.; Formal analysis, Investigation, T.Y.; Writing—review and editing, T.W.; Writing—review and editing, Y.N. All authors have read and agreed to the published version of the manuscript.

Funding: This work was supported by the National key research and development program [Grant No. 2021YFD2000503]; Special fund project for the construction of modern agricultural industrial technology system [Grant No. CARS-04-PS29].

Institutional Review Board Statement: Not applicable.

Informed Consent Statement: Not applicable.

Data Availability Statement: Data are contained within the article.

Conflicts of Interest: The authors declare no conflict of interest.

References

1. He, Y.; Zhou, J.; Sun, J.; Jia, H.; Liang, Z.; Awuah, E. An adaptive control system for path tracking of crawler combine harvester based on paddy ground conditions identification. *Comput. Electron. Agric.* **2023**, *210*, 107948. [CrossRef]
2. Hossain, M.; Hoque, M.; Wohab, M.; Miah, M.M.; Hassan, M. Technical and economic performance of combined harvester in farmers field. *Bangladesh J. Agric. Res.* **2015**, *40*, 291–304. [CrossRef]
3. Lv, Y.; Wu, X.; Fu, Y.; Li, Z. Research on the Influence of Four-wheel Chassis 'Working Stroke Rate. *J. Agric. Mech. Res.* **2015**, *37*, 23–26+30. [CrossRef]
4. Chen, T.; Xu, L.; Ahn, H.S.; Lu, E.; Liu, Y.; Xu, R. Evaluation of headland turning types of adjacent parallel paths for combine harvesters. *Biosyst. Eng.* **2023**, *233*, 93–113. [CrossRef]
5. Guan, Z.; Zhang, M.; Jin, M.; Li, H.; Jiang, T.; Mu, S. Research progress of mechanical damage to soil bywalking mechanism of tracked agricultural equipment. *J. Intell. Agric. Mech.* **2022**, *3*, 62–70.
6. Zhang, L.; Liu, G.; Qi, Y.; Yang, T.; Jin, C. Research progress on key technologies of agriculturalmachinery unmanned driving system. *J. Intell. Agric. Mech.* **2022**, *3*, 27–36.
7. Munir, F.; Azam, S.; Yow, K.-C.; Lee, B.-G.; Jeon, M. Multimodal fusion for sensorimotor control in steering angle prediction. *Eng. Appl. Artif. Intell.* **2023**, *126*, 107087. [CrossRef]
8. Yang, H.; Zhou, J.; Wang, X.; Wu, Y. Parameter Matching and Tillage Depth Control Method for Electric Crawler Tractor Platforms. *J. Eng. Sci. Technol. Rev.* **2021**, *14*, 83–90. [CrossRef]
9. Inoue, E.; Mitsuoka, M.; Rabbani, M. Investigation of nonlinear vibration characteristics of agricultural rubber crawler vehicles. *AMA-Agric. Mech. Asia Afr. Lat. Am.* **2011**, *42*, 89.
10. Shi, R.; Dai, F.; Zhao, W.; Liu, X.; Wang, T.; Zhao, Y. Optimal design and testing of a crawler-type flax combine harvester. *Agriculture* **2023**, *13*, 229. [CrossRef]
11. Liu, Z.; Zhang, G.; Chu, G.; Niu, H.; Zhang, Y.; Yang, F. Design matching and dynamic performance test for an HST-based drive system of a hillside crawler tractor. *Agriculture* **2021**, *11*, 466. [CrossRef]
12. Tang, Z.; Ren, H.; Li, X.; Liu, X.; Zhang, B. Structure Design and Bearing Capacity Analysis for Crawler Chassis of Rice Combine Harvester. *Complexity* **2020**, *2020*, 7610767. [CrossRef]
13. Adams, B.; Darr, M.; Shah, A. Optimized Chassis Stability Relative to Dynamic Terrain Profiles in a Self-Propelled Sprayer Multibody Dynamics Model. *J. ASABE* **2023**, *66*, 127–139. [CrossRef]
14. Sun, Y.; Xu, L.; Jing, B.; Chai, X.; Li, Y. Development of a four-point adjustable lifting crawler chassis and experiments in a combine harvester. *Comput. Electron. Agric.* **2020**, *173*, 105416. [CrossRef]
15. Shinzato, Y.; Komesu, H.; Akati, T.; Ueno, M. Estimating the performance of small sugarcane harvesters in Okinawa. *Eng. Agric. Environ. Food* **2019**, *12*, 499–504. [CrossRef]
16. Xu, L.; Zhou, Z.; Wang, B. Study on Matching Strategies of Hydro-Mechanical Continuously Variable Transmission System of Tractor. *Int. J. Digit. Content Technol. Its Appl.* **2013**, *7*, 843–849.
17. Shao, X.; Yang, Z.; Mowafy, S.; Zheng, B.; Song, Z.; Luo, Z.; Guo, W. Load characteristics analysis of tractor drivetrain under field plowing operation considering tire-soil interaction. *Soil Tillage Res.* **2023**, *227*, 105620. [CrossRef]
18. Liu, P.; Wang, Z.; Li, H.; Zhang, S.; Wei, W. Design and overcoming obstacles ability research of tracked driving chassis with planetary structure. *Trans. Chin. Soc. Agric. Mach.* **2014**, *45*, 17–23.
19. Sun, S.; Wang, J.; Ke, J. New HST optimal design and test in harvester. *J. Mach. Des.* **2014**, *31*, 34–38. [CrossRef]
20. Previati, G.; Gobbi, M.; Mastinu, G. Farm tractor models for research and development purposes. *Veh. Syst. Dyn.* **2007**, *45*, 37–60. [CrossRef]
21. Letherwood, M.D.; Gunter, D.D. Ground vehicle modeling and simulation of military vehicles using high performance computing. *Parallel Comput.* **2001**, *27*, 109–140. [CrossRef]
22. Du, P.; Ma, Z.; Chen, H.; Xu, D.; Wang, Y.; Jiang, Y.; Lian, X. Speed-adaptive motion control algorithm for differential steering vehicle. *Proc. Inst. Mech. Eng. Part. D J. Automob. Eng.* **2020**, *235*, 672–685. [CrossRef]
23. Tang, S.; Yuan, S.; Hu, J.; Li, X.; Zhou, J.; Guo, J. Modeling of steady-state performance of skid-steering for high-speed tracked vehicles. *J. Terramech.* **2017**, *73*, 25–35. [CrossRef]
24. Soodmand, I.; Heidari Shirazi, K.; Moradi, S. Analysis of ride comfort of a continuous tracked bogie system with variable configuration. *Proc. Inst. Mech. Eng. Part. D J. Automob. Eng.* **2020**, *234*, 3429–3439. [CrossRef]

25. Xin, Z.; Jiang, Q.; Zhu, Z.; Shao, M. Design and optimization of a new terrain-adaptive hitch mechanism for hilly tractors. *Int. J. Agric. Biol. Eng.* **2023**, *16*, 134–144. [CrossRef]
26. Matsuda, S.; Mukai, H. Effects of the Combination of Sloped Farm Field, Crawler Compaction and Open Channels on Moisture and Hardness and Temperature of the Surface Soil Layer after Disappearance of the Snow Cover. *Trans. Jpn. Soc. Irrig.* **2010**, *77*, 411–416.
27. Fu, J.; Li, J.; Tang, X.; Wang, R.; Chen, Z. Optimization of Structure Parameters of the Grouser Shoes for Adhesion Reduction under Black Soil. *Agriculture* **2021**, *11*, 795. [CrossRef]
28. Liu, Z.; Guo, J.; Ding, L.; Gao, H.; Guo, T.; Deng, Z. Online estimation of terrain parameters and resistance force based on equivalent sinkage for planetary rovers in longitudinal skid. *Mech. Syst. Signal Process.* **2019**, *119*, 39–54. [CrossRef]
29. Yokoyama, A.; Nakashima, H.; Shimizu, H.; Miyasaka, J.; Ohdoi, K. Effect of open spaces between grousers on the gross traction of a track shoe for lightweight vehicles analyzed using 2D DEM. *J. Terramech.* **2020**, *90*, 31–40. [CrossRef]
30. Nakashima, H.; Yoshida, T.; Wang, X.L.; Shimizu, H.; Miyasaka, J.; Ohdoi, K. On a Gross Traction Generated at Grouser for Tracked Agricultural Vehicles. *IFAC Proc. Vol.* **2013**, *46*, 311–316. [CrossRef]

Article

Evaluation of the Functional Parameters for a Single-Row Seedling Transplanter Prototype

Virgil Vlahidis [1,*], Radu Roșca [1] and Petru-Marian Cârlescu [2]

[1] Department of Pedotechnics, Faculty of Agriculture, "Ion Ionescu de la Brad" Iasi University of Life Sciences, 700490 Iași, Romania; rrosca@uaiasi.ro

[2] Department of Food Technologies, Faculty of Agriculture, "Ion Ionescu de la Brad" Iasi University of Life Sciences, 700490 Iași, Romania; pcarlescu@uaiasi.ro

* Correspondence: vlahidis@uaiasi.ro

Abstract: The development of an automatic seedling planting system for micro-farms requires testing under laboratory conditions to verify the theoretical relationships between essential functional parameters, working speed and planting time. The constructive dimensional values of the prototype, the results measured in stationary mode directly on the transplanter and the auxiliary equipment and the direct determinations of the working parameters on the soil bin are used. Depending on the characteristics of the soil bin trolley, a range of speeds is chosen at which the machine is tested. The data obtained validate the correct operation of the prototype at speeds close to those determined theoretically for the following indicators: distance between plants per row, planter wheel slippage, misplanted seedlings rate and seedling frequency, with results comparable to existing agronomic standards. Once the appropriate operating speeds of the machine have been obtained, between 0.304 and 0.412 m/s, with planting frequencies between 0.899 and 1.157 s^{-1} (respectively, 53.94 and 69.42 seedlings per minute), optimizations and adjustments of some machine components can be made, for subsequent testing in real field conditions.

Keywords: planting machine; seedling; working speed

Citation: Vlahidis, V.; Roșca, R.; Cârlescu, P.-M. Evaluation of the Functional Parameters for a Single-Row Seedling Transplanter Prototype. *Agriculture* **2024**, *14*, 388. https://doi.org/10.3390/agriculture14030388

Academic Editors: Massimiliano Varani and Jin Yuan

Received: 15 December 2023
Revised: 21 February 2024
Accepted: 24 February 2024
Published: 28 February 2024

1. Introduction

All over the world, the automation of seedling planting aims to reduce the shortage of skilled labor, to fit the planting operation into the optimal period and to obtain higher quality indicators than manual planting.

The trend is underlined by intensive research and planting machines introduced on horticultural farms in the last 20 years.

There are currently several directions for automated seedling planting. The first is the development of complex, high-productivity, towed or self-propelled automatic machines by companies from Australia, the US, Great Britain, Italy and Sweden. Characteristic of these machines is their low service staff, high travel speed, long autonomy between feedings and adaptability to different planting schemes; purchase and operating prices are high.

The underlying principles of these machines are air-pruning technology for growing seedlings [1,2] and extraction based either on vacuum from the growth cells of the trays [3] or by finger, needle, or combination systems [4,5].

Another direction of research was initiated in Japan by the agricultural divisions of Yanmar Noki, Co., (Osaka, Japan) Kubota Co., Ltd. (Osaka, Japan) and Iseki Noki Co., Ltd. (Matsuyama-shi, Japan), which developed self-propelled automatic seedling planters derived from rice planters for small and medium-sized Japanese farms. They were designed to be serviced by the machine's sole operator, have a stock of pulp mold cell seedlings trays, use a holding claw seedling extractor and perform a reduced range of planting schemes [6].

The latest approach to plantation automation can be found in research from China and India. Basically, the aim is to use robotic technology to perform the various operations required for planting: seedling extraction from the growing cell [7], seedling distribution [8], selection before planting [9] and guidance of planting machines in greenhouses [10]. Characteristically, in parallel with equipping these planting machines with high-tech components that can be integrated into precision farming, the primary goal remains the trend to develop small-farm-oriented planting systems to achieve better quality and productivity indicators, as illustrated in the mentioned research in India [8] and China [10].

Meanwhile, vegetable farms in Romania are characterized by a preponderance of subsistence micro-farms, of which more than 70% have areas of less than 5 ha; they face a decline in specialized labor and insufficient financial resources for investments on the scale of automatic planting machines and the related standardized seedling production system.

These considerations argue for the need to obtain a seedling technology accessible to small farmers in terms of price and complexity of use, with low and unskilled labor, comprising a simple construction transplanting machine with gravitational extraction, transport and distribution of seedlings with prefabricated Jiffy-type substrate, grown in rigid original plastic trays, in protected shelters. The planting machine is set up as a mono-section, fed with trays of seedlings by an operator, who also drives the aggregate formed by the machine with a walking-behind tractor.

Although it is still aimed at small farms, in comparison with the other approaches presented above, the main feature of the prototype, constructive and functional simplicity, is characterized by certain particularities in all the segments that define the interdependent system of an automatic planting technology: seedling with nutrient substrate, growing medium and automatic planting machine.

1. Jiffy pellets seedling growing media have been used for decades for a diverse range of plants for the definite advantages of good workability, compatibility with almost all growing trays, high germination rate and shock-free transplanting. They have been utilized for automated transplanting of seedlings into pots or other growing media in greenhouses, but in this technology, they are intended for obtaining seedlings directly into the tray, used then for direct transplanting in the field. In addition to the constancy of dimensional parameters, an essential advantage for the horticultural micro-farm is the elimination of obtaining different recipes for substrate mixture [8,11,12], with all the inherent difficulties: disinfection of the components, dosing, substrate formation; the first stage is more difficult, due to the thermal or chemical operations required, which sometimes also affect the viability of the seedlings [13].

2. The use of the aforementioned mechanically, pneumatically or electrically driven robotic systems for the extraction of seedlings with needles and push forks [7], with L-shaped rotating fingers [8], with clamp-type devices [14], as well as distribution devices with belt-conveyor type with cups [9,15], controlled by an electronic module [16], lead to high qualitative and functional parameters of planting, but with difficulties in operation and maintenance. In the case of the prototype intended for micro-farms in Romania, the solution to eliminate complicated and error-prone seedling extraction and transport systems was the use of exclusively gravitational extraction and multi-stage transport of seedlings with nutrient substrate. Due to the limits to which the substrate can withstand different stresses [17], the verification of Jiffy substrates to transport shocks validates the choice of this simplifying solution.

3. Compared to feeding systems from existing high-productivity automatic planters with extraction from universal (Ferrari Growtech, Guidizzolo, Italy) [18] or specialized (Williames Pty Ltd., Warragul, VIC, Australia) [19] trays and transport with different types of conveyors belt types [20,21], the prototype under study comprises a rigid alveolar tray, constructively correlated with the distribution apparatus for simultaneously gravitational discharge of a whole row of seedlings. Thus, intermediate operations performed by third-party systems between the tray and the distribution device are eliminated.

4. Existing automatic planters have either a conventional wheel-linked distribution mechanism [7] or an electronically controlled and electrically driven seedling feeding

system, independent of the machine's movement relative to the ground, based on RTK-GPS technology [22]. In both cases, a slip of the machine's transport system relative to the ground occurs [23], which implies a variation in the distance between plants per row, greater in the first category. The slippage results in non-zero seedling velocities at soil entry, resulting in altered planting depth [24,25] tilting or damage. The designed transplanter has a planting apparatus drive system allowing the zero-speed seedling's planting in relation to the soil, correlated with the slipping of the planting wheel [26].

5. Unlike duckbill [27] or dibble [28]-type planters with an external control mechanism, the system designed for the prototype has seedling release flaps from the planting device, controlled by a cam under the action of its own weight force.

The achievement of this integrated technology for planting seedlings comprises five stages, presented in Figure 1:

- Determining the need for automated seedling planting technology adapted for Romania;
- Design and construction of the machine;
- Laboratory verification of its operation and preliminary calibration of the main working parameters (working speed and planting time);
- Testing under real conditions in the field, with determination of qualitative, energy and economic parameters, for comparison with current agronomic standards;
- Technology implementation in practice.

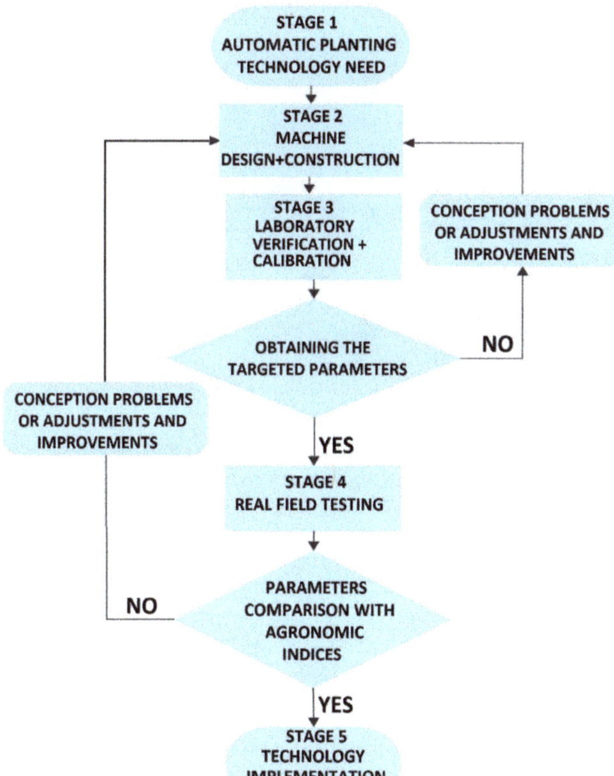

Figure 1. The stages of the automatic seedling planting technology development.

After the design and construction of the prototype, the operating process of the planting machine must be analyzed in terms of two phenomena that take place over interdependent periods of time: the time required to carry out the different operating phases of the feeding system and the time required for the seedling to complete all the stages, from removal from the tray until the entrance into the planting hole. The time intervals involved in the different stages of the planting process are intrinsically linked with the working speed of the machine, which is a defining parameter for machine performance.

These theoretically determined functional characteristics must be verified in practice through calculations and measurements, both at stationary and when operating in the soil bin, to validate the optimum operating values for real operation.

The aim of this paper is to fulfill the second stage of the research:

1. Verification of the actual performance of the main operations of the machine: seedling feeding, distribution, insertion and fixing of the seedlings in the soil;
2. To determine the range of feasible working speeds and to check that certain quality and energy indicators are within the agronomic limits.

2. Materials and Methods

2.1. Seedling Feeding Times

The schematics of the prototype of the planting machine [26], working at speed V_m, which is planting Jiffy nutritive substrate seedlings at distance d_p between the plants on a row, in a time t_p, is shown in Figure 2.

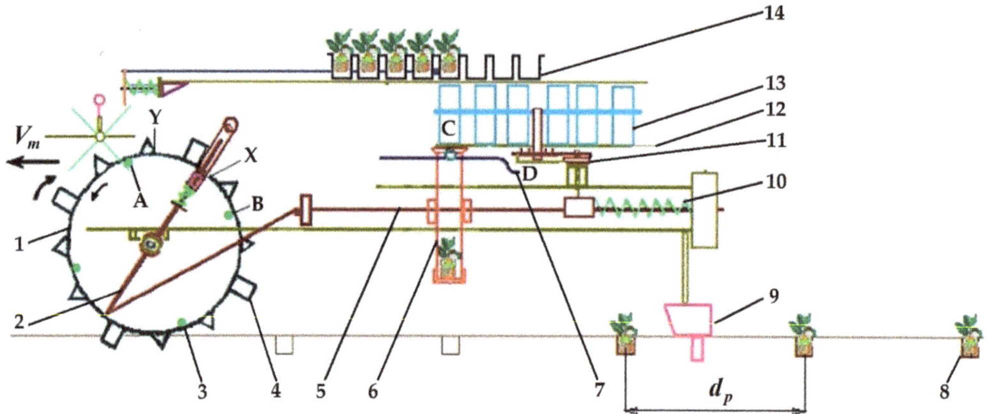

Figure 2. Schematics regarding the constructive and functional parameters of the automatic transplanter [26]: (1) planting driving wheel; (2) oscillating lever; (3) control rod; (4) spur; (5) sliding rods; (6) planting device; (7) cam guide; (8) seedling; (9) fixed winged skids; (10) return springs; (11) ratchet system; (12) holding-release screen; (13) receiving tube of the distribution apparatus; (14) seedlings tray; A—position of the control rod at the previous planting stroke; B—position of the control rod at the next planting stroke; C—position of the planting device on the cam guide at the planting stroke start; D—position of the planting device on the cam guide at the end of the planting stroke; X—oscillating lever position at the start of the planting stroke; Y—oscillating lever position at the end of the planting stroke; d_p—distance between plants on the row; BX and XY circle arcs.

The machine is shown in the operating position when the feed tube of the planting device (6) is at rest and fed with a seedling from the receiving tube of the distribution apparatus (13). At this moment, when the axis of the planting hole formed in the soil by the spur coincides with the axis of the planting device (6), the control rod (3) actuates the oscillating lever (2) on the portion corresponding to the transport stroke of the AB arc; thus, the sliding rods (5) actuate the planting device (6), which is rolling on the CD curve of the cam guide so the seedling is released when it reaches zero relative speed to the ground (Appendix A). The curve *CD* equals the fraction of the arc *AB* corresponding to the transport action (*XY* arc). The rest of the *AB* arc corresponds to the pause (*BX* arc) and return times (*AYX* arc), respectively; these intervals can be determined according to the position of the control rod on the planter wheel.

Figure 3 shows two combined movements: the distribution apparatus (rotational movement of the distribution apparatus disc with receiving tubes) and the planting device (reciprocating rectilinear movement). The seedling planter travels the corresponding distance d_p between two seedlings in a row while the planting device performs two movements:

- Movement I is the return action of the planting device from the seedling release position to the rest position for feeding (under the slot of the holding-release screen of the distribution apparatus), partially deployed during the return time (*YX* arc);
- Movement II is transporting the seedlings by the planting machine over the same distance during transport time and during the period of initiation of the return movement of the planting device (*YA* arc).

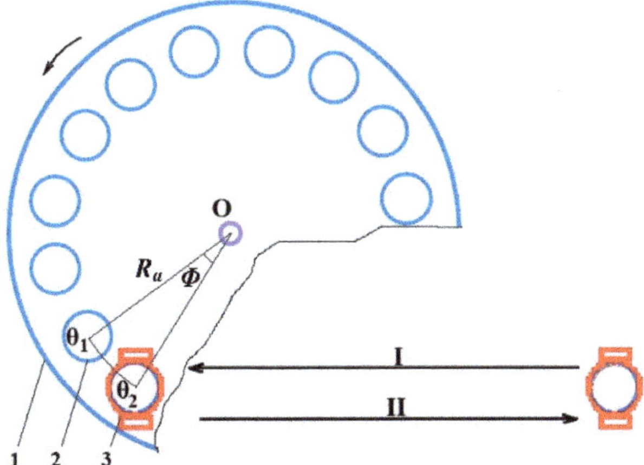

Figure 3. Aspects of distribution and planting movements: (1) distribution apparatus disc; (2) receiving tube; (3) planting device; I—return stroke; II—planting stroke; R_a—the radius of the distribution apparatus at the centers of the receiving tubes Φ—the angle at the center corresponding to the arc $\theta_1\theta_2$ described by the distribution apparatus disc for supplying a seedling to the planting device.

Between these two movements, the planting device comes to rest in relation to the machine frame (pause time) until the planter wheel control rod comes into contact with the oscillating lever drive profile, initiating the transport stroke (planter wheel *XY* arc) (Figure 1).

It should be noted that to avoid damaging the seedling, it is necessary that the seedling complete at least part of the gravity fall trajectory from the receiving tube into the planting device tube by the time the seedling transport stroke starts.

To obtain different planting patterns, the planting machine adjustment by mounting an appropriate equidistantly number of spurs on the planting wheel is based on the fact that the pause time determined by the position of the adjustable control rods on the wheel provides the appropriate time interval for correlating the theoretical distance between plants on ra ow with the interval required for the planting device seedling feeding operation.

These considerations indicate that these times overlap at certain stages of the planting device and distribution apparatus operating process, in varying proportions, depending on the working parameters: distance between plants per row and working speeds.

In conclusion, the following times can be defined as being characteristics of the operation of the planting machine described:

- Return time t_{rev}, in which two separate movements occur: that of the drive rod of the planter wheel, with V_m speed on the YA arc, and the completion of the YX arc by the oscillating driving lever, in the opposite direction relative to the wheel movement;
- Pause time t_0, corresponding to the control rod peripheral movement on the BX arch, with V_m speed;
- Transport time t_t corresponds to the movement of the control rod and oscillating lever on the XY arc, with a peripheral velocity equal to the V_m speed.

As a result, the planting time t_p is given by the relation:

$$t_p = t_{rev} + t_0 + t_t \tag{1}$$

Planting Device Supplying Time

There are two aspects of the seedlings supply to the planting device: the operation of supplying the planting device with a single seedling from the distribution apparatus and, for every four such supplies, the operation of replenishing the distribution apparatus sectors with sets of four seedlings from each row of the tray (Appendix B). Since the second operation overlaps periodically with the first, only this feeding of the planting device is of interest in determining the planting time.

The planting device supplying time t_{al} is given by:

$$t_{al} = t_{per} + t_{cd} \tag{2}$$

where:

- t_{per}—The permutation time of the seedling receiving tube from position θ_1 to θ_2 (Figure 2);
- t_{cd}—Time of gravitational fall of the seedling from the receiving tube into the planting device, s.

The correct supplying condition is as follows: at the end of the permutation stroke of the seedling receiving tube from position θ_1 to θ_2, the planting device should be at rest position under the slot of the holding-release screen of the distributor; i.e., the return time is lower than the permutation time of the receiving tube, and practically, the planter waits for the gravitational feeding of the seedling during the pause time of the transmission system. The conditions imposed by these factors can be expressed by the relations:

$$t_{rev} \leq t_{per} \tag{3}$$

$$t_{cd} \leq t_0 \tag{4}$$

The return t_{rev} and permutation times are determined experimentally by an electronic control block mounted on a planting section operated in stationary conditions.

The gravitational fall time t_{cd} was determined with the same electronic control block mounted on an auxiliary installation [26]. The destination of the auxiliary installation is the measurement of the gravitational feed times for the stages of operation of the planting equipment and for simulating the deformation of the substrates during these operations, thus determining the validity of the choice of such transportation, which is the basis of the simple operation of the machine (Appendix C).

The schematic of the electronic system used to measure the components of planting time is shown in Figure 4.

For these determinations, XNote Stopwatch V 1.63 [29] timing software was used, commanded by the micro-switch of the electronic control block, which also allows the measured times to be recorded.

The return time t_{rev} diagram of the measuring installation, presented in Figure 5, is based on the reciprocating rectilinear movement of the planting device. It is determined by measuring the time from the triggering of the micro-contact by the scotch-yoke device to the shut-off of the infrared beam between the infrared LED and the infrared phototransistor by the same part's lever returns, which has a synchronous movement with the planting device.

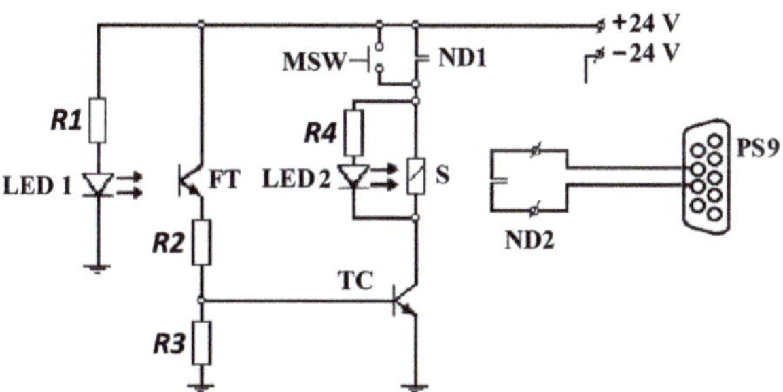

Figure 4. Electronic control block: $R1$—2 kΩ electrical resistance; $R2$—8 kΩ electrical resistance; $R3$—0.560 kΩ electrical resistance; $R4$—2 kΩ electrical resistance; LED 1—light in infrared emission; LED 2—light in infrared phototransistor; MSW—micro-contact; TC—control transistor; ND1—normally open contact; ND2—normally open contact; S—solenoid; PS9—PC nine-pin serial port.

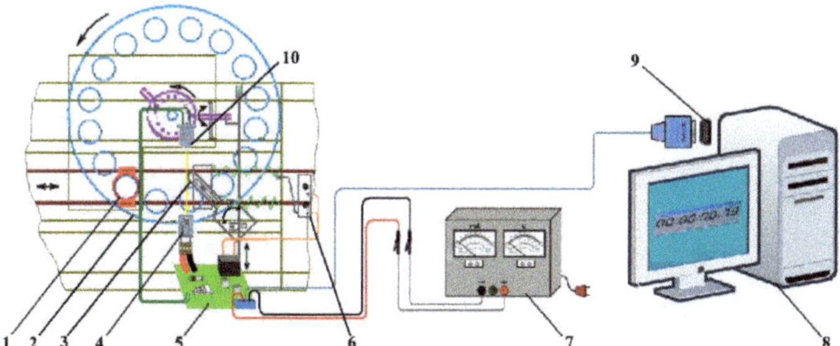

Figure 5. Installation diagram for measuring the return time of the planting device: (1) planting device; (2) distribution apparatus; (3) scotch-yoke device; (4) infrared phototransistor; (5) data acquisition board; (6) micro-switch; (7) DC power supply; (8) computer; (9) serial port; (10) infrared LED.

The measurement of the permutation time t_{per} was based on the step-by-step movement of the distribution apparatus, operated by a ratchet system, as shown in Figure 6. The start of the timer was triggered by the micro-switch through a receiving tube until the infrared beam between the infrared LED and the infrared phototransistor was interrupted by a shutter mounted on the periphery of the distribution apparatus disc.

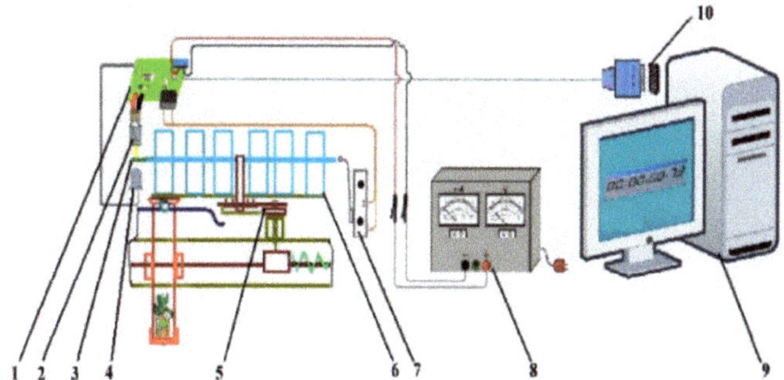

Figure 6. Installation diagram for measuring the distribution device permutation time: (1) data acquisition board; (2) infrared phototransistor; (3) shutter; (4) infrared LED; (5) ratchet system; (6) distribution apparatus; (7) micro-switch; (8) DC power supply; (9) computer; (10) serial port.

The seedling's fall time was measured by triggering a computer timer with a micro-switch when the lever system was actuated, releasing the seedling simulacrum [3]; the timer was stopped when the substrate passed between the infrared LED and photoreceptor, thus interrupting the IR beam. The auxiliary installation is presented in Figure 7.

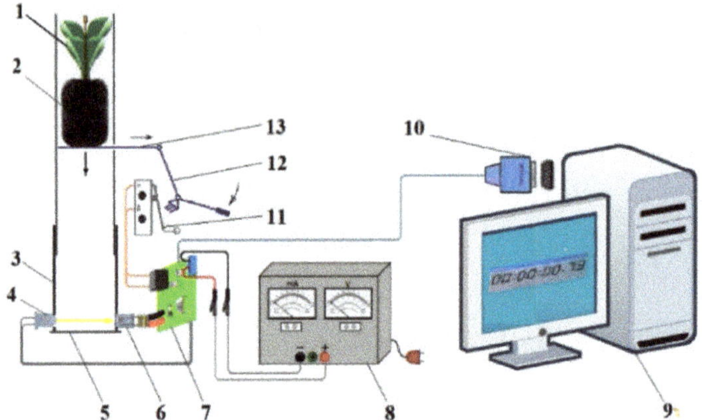

Figure 7. Test rig for measuring fall times for gravitational transport and substrate deformation: (1) seedling simulacrum foliage; (2) seedling substrate; (3) guide tube; (4) infrared LED; (5) impact surface; (6) infrared phototransistor; (7) data acquisition board; (8) DC power supply; (9) computer; (10) serial port; (11) micro-switch; (12) lever system; (13) trap door; ↓—the dropping direction of the seedling simulacrum; →—the trap door direction of opening; ⌒—the lever system direction of rotation.

2.2. Working Speed

Expressing the pause time t_0 as a function of the arc BX leads to:

$$t_0 = \frac{\overline{BX}}{V_m} \Rightarrow V_m = \frac{\overline{BX}}{t_0} \tag{5}$$

where V_m is working speed, m/s, t_0 is pause time, s and the arc length BX, m.

Operating condition (5) allows the determination of the planter's working speed limit:

$$V_{\lim} \leq \frac{\overline{BX}}{t_{cd}} \tag{6}$$

where V_{lim} is the theoretical maximum working speed of the machine, m/s.

To calculate the actual range of working speeds of the planting machine, bounded by the theoretically determined speed limit, the definitions of the times mentioned in Section 2.1 were corroborated with the experimentally determined values on the said installations and with the measured constructive parameters of the planter wheel equipped with four spurs.

For the last-mentioned measurement, the arcs BX and XY (Figure 2) were determined with the machine at rest and the wheel immobilized immediately after the control rod released the oscillating lever, which reached the end of the return stroke.

2.3. Soil Bin Studies

The prototype was tested on a soil bin in order to verify the correct functioning of the component systems and to evaluate the constructive, functional and energetic parameters; one of the parameters investigated was the working speed of the equipment. The speed values for which planting of the seedlings is performed properly can be compared with the current values for existing automatic and semi-automatic planting machines; in particular, the maximum value at which proper planting can be performed may provide information regarding the assumptions that led to the theoretical determination of the maximum working speed.

The soil bin shown in Figure 8 [30] was equipped with a trolley on which the tested prototype was mounted. The horizontality and working depth of the planting equipment may be adjusted using screw mechanisms.

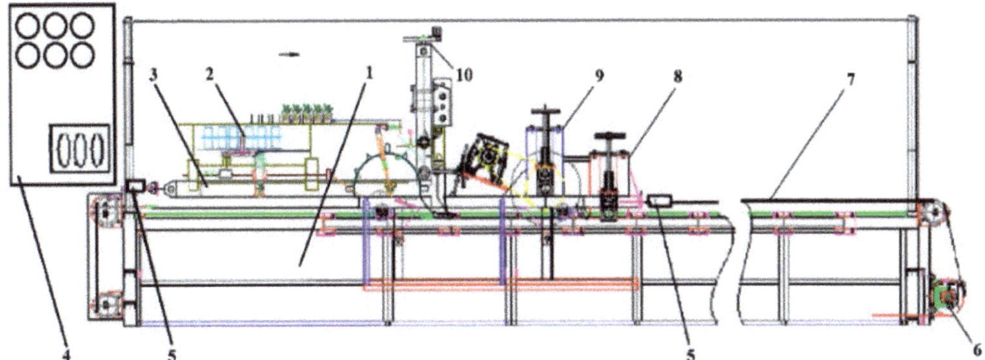

Figure 8. Soil bin test rig: (1) soil channel frame; (2) planting machine; (3) trolley; (4) control panel; (5) sensors for measuring the pulling force; (6) electric motor for driving the trolley; (7) pulling cable; (8) vertical loading mechanism on the roller; (9) vertical loading mechanism on the wheel; (10) mechanism for vertical adjustment of the machine's clamping support.

The towing cable is coupled to the trolley by two LAUMAS Model SL C3 1000 daN tension load cells, allowing for an average tensile force value to be displayed by a controller.

The rotation speed of the electric motor (5.5 kW and 1000 rpm) may be adjusted by means of a frequency converter (type OMRON Hitachi JX Inverter) to vary the frequency of the supply current between 3 Hz and 50 Hz, allowing the working speed to be modified within wide limits, from 0.1 m/s to 1.55 m/s (0.36. . .5.58 km/h); it is also possible to control the return movement of the trolley for a new test run on the channel [30].

All these adjustment possibilities will allow the determination of preliminary speeds for prototype testing.

Four indicators were analyzed, with a direct correlation with working speed:

1. A first quality indicator, the relative deviation of the distance between seedlings per row, by measuring the distance between the seedlings on the row's axis, between the stems points of insertion in the substrate, with a roller; the arithmetic mean was taken:

$$d_{pm} = \frac{\sum_{i=1}^{y} d_{pi}}{y} \tag{7}$$

where:

- $y = z - 1$ is the number of intervals between z planted seedlings.
- d_p—Distance between plants per row, m.

The relative deviation of the distance between plants from the arithmetic mean A_d was calculated with the usual relation:

$$A_d = \frac{\sqrt{\dfrac{\sum_{i=1}^{y} (d_{pm} - d_{pi})^2}{y-1}}}{d_{pm}} \cdot 100 \tag{8}$$

2. A second qualitative index, the misplanted rate; seedlings that do not survive after planting are in one of the following situations: broken at the stem insertion into the substrate, with the substrate not inserted into the soil, buried in the soil and sloped. Regarding the simulacrum seedlings, these situations are similar, except that the real stem breaking is replaced by stem wire bending. The percentage of misplanted M_s seedlings is:

$$M_s = \frac{s_w}{s} \cdot 100 \tag{9}$$

where:

- s_w is the number of the misplanted seedlings.
- s is the total number of planted seedlings.

3. An energetic indicator, the slip coefficient of the planter wheel; knowing the theoretical value between plants per row d_{pt} (for four spurs on the planter wheel), the wheel slip α was calculated with the relation:

$$\alpha = \frac{d_{pm} - d_{pt}}{d_{pm}} \cdot 100 \tag{10}$$

4. A qualitative economic exploitation indicator, planting frequency; F_s represents the number of seedlings distributed per second and is determined by the speed of the machine traveling the average distance between seedlings per row:

$$F_s = \frac{V_m}{d_{pm}} \tag{11}$$

The agronomic indicators of automatic seedling planting to be met by the four mentioned indicators are taken from the literature and are presented in Table 1.

Table 1. Agronomic requirements for seedling planting.

Agronomic Planting Indicator	Relative Deviation A_d (%)	Misplanted Rate M_s (%)	Planter Wheel Slippage α (%)	Planting Frequency F_s (s^{-1})
Value	5–10 [1]	6–10 [2]	10–15 [3]	0.83–5 [4]

[1] Maximum values [8,10,31–35]; [2] Maximum values [24,35–37]; [3] Maximum values [36,38]; [4] Extended values range [10,24,31,33,35,39].

Due to the relatively small number of seedlings planted along the active length of the soil channel (10 seedlings), three repetitions were performed for each selected speed.

3. Results

3.1. Stationary Collected Data

In order to determine the return time of the planting device and the permutation time of the distribution apparatus, the transplanter prototype was mounted on a suitable stationary stand, which allowed the planter wheel to be driven, with the rotational speed measured by means of a version of the same electronic control block, with the microswitch triggered by the control rods of the wheel. The rotational speeds of the planter wheel, ranging from 0.47 to 2.38 rev/s, correspond to the calculated working speeds of the transplanter between 0.1 and 0.5 m/s, by the relation:

$$V = \omega \cdot R \tag{12}$$

where:

- V is the working transplanter speed, m/s;
- ω is the rotational speed, rev/s;

— R is the planter wheel radius, m.

Since the planter wheel has four spurs for making planting holes, eight complete wheel rotations are required to make a number of repetitions equal to the number of seedlings contained in a planting tray (32 seedlings).

3.1.1. The Planting Device Return Time Determination

The return times of the planting device for each machine's working speed and planter wheel rotational speed, determined on installation shown in Figure 9, were processed by calculating the standard deviation from the arithmetic mean (Table 2).

Figure 9. Installation configuration used to determine the return time of the planting device: (1) planting device; (2) oscillating sliding bar; (3) oscillating scotch-yoke lever; (4) micro-switch; (5) infrared LED; (6) infrared phototransistor; (7) data acquisition board.

Table 2. Planting device return time *.

Return Time Speed Rotational/Working	t_{rev1} (s)	t_{rev2} (s)	t_{rev3} (s)	t_{rev4} (s)	t_{rev5} (s)
	$\omega_1 = 0.47$ rev/s $V_1 = 0.1$ m/s	$\omega_2 = 0.95$ rev/s $V_2 = 0.2$ m/s	$\omega_3 = 1.42$ rev/s $V_3 = 0.3$ m/s	$\omega_4 = 1.90$ rev/s $V_4 = 0.4$ m/s	$\omega_5 = 2.38$ rev/s $V_5 = 0.5$ m/s
$X = \bar{x} \pm \sigma_{\bar{x}}$	0.3506 ± 0.0025	0.3496 ± 0.0035	0.3478 ± 0.0021	0.3406 ± 0.0018	0.3328 ± 0.0027

* The table with all measured data can be found in Appendix D, Table A4.

3.1.2. The Distribution Apparatus Permutation Time Determination

The distribution apparatus permutation time was determined, as shown in Figure 10, at the same working and rotational speeds; the measured and processed values are shown in Table 3.

Table 3. Distribution apparatus permutation time **.

Permutation Time Speed Rotational/Working	t_{per1} (s)	t_{per2} (s)	t_{per3} (s)	t_{per4} (s)	t_{per5} (s)
	$\omega_1 = 0.47$ rev/s $V_1 = 0.1$ m/s	$\omega_2 = 0.95$ rev/s $V_2 = 0.2$ m/s	$\omega_3 = 1.42$ rev/s $V_3 = 0.3$ m/s	$\omega_4 = 1.90$ rev/s $V_4 = 0.4$ m/s	$\omega_5 = 2.38$ rev/s $V_5 = 0.5$ m/s
$X = \bar{x} \pm \sigma_{\bar{x}}$	0.7784 ± 0.0052	0.7762 ± 0.0050	0.7765 ± 0.0067	0.7734 ± 0.0070	0.7718 ± 0.0055

** The table with all measured data can be found in Appendix D, Table A5.

Figure 10. Installation configuration used to determine the distribution apparatus permutation time: (1) distribution apparatus disc; (2) receiving tube; (3) shutter; (4) acquisition data board; (5) Infrared LED; (6) micro-switch; (7) infrared phototransistor.

3.1.3. The Gravitational Fall Time Determination

To avoid growing a large number of seedlings for experiments, Jiffy's substrate seedling simulacrum is used, consisting of a real Jiffy substrate and a wire stem with plastic leaves imitation seedling, presented in Figure 11, made according to the model exposed by Hallonborg, U. [3]. The parameters of a 32-simulacrum set are measured after two-phase watering of the substrate according to the Jiffy manufacturer's instructions [40] and are presented in Table 4.

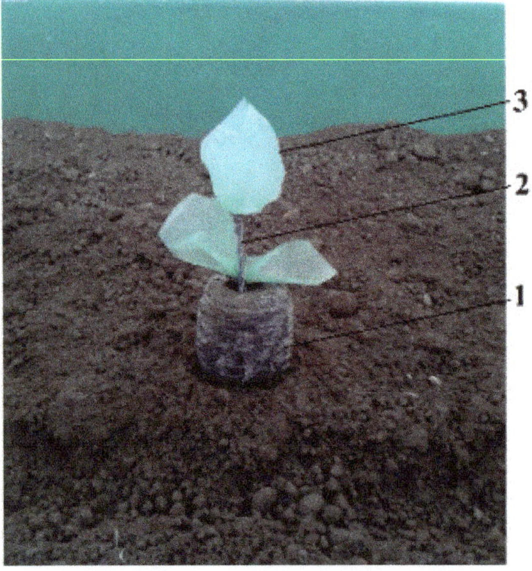

Figure 11. Substrate seedling simulacra: 1. plug; 2. stem; 3. leaves.

Table 4. Seedling simulacra mean parameters *.

Seedling Height [mm]	Stem Height [mm]	Substrate Height [mm]	Substrate Diameter [mm]	Weight [g]	Foliage Diameter [mm]	Leaves Number [pcs.]
157 ± 0.1045	79 ± 0.1254	40.28± 0.1411	41.11 ± 0.2374	58 ± 0.3236	45 ± 1.175	2–4

* The data are the parameters of seedling simulacra after the first gravitational transport presented in Appendix C.

The auxiliary system, shown in Figure 12, was adjusted to measure the gravitational fall time of the seedling simulacrum for the distribution apparatus supply stage from the lower end of the receiving tube (13) into the planting device (6) (as shown in Figure 2), where at the bottom end of the tube, a retention system keeps the substrate seedling until the last operation, the gravitational transport to the planting hole in the soil. The exact height of the tube of the planting device (6), which is 228 mm, represents the actual gravitational fall height for this stage of the planting machine supply operation.

Figure 12. Auxiliary installation for seedling's fall time determination adjusted for the height of 228 mm: (1) receiving tube; (2) trapdoor; (3) planting device tube; (4) plate for stopping the seedling; (5) infrared phototransistor; (6) electronic device (with Infrared LED); (7) micro-switch; (8) lever.

The values determined, and the results regarding the standard deviation from the arithmetic mean are shown in Table 5.

Table 5. Seedlings fall time values.

No.	t_{cd} (s)	No.	t_{cd} (s)	No.	t_{cd} (s)	No.	t_{cd} (s)
1	0.27	9	0.22	17	0.29	25	0.26
2	0.22	10	0.29	18	0.29	26	0.21
3	0.25	11	0.25	19	0.28	27	0.25
4	0.22	12	0.27	20	0.31	28	0.28
5	0.22	13	0.28	21	0.29	29	0.28
6	0.29	14	0.27	22	0.22	30	0.29
7	0.29	15	0.24	23	0.25	31	0.24
8	0.24	16	0.22	24	0.28	32	0.23
$X = \bar{x} \pm \sigma_{\bar{x}}$			0.2590 ± 0.045				

3.2. Soil Bin Test Data

After the soil preparation operations in the bin with the plow attached to the trolley and by processing with the mini-tiller, the degree of soil granulation was analyzed using a metric frame, a set of sieves and a precision balance to determine the percentages of soil components, as shown in Figure 13a. The determinations of the degree of soil compaction and soil moisture at the planting depth of 0–0.150 m were made using the handheld Eijkelkamp cone penetrometer, with humidity electrical conductivity soil sensor ThetaProbe type ML2x (using the software mentioned in Supplementary Materials), as illustrated in Figure 13b.

Figure 13. Aspects of the determinations of soil characteristics in the test bin: (**a**) soil particle size analysis: (1) metric scale; (2) soil sieves set; (3) precision balance; (**b**) penetration resistance and soil moisture determinations are cited.

Following the two types of determinations, the soil in the test bin can be described as a chernozem having a loam-clay texture with 38% clay, 32% silt and 30% sand, with aggregates below 20 mm, moisture content of 19% and compaction of 0.28–0.32 MPa.

During several runs of the soil bin trolley with the planting machine mounted and performed at different speeds without seedling planting, time was measured and divided by the active length of the stand; the speeds obtained are listed in Table 6.

Table 6. Soil bin speed selection.

Parameters/Runs	1	2	3	4	5	6	7	8	9
Frequency (Hz)	10	14.6	19.9	21.7	28.3	31.5	35.4	40.3	48.2
Electrical resistance (KΩ)	1.0	3.0	4.5	5.0	6.5	7.0	7.5	8.0	8.5
Speed V_m (m/s)	0.102	0.150	0.208	0.227	0.285	0.304	0.353	0.412	0.528

With the machine mounted on the soil bin trolley, with the planting wheel engaged into the soil and with the same adjustment of the fixing skids, at speeds close to those used to determine the planting time components, planting runs were executed (Figure 14). The distances between the planted seedlings simulacra were then measured, as shown in Figure 15; both raw and processed data are presented in Table 7 and Figure 16.

Figure 14. Planting machine run in the soil bin (video capture).

(a)   **(b)**

Figure 15. Seedlings simulacra in the soil bin: (**a**) planted in a row at the run end; (**b**) measuring the distance between seedlings: (1) planted seedling; (2) measuring tape; (3) towing cable.

Table 7. Measured and calculated indicators.

No.	Working Speed V_m (m/s)	Seedlings Distance Per Row d_p (m)				Relative Deviation A_d (%)	Misplanted Rate M_s (%)	Wheel Slip α (%)	Planting Frequency F_s (s^{-1})
		Min	Max	Mean	St Dev				
1	$V_1 = 0.102$	0.308	0.396	0.333	0.0151	4.534	<u>10</u>	3.60	0.306
2	$V_2 = 0.208$	0.323	0.352	0.335	0.0080	2.388	0	4.179	0.620
3	$V_3 = 0.304$	0.324	0.356	0.338	0.0086	2.544	3.33	5.029	0.899
4	$V_4 = 0.412$	0.331	0.382	0.356	0.0141	3.960	3.33	9.831	1.157
5	$V_5 = 0.528$	0.333	0.394	0.370	0.0160	4.324	<u>13.33</u>	<u>13.24</u>	1.427

<u>Underlined values:</u> exceeded agronomic limits.

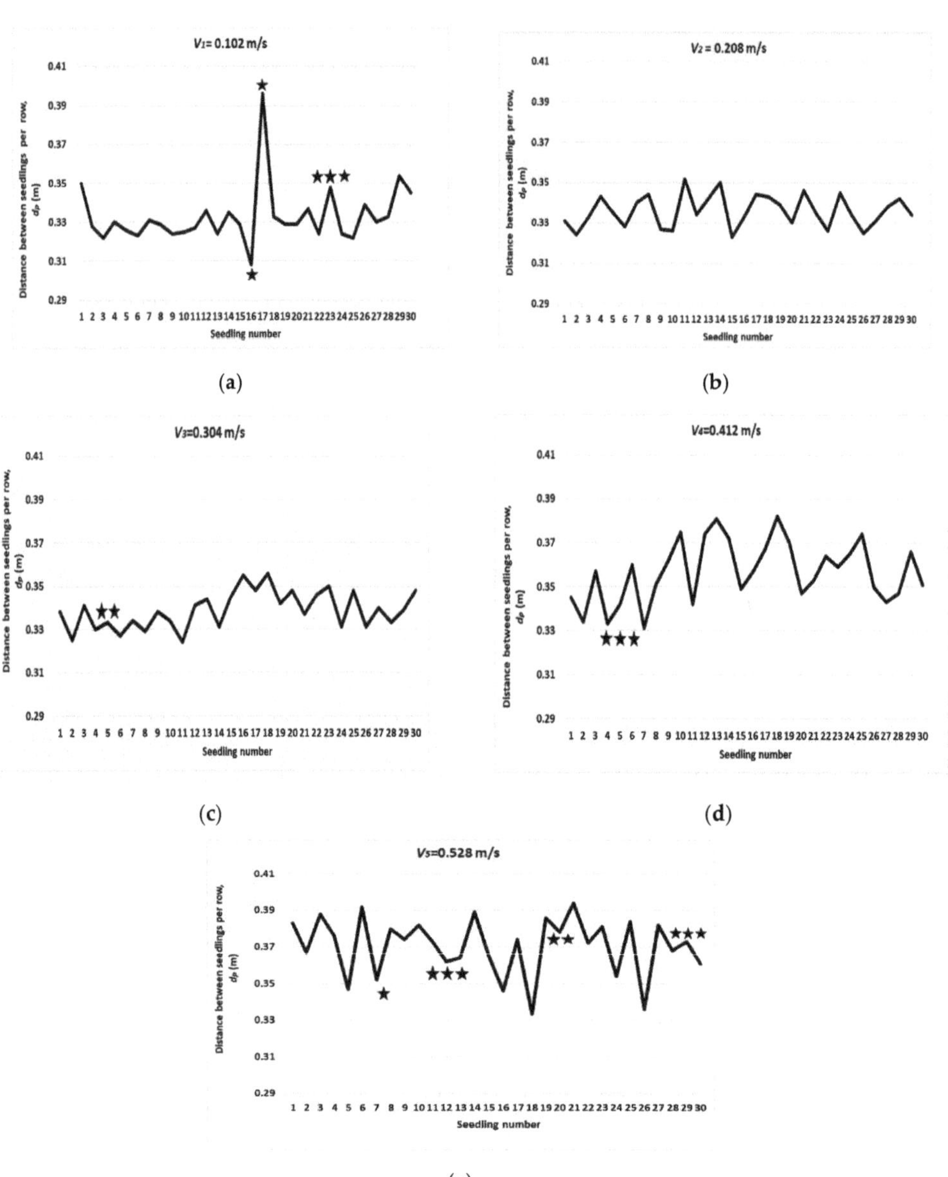

Figure 16. Distances measured between seedlings on the row. (**a**) Results for V_1; (**b**) results for V_2; (**c**) results for V_3; (**d**) results for V_4; (**e**) results for V_5. Errors due to: ★ distribution apparatus blockages, ★★ unfixed seedling, ★★★ wrongly planted seedling.

3.3. Wheel Planter Construction Parameters

Measuring the circle arcs of the planting driving wheel (with diameter $D = 0.420$ m) under the conditions mentioned in Section 2.2 produced the following results:

$$\overline{XY} = 0.085; \quad \overline{BX} = 0.138 \tag{13}$$

where the length of the arc XY, m.

Based on experimental results, it was possible to use the arithmetic mean of the fall time corresponding to transport II from Table 3, i.e.,

$$t_{cd} = t_2 = 0.2590 \tag{14}$$

It therefore follows:

$$V_{\lim} \leq \frac{\overline{BX}}{t_{cd}} \Leftrightarrow V_m \leq 0.533 \tag{15}$$

Thus, for the theoretical maximum limit speed obtained (without slip) of 0.533 m/s (1.918 km/h) and for the theoretical distance between plants per row $d_{pt} = 0.321$ m, determined for the mentioned planting wheel configuration, the planting time t_p can be calculated with the relation:

$$t_p = \frac{d_p}{V_m} \tag{16}$$

The resulting planting time is greater than or at most equal to the minimum limit value, obtained as follows:

$$t_p \geq t_{plim} = 0.602 \tag{17}$$

The relationship (1) of the seedling planting time allows the study of the component times.

Thus, the t_{rev} return times obtained from the experimental tests (Table 1) have very close values, with arithmetic averages differing by a few tenths of a second, indicating that the recovery springs provide sufficient force to overcome resistances that arise during machine operation at different speeds; the fact that the times are inversely proportional to the five rotational speeds of the planting wheel shows that, at very low speeds, the fraction of the return time corresponding to running the YA arc at the planting wheel speed for a correspondingly longer time changes the composition of the planting time. This influence is less pronounced at high speeds, given the small size of the said arc.

For the calculation of the planting time, the measured return time corresponding to the working speed close to the calculated maximum speed V_{lim} shall be chosen, i.e., $t_{rev} = t_{rev5} = 0.3328$ s.

The pause time t_0, corresponding to the XB arc preceding the operation of the planting machine, is given by the relation:

$$t_0 = \frac{\overline{BX}}{V_m} = 0.259 \tag{18}$$

The time corresponding to the transportation of the seedling by the planting device on the curve CD, t_t, can be deduced by measuring its length or by using the size of the arc XY, equal to CD; thus, it results in the following:

$$t_t = \frac{\overline{DC}}{V_m} = 0.159 \tag{19}$$

So, planting time can be expressed as the sum of its components:

$$t_{pmin} = t_{rev} + t_0 + t_t = 0.750 \tag{20}$$

It can be seen that the value obtained, which is the minimum real planting time of a seedling, calculated with the limited maximum working speed of the machine, follows a relation (17), where the theoretical planting time does not take into account the partial overlapping of some components of the planting and distribution times, an essential element in the machine functioning in different operating modes.

4. Discussion

The data measured for different working speeds provide the possibility of evaluating the constructive and functional factors involved in the indices' variations in Table 7.

Thus, for the lowest speed V_1, difficult planting machine operations occur due to component inertia (oscillating lever-driving rod level, distribution apparatus disc), leading to an increased number of wrongly planted seedlings: tilted and insufficiently fixed (Figure 17a) and displaced, as in Figure 17b and, hence, the results in a higher value of the deviation of the distance between plants per row, even if it is well below the permissible limit. Although the misplanted rate is at the upper acceptable limit, and the wheel slip has a low value, what eliminates V_1 working speed is the very low planting frequency, below the accepted range, even for semi-automatic planting machines.

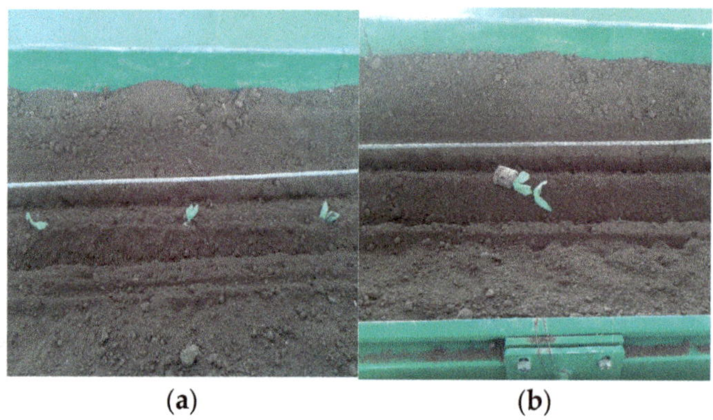

(a)   (b)

Figure 17. Planted seedlings at V_1 working speed. (**a**) Tilted and insufficiently fixed seedlings; (**b**) displaced seedlings.

The V_2 speed ensures a more constant operation of the distribution system, which has resulted in no planting errors and the lowest deviation of planting distances per row. The low travel speed (about 0.75 km/h) leads to reduced slippage but also to a low planting frequency, below the limit mentioned in the literature.

In the case of the V_3 speed, the distribution and planting machine operations are optimal, the only planting fault being the bent seedling simulacrum stem, shown in Figure 18, which would not necessarily compromise planting in a real seedling case. The distance per row deviation, the misplanted rate percentage and the planter wheel slip are within the accepted limits, while the planting frequency, 0.899 s^{-1} or 53.94 seedlings/minute, exceeds the lower accepted value for the first time. Thus, $V_3 = 0.304$ m/s or 1.094 km/h is the first speed at which the operating parameters specific to an automatic seedling planter are reached.

Figure 18. Seedling simulacrum with a bent wire stem.

For the V_4 speed, the working parameters are in normal values for the planting process, with the slip index towards the upper allowed limit. The only planting fault is a buried seedling, presented in Figure 19; the planting frequency has a good value for a process with automatic distribution, 1.157 s^{-1} or 69.42 seedlings per minute, falling well within the range indicated by the literature.

Figure 19. Buried seedling simulacrum.

For the V_5 speed, the relative deviation index of the distance between seedlings per row is in the normal range and the planting frequency is high, 1.427 s^{-1} or 85.62 seedlings per minute; however, the exceeded value of the wheel slip and, in particular, the high percentage of planting errors (such as buried and insufficiently fixed seedlings, as in Figure 20a,b) make this working speed not a viable option for machine operation in this prototype design variant.

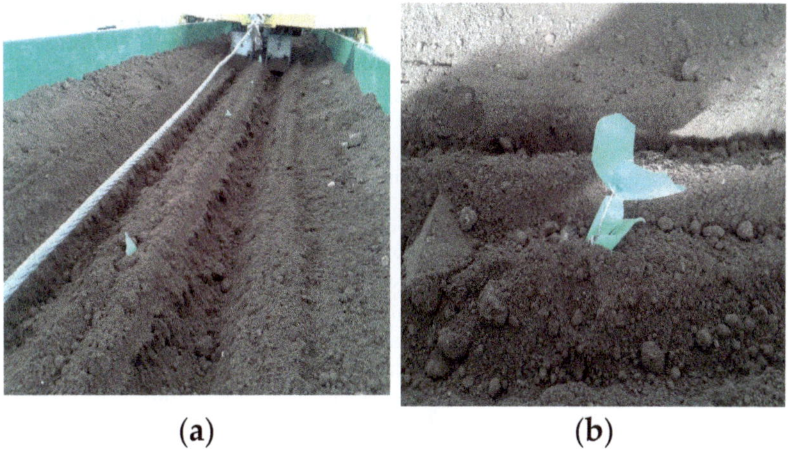

(a)   (b)

Figure 20. Transplanter with V_5 working speed; (**a**) raised planted row with buried seedlings; (**b**) tilted and insufficiently fixed seedling.

The resulting soil profile is noticeable with a raised area on the planted row, bordered by two furrows, a phenomenon responsible for some of the planting defects produced.

The relative deviation of the distance between seedlings per row, although generally increasing slightly with working speed, remains, for all five speeds, below the agronomic limit indicated in the literature.

The exception, for V_1, with the highest value of these indicators among the five variants, is precisely due to the defective distribution, which led to the change in the distance between the two affected seedlings, visible on the graph in Figure 16a as a sudden variation of large amplitude.

The rate of planting faults has a similar variation, increasing with working speed, with the same exception due to planting faults at V_1, where the agronomic upper limit is reached; at V_5, however, there is a net overshoot of this indicator due to multiple planting faults, which are not only due to distribution deviations at extreme speeds but also to seedlings poor fixation by winged skids in the soil.

The most influential energy indicator is the slip coefficient of the planting wheel; in practice, its increase with the working speed is the factor that directly determines the identical variation of the distance between plants per row and indirectly the number of misplanted seedlings by altering the interaction of the fixing skids with the soil and the way the planting wheel spurs makes holes in the soil.

Planting frequency is the main indicator that selects the working speeds for future field trials, among those where planting parameters fall within the agronomic limits for automatic transplanters. From the processing of the measured data, the speeds V_1 and V_2 will not be selected for further testing, even if in V_1 tests there was only one indicator at the upper limit, but not exceeded, and there were no misplanted seedlings in V_2; the planting frequency for these two speeds is too low, even for manually fed machines.

In the case of speed V_5, there is a net overshoot of the slip value, and hence, as pointed out, a high misplanted rate, which also eliminates this speed, even if a very good planting frequency is achieved.

Speeds V_3 and V_4, with good values of the other indicators, are selected as guideline speeds for field tests precisely because of the planting frequency, acceptable for V_3 (according to some cited expert sources, others indicating as a minimum limit a frequency of $1\ \text{s}^{-1}$, i.e., 60 seedlings per minute) and good for V_4.

Future Research Directions

Looking at these results, it can be seen that the range of eligible speeds is relatively narrow, and ways need to be found to expand it. While we have shown that it is not feasible from qualitative and economic points of view to choose low speeds, it is logical to eliminate the shortcomings found to achieve higher working speeds.

As shown in relation 15, the theoretical maximum working speed of the machine $V_{lim} = 0.533\ \text{m/s}$ and the theoretical maximum planting frequency deduced is $1.66\ \text{s}^{-1}$ or 99.62 seedlings/minute (neglecting both aspects related to the real deployment of some feeding times and wheel slip). The V_5 speed has a value close to the limit speed, and obviously, this will be the target working speed if some adjustments or constructive and functional aspects of the prototype are modified.

For this last working speed V_5, it should be noted that the process of supplying and transporting the seedlings works correctly; the planting faults are due to the process of introduction and fixing into the soil: incorrectly formed hole, altered (tilted) fixation, and buried seedlings.

Thus, four systems need to be modified to remove or mitigate these shortcomings to achieve higher working speeds in the future.

The first aspect, even if it has mainly occurred at low speeds, the operating deviations due to the high inertia of the distribution apparatus disc, behaving like a flywheel when actuated by the ratchet mechanism, with intermittent movement at low rotational speeds, require its modification or a more efficient immobilization system.

The second issue concerns the shape of the spurs on the wheel (Figure 21a) for forming planting holes, which at high speeds with high slippage tend to form elongated holes, shown in Figure 21b, which can favor the tilting of the seedling in the direction of the row.

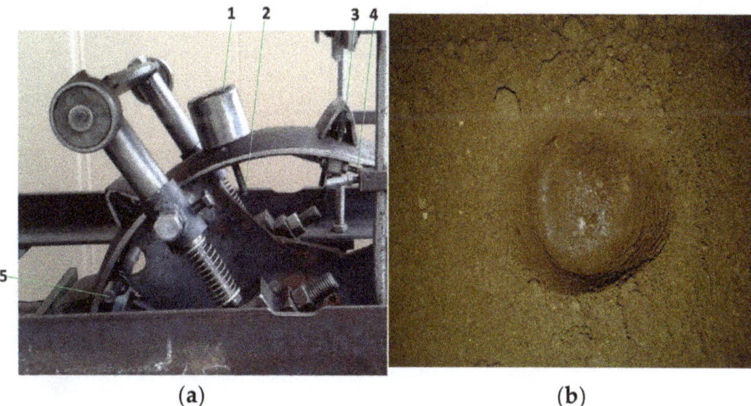

Figure 21. Influence of the spur's shape on the planting hole profile. (**a**) Detail of the planter wheel: (1) spur; (2) fixing and adjusting screw; (3) anti-slip profile; (4) threaded adjusting rod; (5) control rod. (**b**) Elongated planting hole.

The third aspect relates to the fixation of seedlings in the soil by winged fixing skids, presented in Figure 22. It was noted that to standardize the conditions for determining working speeds, the distance and angle of the skid wings are kept constant.

Figure 22. Winged fixing skids-rear view: (1) support; (2) active surface of the skid; (3) skid wing; (4) rotating hub for adjusting the distance between the skid wings: (5) the angle made by the wing with the forward direction; (6) device for adjusting the wing inclination angle with the direction of travel.

It is obvious that, in addition to changing these parameters according to the soil characteristics in field tests, even at the level of the soil bin where the soil has relatively constant properties, adjustments are necessary because soil flow behavior changes between the skid wings at different working speeds. Thus, because increasing the working speed leads to an increase in the misplanted seedlings number, as shown in Figure 16, it is necessary to increase the distance and decrease the angle between the wings to avoid displacing too vigorously and too much soil, as happened at V_5 speed, Figure 20a. In this way, it can be reduced the process of tilting or burial of seedlings already planted in the soil in the fixing process, as shown in Figure 23.

Figure 23. Tilted and partially buried seedlings in the winged skid-fixing process.

The elongated holes can also favor tilting the seedling in the direction of the row.

The fourth direction is aimed at reducing the wheel slip to the maximum—this factor is the most influential on the behavior of the planting machine. Given that control rods of the distribution system are at the base of some of the anti-slip profiles shown in Figure 21a, the change in their number and profile must be made in accordance with the normal operation of the distribution and, also with the new shape of the spurs, in order not to create variations of wheel adhesion to the ground.

In addition to the prototype testing environment, it should be noted that when testing on the soil bin, there are issues that could interfere with the functioning of the prototype and, therefore, with some of the calculated indicator values. The most important considerations in this situation are as follows:

- The slippage is exacerbated due to the soil structure damage after repeated crossings;
- Some events, such as misplanted seedlings, have a higher frequency at the ends of the run (where the soil has more unevenness: moisture, compaction, different grain sizes), observable in the context of a bin run with 10 seedlings planted.

The weight of these influences can only be assessed in actual field tests.

5. Conclusions

The paper presents the test on the soil bin of an automatic seedling planting machine prototype, designed to be attached to a walk-behind tractor, intended for vegetable micro-farms. The work aims at how the transplanter works and the optimal range of speeds for future field tests.

The most important aspects of this research stage are:

1. The results indicate that the chosen speeds, which are below the upper limit of the calculated theoretical working speed, allow for the prototype's testing, which provides indications of the correctness of the planting operation under near-real conditions.
2. The chosen speeds at which the planting parameters comply with agronomic indicators are comparable to those of existing automatic planting machines.

3. The working speeds suitable for this configuration of the automatic seedling transplanter prototype are within the range given by the speeds V_3 and V_4, respectively, 0.304 and 0.412 m/s, which have demonstrated the fulfillment of the agronomic indicators imposed by the literature.

4. The prototype achieves seedling planting frequencies of 0.899 and 1.157 s^{-1}, 53.94 and 69.42 seedlings per minute at the mentioned speeds.

At this point, it should be noted that there are no viable working speeds below 0.304 m/s, given the planting frequency towards the lower agronomic limit obtained for V_3, which is 0.899 s^{-1} (nearly 54 seedlings per minute). Still, there are probably working speeds higher than V_4 = 0.412 m/s in accordance with the fact that theoretical calculated speed limit (with real planting and distribution times taken into account), given by the relations (16) and (20), is 0.428 m/s.

In order to achieve higher working speeds and thus higher quality and economic indicators, in particular higher planting frequencies, following the analysis of the deficiencies found, directions for improvement of the seedling planter prototype were established for future field trials:

- Some low mass inertia components, such as the distribution apparatus disc, should be made to avoid malfunctions at low speeds;
- The shape of the spurs should be modified (e.g., by adopting a double-circular arc profile design) to obtain appropriate openings over the whole speed range;
- The anti-slip profiles need resizing and reconfiguration to minimize planter wheel slip and improve self-cleaning of soil that tends to stick;
- The setting of the fixing skid wings must be correlated with the soil type, moisture content, and working speed, a priority being establishing an appropriate relationship for effective adjustment.

6. Patent

Applicant:
Universitatea de Științe Agricole și Medicină Veterinară "Ion Ionescu de la Brad" [Ro]-Iasi University of Life Sciences (IULS).
Inventor:
Vlahidis Virgil, "Ion Ionescu de la Brad"Iasi University of Life Sciences (IULS), Faculty of Agriculture, Department of Pedotechnics, Romania [RO]
Title:
Equipment for automatically planting seedlings with nutritive pots
Registered by:
OSIM-State Office for Inventions and Trademarks Romania
Priorities:
RO201500134A·2015-02-24
Published as:
RO130295 on 30.03.2022

Supplementary Materials: The following software was accessed and used in this paper: Eijkelkamp Penetrologger, The following supporting information can be downloaded at: https://eijkelkamp-penetroviewer.software.informer.com (accessed on 10 December 2023).

Author Contributions: Writing—original draft preparation, V.V.; review and editing, R.R. and P.-M.C. All authors have read and agreed to the published version of the manuscript.

Funding: This research received no external funding.

Institutional Review Board Statement: Not applicable.

Data Availability Statement: Supplementary data regarding this article will be made available by the corresponding author on reasonable request.

Conflicts of Interest: The authors declare no conflicts of interest.

Appendix A

Avoiding mechanical, pneumatic or hydraulic drive variants with electronic control of the planting device led to the simple design variant of drive from the wheel in contact with the ground. The novelty lies in the lever system (Figure 2), which consists of oscillating levers and sliding rods controlled by the planter wheel that drives the planting device in the two movements shown in Figure 3.

To highlight this aspect, three captures are taken from a video of the planting equipment working on the soil channel, shown in Figure A1.

Figure A1a shows the movement imparted by the planter wheel (1) via the control rod (3) to the oscillating lever (2), which, together with the other components of the lever system, the link bars (4) and the sliding rods (7), drive the planting device (8) in stroke II (Figure 3) on the cam guide (6), with equal speed and opposite direction to the working speed. Sliding rods drive the scotch-yoke mechanism (10) during this time and compress the return springs (12), which will provide the stroke I return movement (Figure 3). The hole (9) is already formed by the spur (5) of the planter wheel and is ahead of the planter; their positions will synchronize at the end of stroke II. The distribution apparatus has a seedling (15) in the receiving tube, supported by the holding-release screen (13) for the future feeding of the planter.

Figure A1b shows the moment when the oscillating lever is towards the end of its rod control action (area A), the planting device has reached the end of stroke II, descends on the downward side of the cam guide (area B), and under the action of its weight, the flaps (16) are opened. The seedling (17) is released and falls into the hole in the ground.

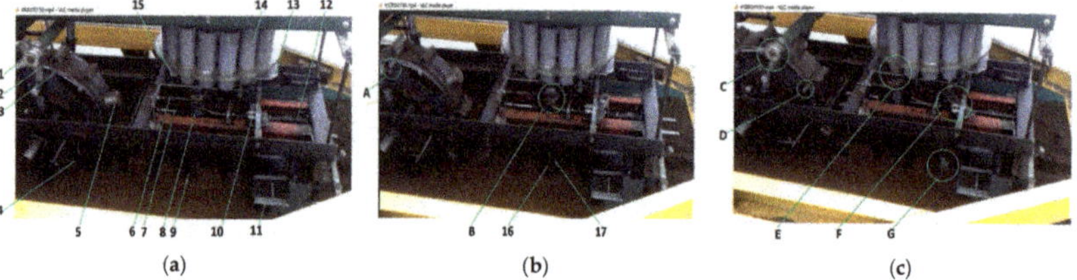

Figure A1. Aspects regarding the operation of the planting device drive system: (**a**) planting stroke II; (**b**) Seedling planting time; (**c**) Operations carried out after stroke I, during pause time t_0; (1) planting driving wheel; (2) oscillating lever; (3) control rod; (4) link bars; (5) spur; (6) cam guide; (7) sliding rods; (8) planting device; (9) planting hole; (10) ratchet system; (11) fixing skids; (12) return springs; (13) holding-release screen; (14) receiving tube of the distribution apparatus; (15) seedling ready for feeding; (16) planting device flaps; (17) Seedling in the ground hole; A—oscillating lever-rod control interaction; B—planting device on cam guide downward side; C—released oscillating lever; D—next control rod; E—new seedling for planting device feeding; F—scotch-yoke mechanism position after distribution apparatus permutation movement; G—seedling in the hole in the ground before skids soil tamping.

At this point, the relative speed of the seedling in the machine's working direction is 0, the only speed being that in the vertical direction, imparted by the last (third) gravitational transport movement made on the machine.

In Figure A1c, stroke I is completed during the duration of t_{rev} under the action of the return springs on the sliding rods. The scotch-yoke mechanism (zone F) ensured the permutation movement of the distribution apparatus, the new seedling (zone E) starting to enter the planting device tube (second gravitational transport movement), all this happening in time pause t_0, during which the planting machine will be at rest relative to the machine frame, under the seedling discharge hole of the holding-release screen. In addition, during this period of time, the oscillating lever, having escaped from the action of the previous control rod (zone C), will enter under the action of the next control rod (zone D). The seedling from the ground hole (zone G) enters under the action of the fixing skids (11).

Each variation in the transplanter linear speed due to various factors (micro-leveling, wheel slippage, areas with different soil compaction or friction coefficients) is copied into the movement of the planting device so that the relative speed in the working direction relative to the ground is 0 when the seedling is released into the planting hole.

Appendix B

In order to eliminate intermediate seedling transport systems on the prototype planter, an integrated system consisting of a tray and a distribution apparatus was designed.

The alveolar tray used is constructed to feed the distribution apparatus with groups of gravity-transmitted substrate seedlings in coordination with the operation of the planting device. The tray contains the growing cells arranged in straight rows lengthways and in rows in a circle arc shape on widthways, congruent (in terms of the number and arrangement of the cells) with a whole number of corresponding sectors of the circle of the distribution apparatus disc, as shown in Figure A2.

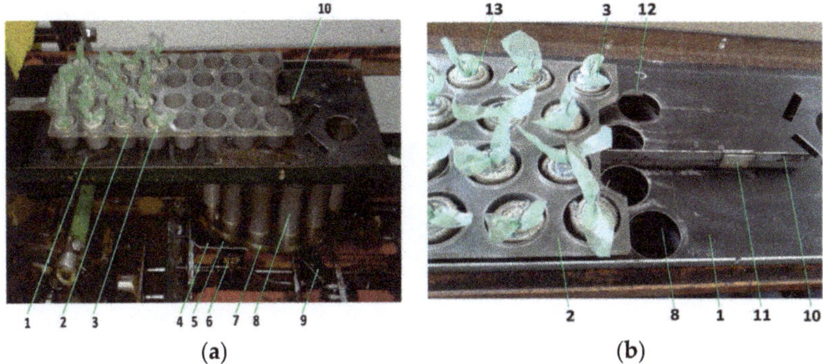

Figure A2. Seedling feeding system: (**a**) General view; (**b**) Alveolar seedling tray on feed table; (1) feed table; (2) alveolar tray; (3) current row of simulacra seedlings; (4) holding-release screen; (5) slot of the holding-release screen; (6) planting device; (7) distribution apparatus; (8) receiving tube of the distribution apparatus; (9) scotch-yoke mechanism; (10) push rod; (11) rod guide; (12) slots of the feed table; (13) simulacra seedling in growing cell.

The distribution apparatus is of the horizontal rotating disc type with receiving tubes and a holding-release screen with a slot for each seedling to fall into the tube of the planting device. The distribution device is mounted on the shaft of the ratchet mechanism, which drives it in a stepwise permutation movement. The role of this apparatus is to allocate the seedlings, one by one, from the gravitational unloaded row from the alveolar tray to the planting device.

When the last seedling of the previous series falls from the receiving tube (8) through the slot (5) of the holding-release screen (4) into the planting device tube (6), the push rod (10) pushes the alveolar tray (2) with the growing cell with seedling (13) forward with the distance between two rows of trays.

Through the slots of the feed table (1), the current row (3) of the tray is discharged into the receiving tubes (8) of the corresponding sector of the distribution apparatus disc (7) (Figure A2b). The seedlings with the nutrient substrate are stopped at the bottom of the tubes by the holding-release screen and slide on it at the step-by-step rotation of the distribution apparatus, printed by the ratchet device driven by scotch-yoke mechanism (9); when the slot in the hold-release screen is reached, the first seedling in the group drops gravitationally into the planting device tube.

The push rod retracts and is in position by the time the last seedling in the group is delivered to the planting device; one of the control rods of the planter wheel drives the arms of the feeding system and ensures a new forward movement of the tray.

This feeding system allows the seedlings to drop from the growing tray through the distribution unit to the planting device by two successive gravitational falls in a simple and compact design.

Appendix C

The auxiliary installation was designed to reproduce both the deformation of the seedling substrate and the fall times corresponding to the three different stages of gravitational transport [26]. If the fall time t_{cd} is used to determine the feed time for the planting device, the fall times for the other two gravitational transports are used for the calculation of the alveolar tray driving system, respectively, for the dimensioning of the receiving tubes of the distribution apparatus and the cam guide.

The gravitational type of transport simplifies the transmission and seedling supply of the prototype planter but is limited by the degree of deformation that the seedling substrate can withstand, especially at multiple stages of transport and by the dynamics of falling [3], and can be adopted only after checking the seedling deformation of the Jiffy Pellets substrates after successive fall from the real heights in the working process.

The auxiliary installation is dimensionally adjusted for each fall height that reproduces the targeted gravitational transport stage, is loaded with the seedling simulacrum, and is manually operated to trigger the fall; the hatch mimics the sliding on the real supports in operation (feeding table, hold-release screen, planting device flaps).

The fall times are measured using the electronic block mentioned above, and the dimensions after the deformation of the substrates are measured with the caliper. Determinations start with the expansion of Jiffy substrates by watering according to the manufacturer's instructions; the prefabricated nutrient substrate in its initial state has the following dimensions: a diameter of 36 mm and height of 8 mm. Measure the mass and the lower and upper diameter on a batch of 32 substrates using a scale and a caliper; the values are listed in Table A1.

Table A1. Mean parameters of expanded Jiffy Seedling substrate *.

Lower Diameter [mm]	Upper Diameter [mm]	Height [mm]	Weight (Wet) [g]
39.86 ± 0.1160	38.64 ± 0.1973	41.02 ± 0.1205	44.36 ± 0.2093

* Data processed by calculating the standard deviation from the arithmetic mean.

After expansion, make the simulacrum seedling by adding the metal stem and plastic leaves. The initial parameters of the simulacra are given in Table A2.

Table A2. Seedling simulacra mean parameters before transport **.

Seedling Height [mm]	Stem Height [mm]	Substrate Height [mm]	Substrate Diameter [mm]	Weight [g]	Foliage Diameter [mm]	Leaves Number [pcs.]
158 ± 0.0345	80 ± 0.032	41.02 ± 0.1205	39.86 ± 0.1160	58.6 ± 0.6432	45 ± 1.894	2–4

** Data processed by calculating the standard deviation from the arithmetic mean.

Next, using the installation in Figure 12 for Transport II from the planting device tube and the variants for Transports I and III, shown in Figure A3, the substrate deformations corresponding to the parameters: substrate lower diameter and substrate height are determined; the first parameter serves for the actual dimensioning of the drop tubes, while the second provides information on the degree of deformation after successive transports. The values are presented in Table A3.

The mass of the simulacra was constant during the short time of the determinations; there were also no variations regarding stem and leaf simulacrums. The variations of the measured parameters are relatively small, and the most important aspect is that the seedling simulacrums maintained their integrity, which demonstrates the viability of the chosen gravity transport system.

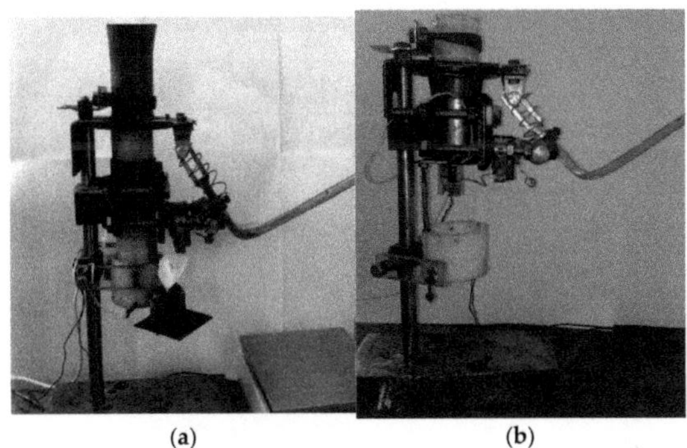

(a)   (b)

Figure A3. Auxiliary installation for gravity transport simulation. (**a**) Height adjustment for Transport II; (**b**) height adjustment for Transport III.

Table A3. Seedling simulacra parameters after transport stages ***.

Initial Values		Transport I $h_1 = 185$ mm		Transport II $h_2 = 228$ mm		Transport III $h_3 = 80$ mm	
Lower Diameter [mm]	Height [mm]	Lower Diameter [mm]	Height [mm]	Lower Diameter [mm]	Height [mm]	Lower Diameter [mm]	Height [mm]
39.86 ± 0.1160	41.02 ± 0.1205	41.11 ± 0.2374	40.28 ± 0.1411	41.53 ± 0.1129	38.88 ± 0.1569	42.80 ± 0.1039	37.10 ± 0.1767

*** Data processed by calculating the standard deviation from the arithmetic mean.

Appendix D

Table A4. Planting device return time.

Return Time Speed Rotational/Working	t_{rev1} (s) $\omega_1 = 0.47$ rev/s $V_1 = 0.1$ m/s	t_{rev2} (s) $\omega_2 = 0.95$ rev/s $V_2 = 0.2$ m/s	t_{rev3} (s) $\omega_3 = 1.42$ rev/s $V_3 = 0.3$ m/s	t_{rev4} (s) $\omega_4 = 1.90$ rev/s $V_4 = 0.4$ m/s	t_{rev5} (s) $\omega_5 = 2.38$ rev/s $V_5 = 0.5$ m/s
1	0.37	0.36	0.35	0.35	0.34
2	0.37	0.32	0.35	0.33	0.34
3	0.35	0.38	0.37	0.35	0.36
4	0.34	0.38	0.32	0.33	0.35
5	0.37	0.32	0.35	0.34	0.32
6	0.35	0.36	0.34	0.35	0.33
7	0.33	0.35	0.34	0.34	0.32
8	0.37	0.37	0.33	0.34	0.31
9	0.36	0.34	0.33	0.34	0.35
10	0.35	0.33	0.34	0.33	0.34
11	0.35	0.33	0.36	0.36	0.35
12	0.34	0.32	0.35	0.34	0.33
13	0.37	0.35	0.34	0.33	0.32
14	0.37	0.34	0.35	0.33	0.32
15	0.34	0.37	0.36	0.34	0.31
16	0.36	0.33	0.35	0.33	0.32
17	0.33	0.37	0.34	0.33	0.35
18	0.32	0.33	0.36	0.35	0.34
19	0.34	0.35	0.35	0.35	0.32
20	0.33	0.37	0.35	0.33	0.31
21	0.36	0.33	0.36	0.33	0.32
22	0.33	0.36	0.37	0.35	0.35
23	0.34	0.33	0.33	0.35	0.33
24	0.35	0.34	0.36	0.34	0.34
25	0.35	0.38	0.35	0.36	0.36
26	0.37	0.36	0.36	0.34	0.35
27	0.36	0.38	0.33	0.32	0.34
28	0.35	0.33	0.36	0.34	0.33
29	0.36	0.35	0.35	0.33	0.31
30	0.34	0.33	0.33	0.34	0.32
31	0.36	0.38	0.35	0.35	0.32
32	0.34	0.35	0.35	0.36	0.35
$X = \bar{x} \pm \sigma_{\bar{x}}$	0.3506 ± 0.0025	0.3496 ± 0.0035	0.3478 ± 0.0021	0.3406 ± 0.0018	0.3328 ± 0.0027

Table A5. Distribution apparatus permutation time.

Permutation Time Speed Rotational/Working	t_{per1} (s) $\omega_1 = 0.47$ rev/s $V_1 = 0.1$ m/s	t_{per2} (s) $\omega_2 = 0.95$ rev/s $V_2 = 0.2$ m/s	t_{per3} (s) $\omega_3 = 1.42$ rev/s $V_3 = 0.3$ m/s	t_{per4} (s) $\omega_4 = 1.90$ rev/s $V_4 = 0.4$ m/s	t_{per5} (s) $\omega_5 = 2.38$ rev/s $V_5 = 0.5$ m/s
1	0.75	0.72	0.78	0.78	0.74
2	0.75	0.74	0.83	0.73	0.80
3	0.81	0.77	0.76	0.81	0.79
4	0.74	0.74	0.75	0.71	0.80
5	0.79	0.77	0.73	0.79	0.79
6	0.74	0.71	0.79	0.75	0.76
7	0.78	0.77	0.77	0.83	0.71
8	0.74	0.80	0.72	0.79	0.81
9	0.78	0.76	0.77	0.83	0.79
10	0.81	0.82	0.75	0.72	0.77
11	0.83	0.79	0.83	0.83	0.80
12	0.82	0.77	0.78	0.82	0.76
13	0.75	0.82	0.84	0.71	0.75
14	0.77	0.77	0.75	0.81	0.73
15	0.77	0.75	0.83	0.74	0.76
16	0.80	0.80	0.79	0.74	0.82
17	0.82	0.80	0.81	0.78	0.74
18	0.81	0.75	0.78	0.76	0.79
19	0.83	0.81	0.71	0.73	0.76
20	0.79	0.81	0.82	0.80	0.81
21	0.77	0.77	0.73	0.81	0.76
22	0.75	0.82	0.81	0.79	0.79
23	0.75	0.80	0.75	0.81	0.79
24	0.73	0.78	0.83	0.70	0.81
25	0.80	0.79	0.74	0.79	0.71
26	0.76	0.80	0.84	0.78	0.81
27	0.80	0.76	0.73	0.78	0.76
28	0.74	0.77	0.80	0.77	0.72
29	0.80	0.80	0.74	0.81	0.75
30	0.79	0.78	0.78	0.70	0.75
31	0.77	0.75	0.76	0.78	0.81
32	0.77	0.75	0.75	0.77	0.76
$X = \bar{x} \pm \sigma_{\bar{x}}$	0.7784 ± 0.0052	0.7762 ± 0.0050	0.7765 ± 0.0067	0.7734 ± 0.0070	0.7718 ± 0.0055

References

1. Huang, B.K.; Ai, F. Air-pruned transplant production system for fully automated transplanting. *Acta Hortic.* **1992**, *319*, 523–528. [CrossRef]
2. Gao, J.H.; Huang, B.K. Computer-Controlled Transplanting of Air-Pruned Seedlings/Plugs. *SNA Res. Conf.* **1993**, *38*, 229–234.
3. Hallonborg, U. Wobbling in Pneumatic Systems for Transporting Seedlings. *J. For. Eng.* **1998**, *9*, 7–14. [CrossRef]
4. Zhang, N.; Zhang, G.; Liu, H.; Liu, W.; Wei, J.; Tang, N. Design of and Experiment on Open-and-Close Seedling Pick-Up Manipulator with Four Fingers. *Agriculture* **2022**, *12*, 1776. [CrossRef]
5. Jorg, O.J.; Sportelli, M.; Fontanelli, M.; Frasconi, C.; Raffaelli, M.; Fantoni, G. Design, Development and Testing of Feeding Grippers for Vegetable Plug Transplanters. *AgriEngineering* **2021**, *3*, 669–680. [CrossRef]
6. Tsuga, K. Development of fully automatic vegetable transplanter. *Jarq-Jpn. Agric. Res. Q.* **2000**, *34*, 21–28.
7. Ye, B.L.; Li, L.; Yu, G.H.; Liu, A.; Cai, D. Design and test on cam mechanism of seedling pick-up arm for vegetable transplanter for pot seedling. *Trans. CSAE* **2014**, *30*, 21–29.
8. Khadatkar, A.; Mathur, S.M.; Dubey, K.; BhusanaBabu, V. Development of embedded automatic transplanting system in seedling transplanters for precision agriculture. *Artif. Intell. Agric.* **2021**, *5*, 175–184. [CrossRef]
9. Jin, X.; Yuan, Y.; Ji, J.; Zhao, K.; Li, M.; Chen, K. Design and implementation of anti-leakage planting system for transplanting machine based on fuzzy information. *Comput. Electron. Agric.* **2020**, *169*, 105204. [CrossRef]
10. Yang, Q.; Huang, G.; Shi, X.; He, M.; Ahmad, I.; Zhao, X.; Addy, M. Design of a control system for a mini-automatic transplanting machine of plug seedling. *Comput. Electron. Agric.* **2020**, *169*, 105226. [CrossRef]
11. Jin, X.; Li, D.Y.; Ma, H.; Ji, J.T.; Zhao, K.X.; Pang, J. Development of single row automatic transplanting device for potted vegetable seedlings. *Int. J. Agric. Biol. Eng.* **2018**, *11*, 67–75. [CrossRef]
12. Han, L.; Mo, M.; Gao, Y.; Ma, H.; Xiang, D.; Ma, G.; Mao, H. Effects of New Compounds into Substrates on Seedling Qualities for Efficient Transplanting. *Agronomy* **2022**, *12*, 983. [CrossRef]
13. Parladé, J.; Pera, J.; Álvarez, I.F.; Bouchard, D.; Généré, B.; Le Tacon, F. Effect of inoculation and substrate disinfection method on rooting and ectomycorrhiza formation of Douglas fir cuttings. *Ann. For. Sci.* **1999**, *56*, 35–40. [CrossRef]
14. Iqbal, M.Z.; Islam, M.N.; Ali, M.; Kabir, M.S.N.; Park, T.; Kang, T.G.; Park, K.-S.; Chung, S.-O. Kinematic analysis of a hopper-type dibbling mechanism for a 2.6 kW two-row pepper transplanter. *J. Mech. Sci. Technol.* **2021**, *35*, 2605–2614. [CrossRef]
15. Han, C.; Hu, X.; Zhang, J.; You, J.; Li, H. Design and testing of the mechanical picking function of a high-speed seedling auto-transplanter. *Artif. Intell. Agric.* **2021**, *5*, 64–71. [CrossRef]

16. Jin, X.; Li, M.; Li, D.; Ji, J.; Pang, J.; Wang, J.; Peng, L. Development of automatic conveying system for vegetable seedlings. *EURASIP J. Wirel. Commun. Netw.* **2018**, *2018*, 178. [CrossRef]
17. Zeng, F.; Cui, J.; Li, X.; Bai, H. Establishment of the Interaction Simulation Model between Plug Seedlings and Soil. *Agronomy* **2024**, *14*, 4. [CrossRef]
18. Ferrari Growtech. Available online: https://ferrarigrowtech.com/en/tray-transplanters/8-futura-automated-transplanter.html. (accessed on 10 February 2024).
19. Williames Pty Ltd. Available online: http://www.williames.com/nursery-machinery/williames-trays-flats/ (accessed on 10 February 2024).
20. Wen, Y.; Zhang, J.; Tian, J.; Duan, D.; Zhang, Y.; Tan, Y.; Yuan, T.; Li, X. Design of a traction double-row fully automatic transplanter for vegetable plug seedlings. *Comput. Electron. Agric.* **2021**, *182*, 106017. [CrossRef]
21. Han, L.; Mao, H.; Hu, J.; Kumi, F. Development of a Riding-Type Fully Automatic Transplanter for Vegetable Plug Seedlings. *Span. J. Agric. Res.* **2019**, *17*, e0205. [CrossRef]
22. Sun, H.; Slaughter, D.C.; Ruiz, M.P.; Gliever, C.; Upadhyaya, S.K.; Smith, R.F. RTK GPS mapping of transplanted row crops. *Comput. Electron. Agric.* **2010**, *71*, 32–37. [CrossRef]
23. Pérez-Ruiz, M.; Slaughter, D.C. Development of a precision 3-row synchronised transplanter. *Biosyst. Eng.* **2021**, *206*, 67–78. [CrossRef]
24. Reza, M.N.; Ali, M.; Habineza, E.; Kabir, M.S.; Kabir, M.S.N.; Lim, S.-J.; Choi, I.-S.; Chung, S.-O. Analysis of operating speed and power consumption of a gear-driven rotary planting mechanism for a 12-kW six-row self-propelled onion transplanter. *Span. J. Agric. Res.* **2023**, *21*, e0207. [CrossRef]
25. Manilla, R.D.; Shaw, L.N. A high-speed dibbling transplanter. *Trans. ASAE* **1987**, *30*, 904–908. [CrossRef]
26. Vlahidis, V. Contributions to Improving the Seedling Planting Process. Ph.D. Thesis, "Gh.Asachi" Technical University Iași, Faculty of Mechanics, Specialisation Mechanical Engineering, Iași, Romania, 2014.
27. He, T.; Li, H.; Shi, S.; Liu, X.; Liu, H.; Shi, Y.; Jiao, W.; Zhou, J. Preliminary Results Detailing the Effect of the Cultivation System of Mulched Ridge with Double Row on Solanaceous Vegetables Obtained by Using the 2ZBX-2A Vegetable Transplanter. *Appl. Sci.* **2023**, *13*, 1092. [CrossRef]
28. Wang, Y.; He, Z.; Wang, J.; Wu, C.; Yu, G.; Tang, Y. Experiment on transplanting performance of automatic vegetable pot seedling transplanter for dry land. *Trans. Chin. Soc. Agric. Eng.* **2018**, *34*, 19–25. [CrossRef]
29. Xnote Stopwatch V 1.63. Available online: http://www.xnotestopwatch.com/download.html (accessed on 11 December 2023).
30. Țenu, I.; Roșca, R.; Cârlescu, P.; Vlahidis, V. *Laboratory Stand for Impact Study Agricultural Traffic and Technological Works on Soil Physical Properties*; Iasi Polytechnic Institute Bulletin, Machine Constructions Section: Iași, Romania, 2012; Volume LVIII (LXII), pp. 255–262.
31. Stubbs, S.; Colton, J. The Design of a Mechanized Onion Transplanter for Bangladesh with Functional Testing. *Agriculture* **2022**, *12*, 1790. [CrossRef]
32. Khadatkar, A.; Mathur, S.M.; Dubey, K.; Magar, A.P. Automatic Ejection of Plug-type Seedlings using Embedded System for use in Automatic Vegetable Transplanter. *J. Sci. Ind. Res.* **2021**, *80*, 1042–1048. [CrossRef]
33. Prasanna Kumar, G.V.; Raheman, H. Development of a walk-behind type hand tractor powered vegetable transplanter for paper pot seedlings. *Biosyst. Eng.* **2011**, *110*, 189–197. [CrossRef]
34. Brewer, H.L. Experimental Automatic Feeder for Seedling Transplanter. *Appl. Eng. Agric.* **1988**, *4*, 24–29. [CrossRef]
35. Liu, J.; Zhao, S.; Li, N.; Faheem, M.; Zhou, T.; Cai, W.; Zhao, M.; Zhu, X.; Li, P. Development and field test of an autonomous strawberry plug seeding transplanter for use in elevated cultivation. *Appl. Eng. Agric.* **2019**, *35*, 1067–1078. [CrossRef]
36. Yang, L.; He, X.T.; Cui, T.; Zhang, D.X.; Shi, S.; Zhang, R.; Wang, M. Development of mechatronic driving system for seed meters equipped on conventional precision corn planter. *Int. J. Agric. Biol. Eng.* **2015**, *8*, 1–9. [CrossRef]
37. Iqbal, M.Z.; Islam, M.N.; Chowdhury, M.; Islam, S.; Park, T.; Kim, Y.-J.; Chung, S.-O. Working Speed Analysis of the Gear-Driven Dibbling Mechanism of a 2.6 kW Walking-Type Automatic Pepper Transplanter. *Machines* **2021**, *9*, 6. [CrossRef]
38. Garibaldi-Márquez, F.; Martínez-Reyes, E.; García-Hernández, R.V.; Galindo-Reyes, M.A. Planter to distribute seeds in four-row beds and harvest rainwater. *Ing. Agrícola Y Biosist.* **2020**, *12*, 21–40. [CrossRef]
39. Wen, Y.; Zhang, L.; Huang, X.; Yuan, T.; Zhang, J.; Tan, Y.; Feng, Z. Design of and Experiment with Seedling Selection System for Automatic Transplanter for Vegetable Plug Seedlings. *Agronomy* **2021**, *11*, 2031. [CrossRef]
40. Jiffy Pellets Products. Available online: https://jiffygroup.com/products/jiffy-pellets/#jiffypellets (accessed on 3 January 2024).

Article

Study on the Hole-Forming Performance and Opening of Mulching Film for a Dibble-Type Transplanting Device

Xiaoshun Zhao [1,2], Zhuangzhuang Hou [2], Jizong Zhang [3,*], Huali Yu [2], Jianjun Hao [2] and Yuhua Liu [3]

[1] State Key Laboratory of North China Crop Improvement and Regulation, Baoding 071001, China; zxs@hebau.edu.cn
[2] College of Mechanical and Electrical Engineering, Hebei Agricultural University, Baoding 071001, China; 20232090224@pgs.hebau.edu.cn (Z.H.); yhl@hebau.edu.cn (H.Y.); hjj@hebau.edu.cn (J.H.)
[3] College of Agronomy, Hebei Agricultural University, Baoding 071001, China; liuyuhua@hebau.edu.cn
* Correspondence: zhjz@hebau.edu.cn

Abstract: In order to improve the quality of transplanting devices and solve the problems of the poor effect on soil moisture conservation and more weeds easily growing due to the high mulching-film damage rate with an excessive number of hole openings, we developed a dibble-type transplanting device consisting of a dibble-type transplanting unit, a transplanting disc, and a dibble axis. The ADAMS software Adams2020 (64bit) was used to simulate and analyze the kinematic track of the transplanting device. The results of the analysis show that, when the hole opening of the envelope in the longitudinal dimension was the smallest, the transplanting characteristic coefficient was 1.034, the transplanting angle was 95°, and the transplanting frequency had no influence. With the help of the ANSYS WORKBENCH software Ansys19.2 (64bit), an analysis of the process of the formation of an opening in the mulching film and a mechanical simulation of this process were completed. The results indicate that, when the maximum shear stress of the mulching film was the smallest, the transplanting characteristic coefficient was 1.000, the transplanting frequency was 36 plants·min^{-1}, and the transplanting angle was 95°. In addition, the device was tested in a film-breaking experiment on a soil-tank test bench to verify the hole opening in the mulching film. The bench test showed that, when the longitudinal dimension was the smallest, the transplanting characteristic coefficient was 1.034, the transplanting frequency was 36 plants·min^{-1}, and the transplanting angle was 95°. When the lateral dimension was the smallest, the transplanting characteristic coefficient was 1.034, the transplanting frequency was 36 plants·min^{-1}, and the transplanting angle was 90°. The theoretical analysis, kinematic simulation, and soil-tank test results were consistent, verifying the validity and ensuring the feasibility of the transplanting device. This study provides a reference for the development of transplanting devices.

Keywords: dibble-type transplanting device; transplanting characteristic coefficient; ADAMS; kinematic analysis; film-breaking experiment; ANSYS WORKBENCH

Citation: Zhao, X.; Hou, Z.; Zhang, J.; Yu, H.; Hao, J.; Liu, Y. Study on the Hole-Forming Performance and Opening of Mulching Film for a Dibble-Type Transplanting Device. *Agriculture* **2024**, *14*, 494. https://doi.org/10.3390/agriculture14030494

Academic Editor: Massimiliano Varani

Received: 27 January 2024
Revised: 7 March 2024
Accepted: 11 March 2024
Published: 18 March 2024

1. Introduction

At present, vegetable seedling transplanting operations are performed either manually or by semi-automatic transplanting machines. Manual transplantation on a large commercial scale remains labor-intensive and time-consuming and has high operating costs and low work efficiency [1,2]. This places a serious restriction on the large-scale industrialization of vegetables. However, mulching-film transplanting machines can effectively cut costs, improve the transplanting efficiency, and provide economic benefits [3–7]. However, the high mulching-film damage rate with an excessive number of hole openings can result in poor soil moisture retention and more weed growth, which have become common problems for dibble-type transplanters. As the most important part of the transplanting machine, the transplanting device has a direct impact on the hole-forming performance and transplant quality of the transplanting machine. Thus, the transplanting device also affects the survival rate of vegetable seedlings.

Research on improving the transplant quality and optimizing performance has attracted more and more attention from scholars. Hu et al. [8] designed a new automatic transplanting mechanism based on a clamping stem. Quan et al. [9] proposed a scheme for eliminating the forward speed of the entire machine by the horizontal linear velocity of the reverse rotation of the hole-forming mechanism. Vlahidis et al. [10] evaluated the functional parameters of a prototype single-row seedling transplanter. Fu et al. [11] simulated and performed an experiment on the compression molding behavior of a substrate block suitable for mechanical transplanting based on the Discrete Element Method (DEM). Du et al. [12] designed a mathematical model, based on the movement trajectory formula of a dibble-type transplanting device, that considers the planting depth, zero-speed transplanting, and the smallest amount of mulching-film damage in order to solve the mulching-film damage problem. Feng et al. [13] proposed design criteria for a dibble-type transplanting device and obtained the constraint formulas for the zero-speed-transplanting throwing conditions and the minimum-film-breaking conditions of a dibble-type transplanter through a theoretical analysis. Cui et al. [14] optimized the movement trajectory of a dibble-type transplanting device and found that the best position for planting seedings is the rising stage of the planting devices after reaching the lowest point, and plant spacing can be changed by changing the transmission ratio or the number of planting devices. The above-mentioned scholars have studied the hole-opening dimension of the transplanter. In terms of the kinematic trajectory, Li and Liu et al. [15,16] reduced the degree of tearing in the mulching film by optimizing the dimensions and kinematic parameters of a dibble-type transplanting device. Tian and Zhang et al. [17,18] analyzed the laws governing the opening of the duckbill of a dibble-type transplanter based on Pro/E and obtained the displacement–time characteristic curve of the motion of the transplanter. Jin et al. [19] designed a duckbill-type pot transplanter to solve the problem of the easy tearing of the mulching film by the dibble-type transplanter and determined the influence of the main parameters on the transplanting track based on ADAMS. Xue et al. [20] optimized a fully automated potted cotton seedling transplanting device. The authors found that, during the planting process, the absolute trajectory of the transplanter must maintain an approximately vertical posture at both the entry and excavation stages. Li et al. [21] studied a kind of automatic pick-up device for chili plug seedlings and found that the relative trajectories of the seedling needle and end-point were consistent with the theoretical ones. Zhou et al. [22,23] designed and improved an end effector of a transplanting device to reduce the seedling-carrying phenomenon of flower plug seedlings. Nevertheless, few studies have conducted a film-breaking experiment on a dibble-type transplanting device.

In this study, we fully considered the agronomic requirements of zero-speed transplanting and the minimum damage to the mulching film and the influence of the transplanting characteristic coefficient, the transplanting frequency, and the transplanting angle on the size of the hole openings in the mulching film based on an ADAMS2020 (64bit) (MSC Software, Inc., Newport Beach, CA, USA)kinematic track analysis and an ANSYS19.2 (64bit) WORKBENCH (ANSYS, Inc., New Castle, DE, USA) dynamic simulation. We also carried out a film-breaking experiment on a soil-tank test bench to verify the simulation analysis results and determine the optimal operating parameters. The optimal operating parameters of the transplanting machine can minimize the hole opening in the mulching film so as to achieve moisture preservation and weed control, improve the survival rate of vegetable seedlings, and improve the quality of transplanting operations, which is of great significance for the large-scale production and industrialization of vegetables in China.

2. Materials and Methods

2.1. The Whole Structure and Working Principle of the Transplanting Device

The prototype of the transplanting device is composed of a shell, 6 dibble-type transplanting units, 2 transplanting discs, a sprocket, a spindle, 6 dibble axes, 2 eccentric discs, an eccentric axis, and eccentric connections, as shown in Figure 1.

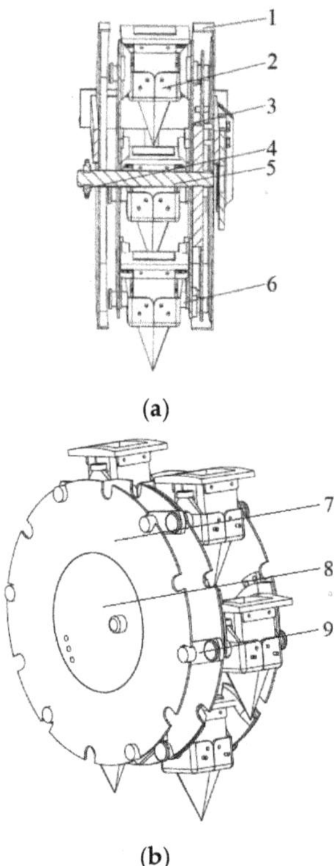

(a)

(b)

Figure 1. The whole structure of the transplanting device. (**a**) A sectional view of the transplanting device. 1. Shell; 2. dibble-type transplanting unit; 3. transplanting disc; 4. sprocket; 5. spindle; 6. dibble axis. (**b**) An axonometric view of the transplanting device without the shell. 7. Eccentric disc; 8. eccentric axis; 9. eccentric connection.

Each transplanting device is installed with 6 transplanting units. Each transplanting unit is evenly mounted between the transplanting discs by means of a dibble axis. The transplanting discs are power discs. Driven by the transplanting disc, the transplanting units rotate around the center of the transplanting discs. The eccentric discs are mounted on the outside of the transplanting discs by the eccentric axes and are connected to the transplanting discs by the eccentric connections. Under the combined action of the transplanting discs and the eccentric discs, the working angles of the transplanting units remain the same, as shown in Figure 1b.

When the transplanter is operating in the field, the transplanting discs are driven by a ground wheel. The 6 transplanting units rotate with the transplanting disc in the direction of the tractor. A vegetable seedling is manually or automatically fed into the topmost transplanting unit. Under the combined action of the transplanting discs and the eccentric discs, the angle between the transplanting unit with the vegetable seedling and the ground remains unchanged. When it moves to the lower end, it breaks through the mulching film, inserts into the soil, and then opens the dibble under the action of the eccentric discs (equivalent to a cam mechanism), and the vegetable seedling naturally falls into the seedling-hole in the ground under its own weight, followed by soil covering and suppression. The above-mentioned steps form the entire transplanting process of the transplanting device shown in Figure 2.

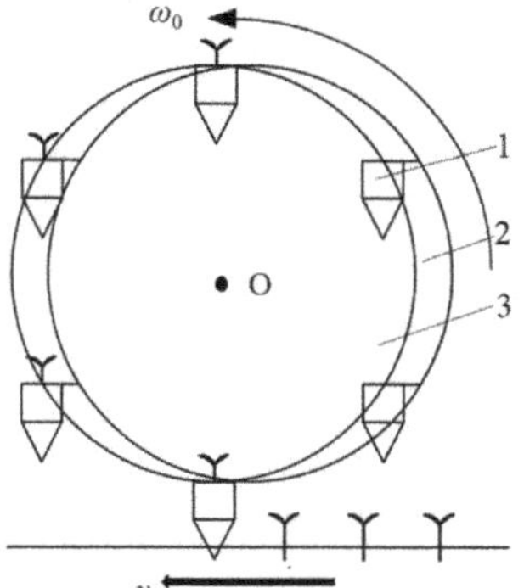

Figure 2. A schematic diagram of the working principle of the dibble-type transplanting device. 1. Dibble-type transplanting unit; 2. eccentric disc; 3. transplanting disc.

2.2. The Kinematic Analysis of the Transplanting Device

While the transplanting unit moves in a straight line along the direction of travel of the transplanter (Figure 3), the transplanting unit moves in a circular motion along the circumference of the transplanting disc on the transplanting device. The transplanting angle φ is between the center axis of the transplanting device and the direction of the forward speed.

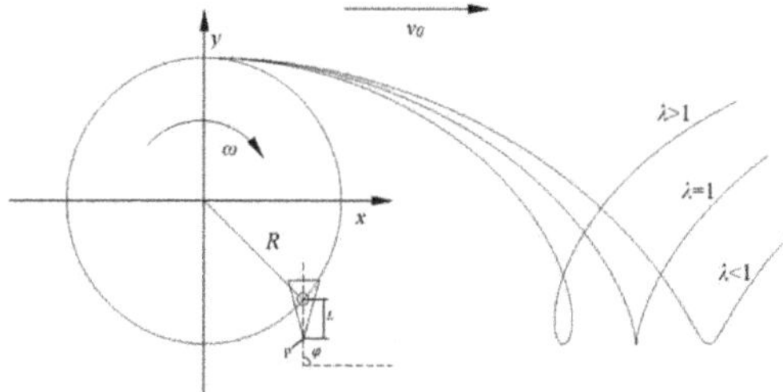

Figure 3. A schematic diagram of the track of the dibble-type transplanting device.

The coordinate system is established with the center of the transplanting disc at the initial position as the coordinate system origin. And then, the kinematic equation for the lower end-point p of the transplanting unit is established, as shown in Equation (1).

$$\begin{cases} x = v_0 t + R\cos\omega t \\ y = -R\sin\omega t - L \end{cases} \tag{1}$$

where v_0 is the forward speed for the transplanting device, m·s⁻¹; R is the rotation radius of the transplanting unit, m; ω is the rotation angular velocity for the transplanting unit, rad·s⁻¹; t is the transplanting time, s; and L is the distance from the center of the transplanting device to the lower end-point P, m.

The first-order derivation of Equation (1) is carried out to obtain the kinematic velocity equation of the lower end-point P of the transplanting unit.

$$\begin{cases} v_x = v_0 - R\omega\sin\omega t \\ v_y = -R\omega\cos\omega t \end{cases} \tag{2}$$

The kinematic track of the transplanting unit is mainly related to the transplanting characteristic coefficient λ. λ can be defined by Equation (3):

$$\lambda = \frac{R\omega}{v_0} \tag{3}$$

The kinematic track of the transplanting unit is greatly affected by λ, as shown in Figure 3. When $\lambda > 1$, the kinematic track of the transplanting unit is a trochoid, and the transplanting unit has two points with a horizontal velocity of 0 on both sides of the trochoid loop. When $\lambda = 1$, the horizontal velocity v_x at the lowest point of the kinematic track is 0. When $\lambda < 1$, there is no point where the horizontal velocity is 0 in the kinematic track of the transplanting unit. When the horizontal velocity of the transplanting unit is 0, the vegetable seedling is put in a seedling-hole in the ground, which is conducive to improving the transplanting uprightness of the vegetable seedling. To achieve zero-speed transplanting, a transplanting characteristic coefficient $\lambda \geq 1$ should be ensured. However, when the transplanting characteristic coefficient is too large, the transplanting unit will form a large hole opening in the mulching film, and the problems of mulching-film hanging and mulching-film tearing will occur in severe cases.

The mulching film is close to the ground, and the smaller the hole opening formed in the mulching film when transplanting, the more helpful in preserving moisture. Meanwhile, it can effectively control weeds, which is more conducive to improving the survival rate of vegetable seedlings. The kinematic tracks of the transplanting unit under different transplanting characteristic coefficients λ are shown in Figure 4. The distance H between the mulching film and the bottom of the hole opening is the transplanting depth of the transplanting unit. When the transplanting unit is moved to the lowermost end, the lowermost end-point is at the bottom of the hole opening. The size of the hole opening in the mulching film mainly depends on the horizontal displacement under the mulching film from the lower end-point p from the soil entry point A to the unearthed point B. The horizontal displacement under the mulching film mainly depends on the horizontal displacement l_{AB} formed by the soil entry point A and the unearthed point B and the dimension of l_{CD} at the maximum trochoid loop. The transplanting characteristic coefficient from track 1 to track 5 gradually decreases, and the l_{AB} of track 3 is equal to l_{CD}. It can be seen from Figure 4 that track 3 is a boundary point. With the increase in λ, the displacement l_{CD} of the lower end-point gradually increases. In addition, with the decrease in λ, the displacement under the mulching film formed by the entry point and unearthed point of the lower end-point of the transplanting unit gradually increases. Therefore, when $l_{AB} = l_{CD}$, the displacement under the mulching film at the lower end-point p of the transplanting unit is the smallest.

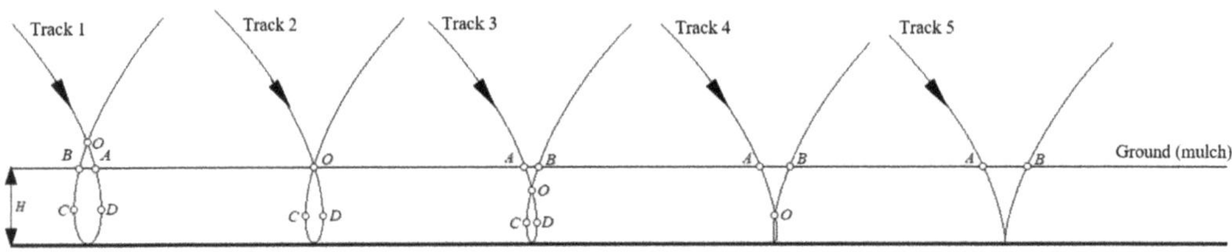

Figure 4. The hole-forming kinematic track of the transplanting unit with different transplanting characteristic coefficients λ.

When the transplanting unit is moved to the soil entry point A, the height of the lower end-point p from the hole opening is H, and the ordinate y_A of the lower end-point p can be defined by Equation (4):

$$y_A = -(R + L) + H \tag{4}$$

In Equation (1), $\theta = \omega t$, and then θ_A is the angle at which the lower end-point P of the dibble-type transplanting unit moves to point A. Then, from Equations (1) and (4), it can be calculated:

$$\theta_A = \arcsin(1 - \frac{H}{R}) \tag{5}$$

The rotating radius of the transplanting unit $R = 240$ mm and the transplanting depth $H = 60$ mm, so the transplanting angle is calculated as $\theta_A = 0.8481$ rad by Equation (5).

According to Equation (1), the abscissa x_A of the lower end-point p can be calculated at point A:

$$x_A = 0.8481\frac{v_0}{\omega} + 0.6614R \tag{6}$$

When the transplanting unit moves to point C, the horizontal velocity will be equal to 0.

$$v_{Cx} = v_0 - \omega R \sin \theta_C = 0 \tag{7}$$

Then, the transplanting angle θ_C from Equation (7) can be calculated, as shown in Equation (8):

$$\theta_C = \pi - \arcsin\frac{v_0}{\omega R} \tag{8}$$

Bringing θ_C into Equation (1), the x_C of the lower end-point p at point C can be calculated by Equation (9):

$$x_C = \frac{v_0}{\omega}(\pi - \arcsin\frac{v_0}{\omega R}) + R\cos(\pi - \arcsin\frac{v_0}{\omega R}) \tag{9}$$

When $x_A = x_B$, Equation (10) can be calculated:

$$x_C - x_A = 0 \tag{10}$$

Equation (11) is calculated from Equations (3), (6), (9) and (10):

$$R\left[\frac{1}{\lambda}(\pi - \arcsin\frac{v_0}{\omega R}) + \cos(\pi - \arcsin\frac{v_0}{\omega R}) - 0.8484\frac{1}{\lambda} - 0.6614\right] = 0 \tag{11}$$

From Equation (11), it can be seen that when $\lambda = 1.068$, the horizontal displacement of the lower end-point p under the mulching film is the smallest during the hole-forming process for the transplanting unit. The transplanting characteristic coefficient is related not only to the structural parameters of the transplanting device but also to the transplanting depth of the vegetable seedling.

2.3. ADAMS-Based Kinematic Simulation and Analysis for Transplanting Unit

2.3.1. The Displacement Curve of the Transplanting Unit

Equation (1) for the kinematic track of the transplanting unit shows that the displacement y in the vertical direction is only related to the rotational speed ω and is not related to the transplanting characteristic coefficient λ. According to the above, the transplanting characteristic coefficient is selected to be $\lambda = 1.068$, and the transplanting frequency is 36 plants·min^{-1} according to the actual artificial transplanting speed. At this transplanting frequency, the operator has enough time to place the vegetable seedling in the dibble. The kinematic simulation of the transplanting unit was carried out using ADAMS, as shown in Figure 5. The displacement curve coordinates of the lower end-point in the vertical direction were calculated by the ADAMS post-processing module, and the curve coordinates were imported into ORIGIN2021 (64bit) (OriginLab, Inc., Northampton, MA, USA) to generate the vertical displacement curve of the lower end-point. The lowest point of the displacement of the lower end-point for the transplanting unit is the bottom of the hole opening; the ground is 60 mm away from the bottom of the hole opening, and the ordinate of the ground is 0. The two points where the kinematic track intersects with the ground are the soil entry point and the unearthed point, respectively, and then we can calculate that the operation time for the mulching film at the lower end-point of the transplanting unit is 3.86~6.16 s.

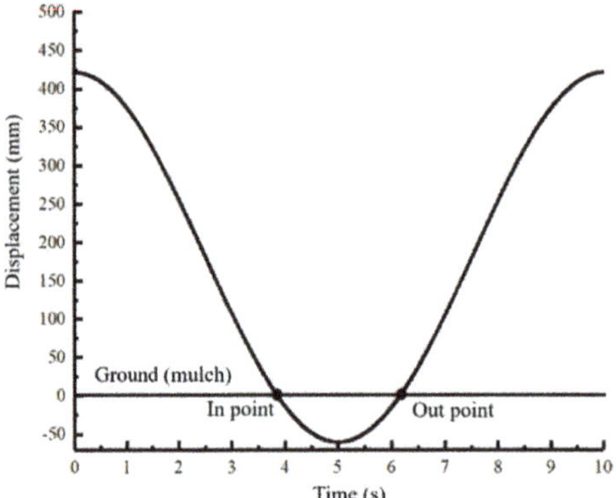

Figure 5. Vertical displacement curve of the bottom end-point of the transplanting unit.

According to the above, the transplanting characteristic coefficient $\lambda \geq 1$, and when the displacement under the mulching film of the lower end-point p is the smallest, the transplanting characteristic coefficient $\lambda = 1.068$. So, the transplanting frequency is 36 plants·min^{-1}. The transplanting characteristic coefficient was set to 1.000, 1.034, 1.068, 1.102, and 1.136 for the kinematic simulation. The influence of different transplanting characteristic coefficients λ on the displacement under the mulching film of the lower end of the transplanting unit was analyzed.

The ADAMS post-processing module was used to obtain the horizontal displacement curve coordinates of the lower end-point under the mulching film for different transplanting characteristic coefficients, and the curve coordinates were imported into ORIGIN to generate the horizontal displacement curve of the lower end-point, as shown in Figure 6. With the increase in the transplanting characteristic coefficient λ, the displacement curve span in the horizontal direction under the mulching film first decreases and then increases. When $\lambda = 1.068$, the displacement curve span is the smallest, and the curve is the most stable.

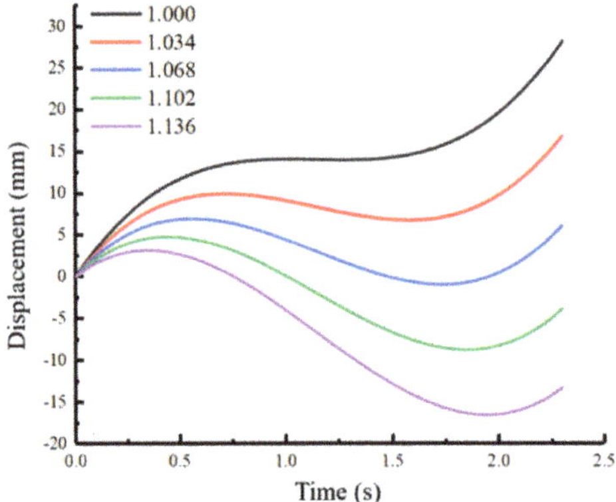

Figure 6. Horizontal displacement curve under the mulching film of the lower end-point of the transplanting unit.

The maximum displacement of the lower end-point for the transplanting unit under the mulching film for different transplanting characteristic coefficients λ is shown in Table 1.

Table 1. The maximum displacement of the lower end-point under the mulching film with different transplanting characteristic coefficients.

λ	Maximum Displacement/(mm)
1.000	28.22
1.034	16.81
1.068	7.91
1.102	13.53
1.136	19.68

As the transplanting characteristic coefficient λ increases, the displacement of the lower end-point first decreases and then increases. When $\lambda = 1.068$, the displacement is a minimum of 7.91 mm, and when $\lambda = 1.000$, the displacement is a maximum of 28.22 mm. To reduce the horizontal displacement of the lower end-point of the transplanting unit under the mulching film, the transplanting characteristic coefficient should be set as close as possible to 1.068.

2.3.2. Kinematic Simulation Analysis

ADAMS was used to simulate the kinematics of the transplanting unit, and the hole-opening formation process was simulated by the envelope of the kinematic track generated by the outline of the duckbill of the transplanting unit, as shown in Figure 7. Seven points were taken from the bottom of the duckbill of the transplanting unit to 60 mm on both sides, and a kinematic track envelope of 14 points was obtained. The dimensions of the hole opening were divided into a longitudinal dimension and a lateral dimension. The dimension of the hole opening formed in the forward direction is the longitudinal dimension. The lateral dimension formed in the direction of the duckbill's opening and closing movement is perpendicular to the forward direction of the transplanter.

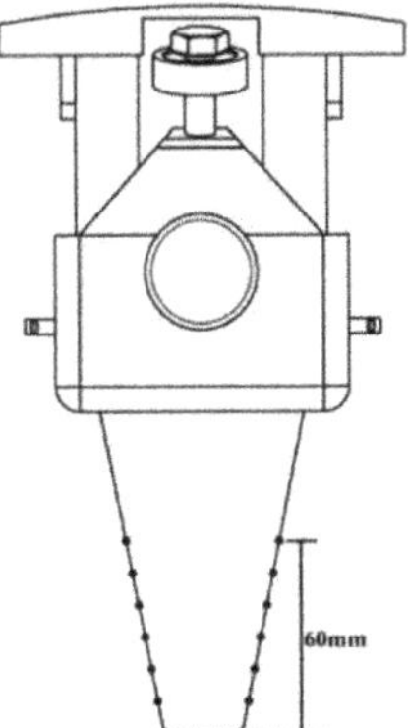

Figure 7. A schematic diagram of the simulated picking point position of the transplanting unit.

The Effect of the Transplanting Characteristic Coefficient on the Longitudinal Dimension of the Hole Opening

The transplanting angle was $90°$, the transplanting frequency was 36 plants·min^{-1}, and the transplanting characteristic coefficient was set to 1.000, 1.034, 1.068, 1.102, and 1.136. The envelope hole openings formed under different transplanting characteristic coefficients λ are shown in Figure 8. The bottom end of the envelope line was set as the hole-opening bottom, and the dimension of the lateral direction of the envelope line at 60 mm from the hole-opening bottom was used to simulate the hole-opening dimension, which is the theoretical hole-opening dimension of the transplanting unit. As the transplanting characteristic coefficients λ increases, the ring buckle of the kinematic envelope line increases, and the dimension of the envelope line first decreases and then increases. When $\lambda = 1.034$, the envelope hole opening formed by the kinematic track of the transplanting unit is the smallest, and the hole-opening dimension is 49.00 mm. When $\lambda = 1.068$, the displacement under the mulching film at the lower end-point of the transplanting unit is the smallest, and the time for the upper end to enter the soil is short because the transplanting unit is wedge-shaped with a wide top and a narrow bottom. The smaller the trochoid loop of the track of the transplanting unit, the smaller the displacement under the mulching film at the upper end, so the dimension of the trochoid loop of the track also affects the dimension of the envelope hole opening.

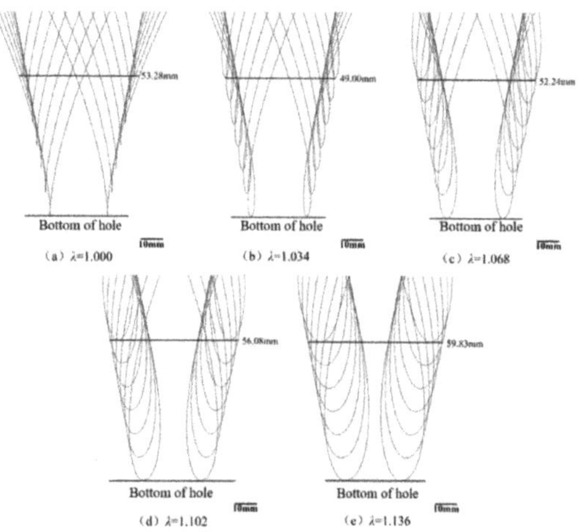

Figure 8. Diagrams of envelope hole openings with different transplanting characteristic coefficients. (**a**) The envelope hole opening at $\lambda = 1.000$. (**b**) The envelope hole opening at $\lambda = 1.034$. (**c**) The envelope hole opening at $\lambda = 1.068$. (**d**) The envelope hole opening at $\lambda = 1.102$. (**e**) The envelope hole opening at $\lambda = 1.136$.

The Effect of the Transplanting Frequency on the Longitudinal Dimension of the Hole Opening

The transplanting angle was 90°, and the transplanting characteristic coefficient λ was 1.068. Three transplanting frequencies of 24, 36, and 48 plants·min^{-1} were selected for the simulation experiments. With the increase in the transplanting frequency, the speed of hole-opening formation is also accelerated, the dimensions of the envelope do not change, and the dimensions of the envelope trochoid are the same, as shown in Figure 9. When the transplanting angle and the transplanting characteristic coefficient λ remain unchanged, the longitudinal hole-opening dimension of the envelope formed by the transplanting unit is independent of the transplanting frequency. So, in the simulation experiment, the transplanting characteristic coefficient λ was 1.068, and the transplanting frequency was 36 plants·min^{-1}. In addition, the transplanting unit is obliquely inserted into the soil at the maximum transplanting angle but does not make an inclined hole opening; it is advisable to take the adjustable maximum limit of the transplanting angle of the transplanting unit as the experimental maximum (95°), also consider the value of the transplanting angle follows the principle of symmetry.

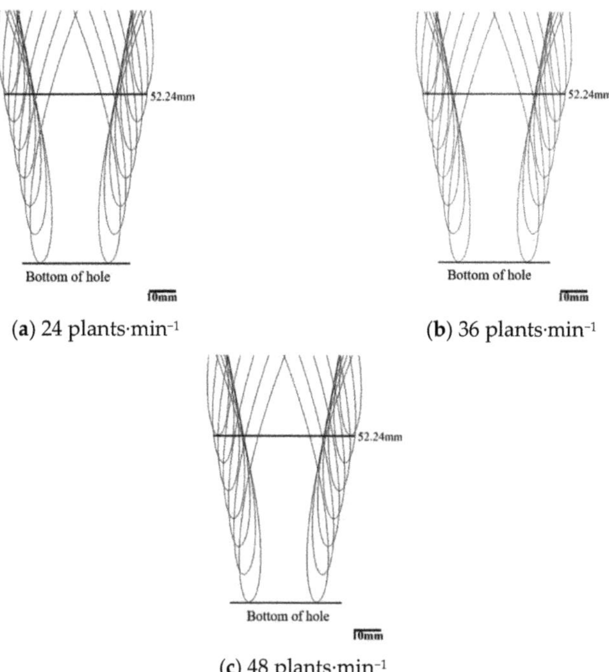

(**a**) 24 plants·min⁻¹ (**b**) 36 plants·min⁻¹

(**c**) 48 plants·min⁻¹

Figure 9. Diagrams of envelope hole openings with different transplanting frequencies. (**a**) The envelope hole opening when the transplanting frequency is 24 plants·min^{-1}. (**b**) The envelope hole opening when the transplanting frequency is 36 plants·min^{-1}. (**c**) The envelope hole opening when the transplanting frequency is 48 plants·min^{-1}.

The Effect of the Transplanting Angle on the Longitudinal Dimension of the Hole Opening

According to the above, in the simulation experiment, the transplanting characteristic coefficient λ was 1.068, and the transplanting frequency was 36 plants·min^{-1}. We selected five transplanting angles, 75°, 80°, 85°, 90°, and 95°, to perform the simulation experiment, as shown in Figure 10. As the transplanting angle increases, the horizontal span of the envelope formed on the left side of the duckbill gradually increases, and the envelope becomes more and more scattered. However, the span of the envelope formed on the right side of the duckbill gradually shrinks, and the envelope becomes more and more compact. Thus, with the increase in the transplanting angle, the displacement on the left side of the duckbill of the transplanting unit increases, the displacement on the right side of the duckbill decreases, and the envelope hole opening decreases. When the transplanting angle is 95°, the envelope hole opening is the smallest, and the longitudinal dimension of the hole opening is 51.82 mm. In the actual transplanting operation, the transplanting angle should be adjusted accordingly to achieve the purpose of reducing the hole opening.

The Lateral Dimension of the Hole Opening

The kinematic envelope formed by the transplanting unit in the lateral direction during the transplanting process is shown in Figure 11. Since the lateral kinematic only is the duckbill opening and closing movement, the formation of the lateral hole opening is only related to the dimension the cam that controls the duckbill opening and closing and is independent of other kinematic parameters. Therefore, the theoretical dimension of the lateral hole opening formed by the transplanting unit during the transplanting process is 62.41 mm.

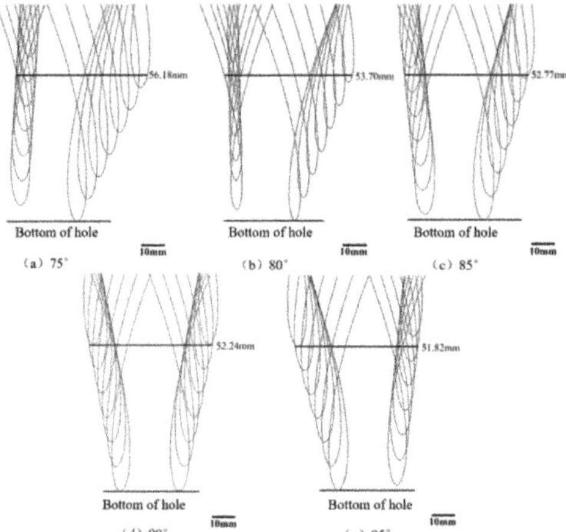

Figure 10. Diagrams of envelope hole openings with different transplanting angles. (**a**) The envelope hole opening when the transplanting angle is 75°. (**b**) The envelope hole opening when the transplanting angle is 80°. (**c**) The envelope hole opening when the transplanting angle is 85°. (**d**) The envelope hole opening when the transplanting angle is 90°. (**e**) The envelope hole opening when the transplanting angle is 95°.

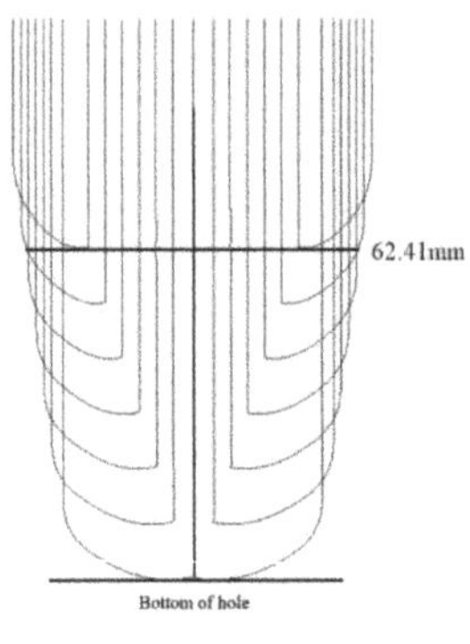

Figure 11. Envelope diagram of the transplanting unit in the lateral direction.

2.4. Dynamic Simulation Analysis Based on ANSYS WORKBENCH

Since the envelope hole opening based on ADAMS is formed by the kinematic track of the transplanting unit, the dimension of the envelope hole opening is the theoretical value of the hole opening. During the transplanting operation process, the plastic mulching film is tough, so the shear stress of the mulching film is also different for different transplanting parameters. In order to obtain the influence of different experimental factors on the maximum shear stress in the process of hole-opening formation, a dynamic simulation of the process of the transplanting unit damaging the mulching film was carried out based on the LS−DYNA module of ANSYS WORKBENCH. In the simulation, only the structural model involved in the film-breaking process was retained, and the model was simplified. The simulation model is shown in Figure 12.

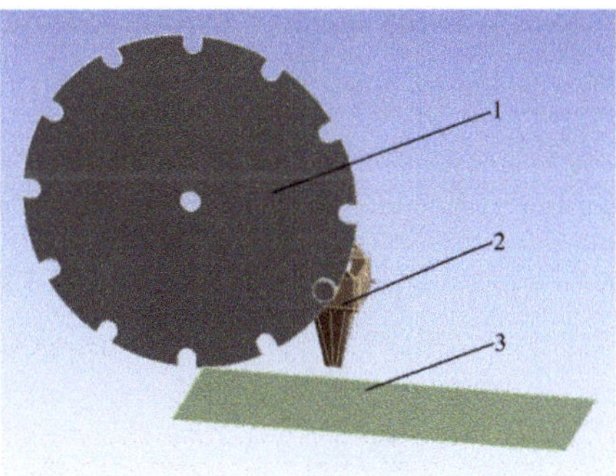

Figure 12. Dynamic simulation model of film-breaking experiment of the transplanting unit. 1. Transplanting disc; 2. transplanting unit; 3. mulching film.

The transplanting unit and the transplanting disc are made of structural steel and were set as rigid bodies. The thickness of the mulching film was set to 1 mm, and it was set as a flexible body. The main material properties of the mulching film are shown in Table 2.

Table 2. Table of mulching-film material properties.

Material Properties	Density ($kg \cdot m^{-3}$)	Young's Modulus (MPa)	Poisson's Ratio	Yield Stress (MPa)
Value	940	0.1	0.45	2×10^{-4}

A tetrahedral mesh shape was selected to mesh the simulation model, as shown in Figure 13. The grid dimensions of the transplanting disc and the transplanting unit were set to 5 mm, and the grid dimension of the mulching film was set to 2 mm. After meshing, a driver was added to the simulation model, a fixed constraint was added to the mulching film, and the simulation time was set to 3 s.

Figure 13. Meshing for the simulation model.

2.5. The Film-Breaking Experiment on the Soil−Tank Test Bench

The film-breaking experiment was carried out on a soil-tank test bench, as shown in Figure 14. The soil-tank test bench is the TCC−II electric four-wheel-drive soil-tank test bench produced by Harbin Bona Technology Co., Ltd., Harbin, Heilongjiang Province, China. Other experimental equipment and tools included a transplanting device, a mulching film (80 cm width), a three-phase asynchronous gear reduction motor, and a steel ruler.

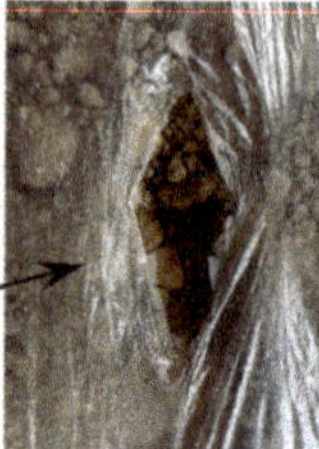

Figure 14. Film-breaking experiment on soil-tank test bench.

The mulching film was first laid in the soil tank, and the laying distance was not less than 5 m. The dibble-type transplanting device moved forward with the soil-tank experiment vehicle, and the transplanting units were driven by the three-phase asynchronous gear reduction motor. Taking the transplanting characteristic coefficient, the transplanting frequency, and the transplanting angle as the experimental factors, 3 rows were used for experiments run under each experimental factor, and the number of hole openings in each row was not less than 20. A straight steel ruler was used to measure the size of every hole opening in the mulching film. The dimensions of the largest hole opening formed in the longitudinal and lateral directions of the mulching film were measured. The experiment was repeated three times, and then the average of the measurements was obtained.

3. Results and Discussion

3.1. The Process of Hole-Opening Formation in Mulching Film

The simulation time for the process of hole-opening formation in a mulching film was 3 s. The pressure nephograms of the process of hole-opening formation in a mulching film at different times are shown in Figure 15.

The pressure nephogram of the mulching film at 0.32 s shows that the transplanting unit has not yet contacted the mulching film (Figure 15a). When the simulation time is 0.79 s, the transplanting unit begins to insert the mulching film (Figure 15b). At this time, the mulching film has not been broken, but it is subjected to the greatest stress in the forward direction of the transplanter. Figure 15c,d show the hole-opening formation stage of the mulching film, and the stress on the mulching film is less than the stress before 0.79 s. Figure 15e,f are the pressure nephograms after the hole opening in the mulching film has formed.

From the pressure nephograms of the mulching film at different times, it can be seen that the stress change on both sides of the longitudinal hole of the mulching film is more obvious, and the shear stress of the mulching film on both sides of the lateral hole is relatively uniform. The stress of the mulching film in the longitudinal direction is greater than that in the lateral direction, and the size of the hole formed in the longitudinal direction is larger than that in the lateral direction.

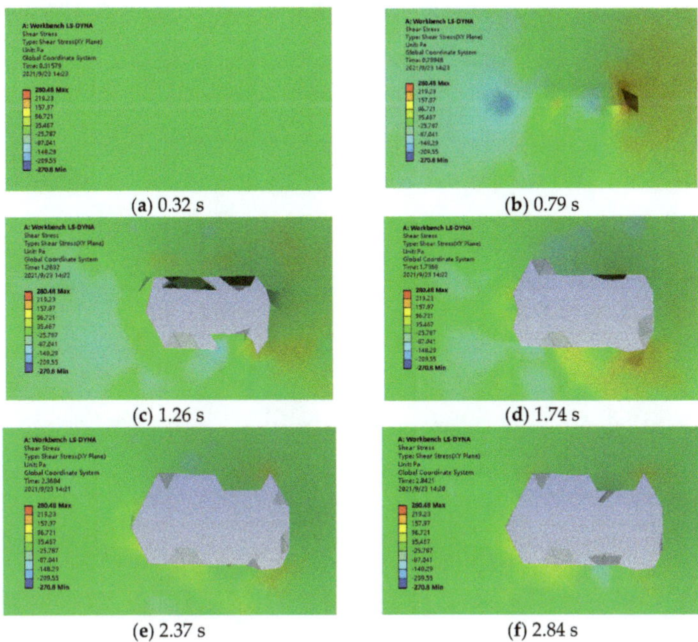

(**a**) 0.32 s   (**b**) 0.79 s

(**c**) 1.26 s   (**d**) 1.74 s

(**e**) 2.37 s   (**f**) 2.84 s

Figure 15. Pressure nephograms of the mulching film during the hole opening forming process.

3.2. The Effect of the Maximum Shear Stress on the Mulching Film under Different Experimental Factors

The statistical curves of the maximum shear stress of the mulching film under different experimental factors are shown in Figure 16. With the increase in the transplanting characteristic coefficient λ, the maximum shear stress shows an increasing trend. When the transplanting characteristic coefficient $\lambda = 1.000$, the maximum shear stress is the smallest. With the increase in the transplanting frequency, the maximum shear stress first decreases and then increases. When the transplanting frequency is 36 plants·min^{-1}, the maximum shear stress is the smallest. With the increase in the transplanting angle, the maximum shear stress first increases and then decreases. When the transplanting angle is 85°, the maximum shear stress is the largest. When the transplanting angle is 95°, the maximum shear stress is the smallest. The smaller the shear stress of the mulching film during the formation of the hole opening, the smaller the damage caused by the transplanting unit to the mulching film, and the smaller the size of the hole opening. Therefore, the transplanting operation should be carried out under operating parameters that subject the mulching film to less shear stress.

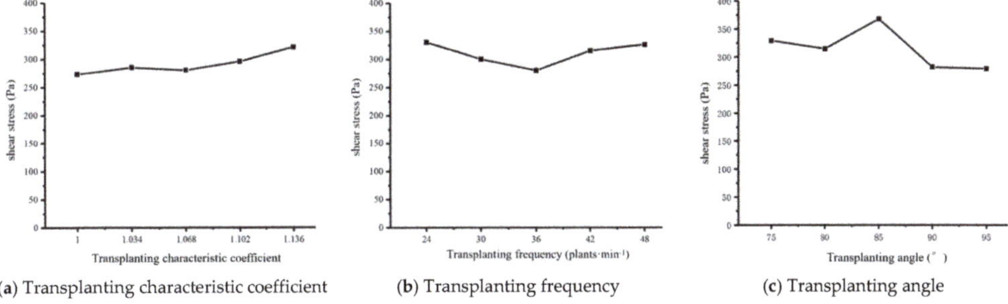

(**a**) Transplanting characteristic coefficient   (**b**) Transplanting frequency   (**c**) Transplanting angle

Figure 16. The effect of the maximum shear stress on the mulching film under different experimental factors.

3.3. The Effect of Different Experimental Factors of the Mulching Film on the Size of the Hole Opening

The effects of different operating parameters on the dimensions of the hole opening of a mulching film are shown in Figure 17.

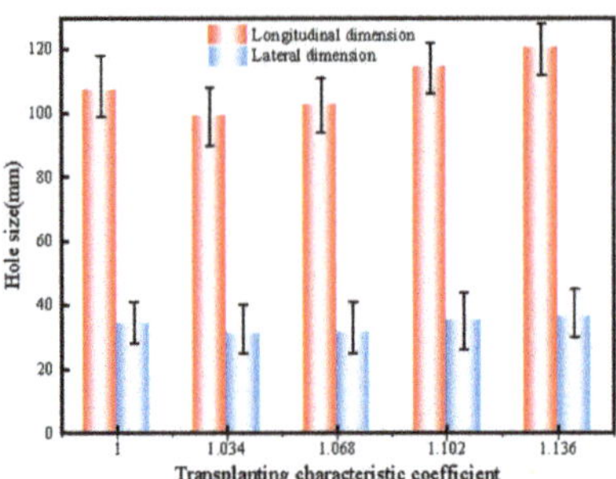

(**a**) Transplanting characteristic coefficient

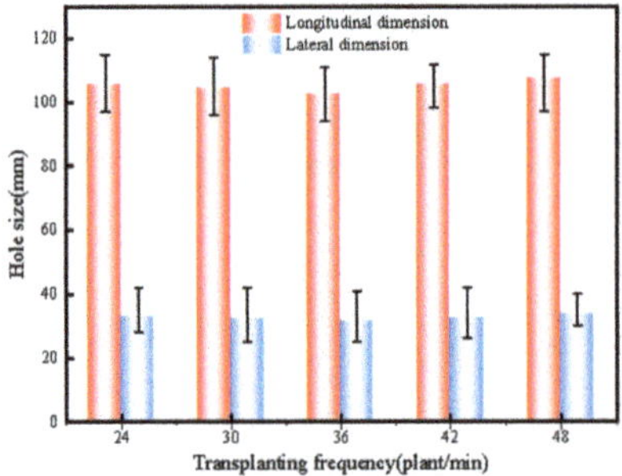

(**b**) Transplanting frequency

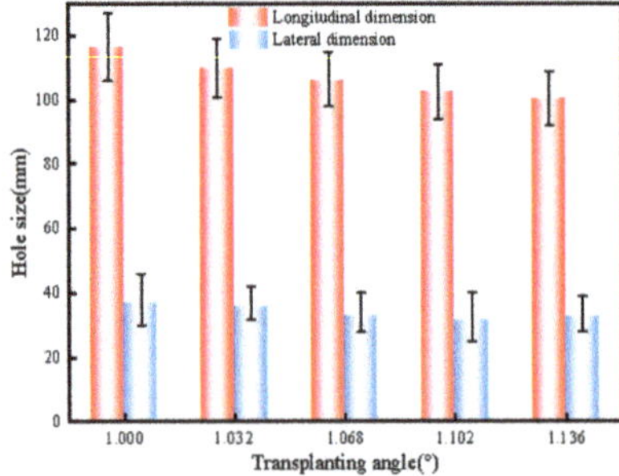

(**c**) Transplanting angle

Figure 17. The effects of different experimental factors on the dimensions of the hole opening of a mulching film.

With the increase in the transplanting characteristic coefficient λ, the longitudinal and lateral dimensions of the hole opening first decrease and then increase. The longitudinal dimension of the hole opening ranges from 99.20 to 120.40 mm, and the lateral dimension ranges from 31.67 to 36.53 mm. When the transplanting characteristic coefficient $\lambda = 1.034$, the dimension of the hole opening is the smallest. When the characteristic coefficient $\lambda = 1.136$, the dimension of the hole opening is the largest. At different transplanting frequencies, the longitudinal dimension of the hole opening ranges from 102.53 to 107.73 mm, and the lateral dimension ranges from 31.67 to 33.87 mm. When the transplanting frequency is 36 plants·min^{-1}, the hole-opening dimension is the smallest. When the transplanting frequency is 48 plants·min^{-1}, the hole-opening dimension is the largest. With different transplanting angles, the longitudinal dimension of the hole opening ranges from 100.47 to 116.33 mm, and the lateral dimension ranges from 31.67 to 37.07 mm. The longitudinal dimension of the hole opening decreases with the increase in the transplanting angle. When the transplanting angle is 95°, the longitudinal dimension is the smallest, while the lateral dimension is the smallest when the transplanting angle is 90°.

From the above-mentioned analysis, it can be seen that the influence of the transplanting characteristic coefficient and the transplanting angle on the longitudinal size of the hole opening is the same as that shown by the kinematic simulation results (Figures 8 and 10). The influence of the transplanting frequency on the longitudinal dimension of the hole opening is related to the shear stress of the mulching film in the dynamic simulation (Figures 15 and 16). The longitudinal dimension of the hole opening is larger than the theoretical dimension, while the lateral dimension of the hole opening is smaller than the theoretical dimension. From the results in Figure 17, we can see that the movement of the transplanting unit in the direction of the forward velocity causes great damage to the mulching film, causing the mulching film to tear in the longitudinal direction and making the actual dimension of the hole opening larger than the theoretical dimension of the hole opening.

4. Conclusions

(1) The kinematic track of the lower end-point of the transplanting unit was analyzed, and the kinematic track equation was established. According to the results calculated by the kinematic track Equation (11), the actual transplanting depth of the vegetable seedling is 60 mm, the transplanting characteristic coefficient λ of the minimum displacement of the lower end of the transplanting unit under the mulching film is 1.068, and the horizontal displacement of the transplanting unit under the mulching film is minimal.

(2) The results of the ADAMS simulation show that the transplanting characteristic coefficient, the transplanting frequency, and the transplanting angle have no effect on the lateral hole-opening dimension of the envelope, and the transplanting frequency has no effect on the longitudinal dimension of the hole opening. But when the longitudinal dimension of the hole opening of the envelope is the smallest, the transplanting characteristic coefficient is 1.034, and the transplanting angle is 95°. Therefore, in the actual transplanting operation, in order to reduce the size of the hole opening in the mulching film, the transplanting characteristic coefficient and the transplanting angle should be reasonably controlled.

(3) The pressure nephogram from the ANSYS WORKBENCH finite element analysis shows that the maximum shear stress of the mulching film in the longitudinal direction is greater than that in the lateral direction. When the transplanting characteristic coefficient, the transplanting frequency, and the transplanting angle are 1.000, 36 plants·min^{-1}, and 95°, respectively, the maximum shear stress of the mulching film in the process of hole-opening formation is the smallest. So, in the process of transplanting, the operating parameters with less shear stress on the mulching film should be selected as much as possible to reduce the occurrence of mulching-film damage and tearing.

(4) The optimal parameters were verified through a film-breaking experiment on a soil-tank test bench. Under the conditions of different transplanting characteristic coefficients, transplanting frequencies, and transplanting angles, the differences between the maximum and minimum sizes of the longitudinal dimension of the hole opening in the mulching film are 21.20 mm, 5.20 mm, and 15.86 mm, respectively, and the differences between the maximum and minimum sizes of the lateral dimension of the hole opening in the mulching film are 4.86 mm, 2.20 mm, and 5.40 mm, respectively.

The larger the difference between the maximum size and the minimum size of the hole opening in the mulching film, the more significant the influence of the experimental factors on the hole-opening dimensions. The transplanting characteristic coefficient has the most significant effect on the longitudinal dimension of the hole opening, the transplanting angle has the most significant effect on the lateral dimension of the hole opening, and the transplanting frequency has no significant effect on the dimensions of the hole opening. The test results show that the influence of each test factor on hole-opening formation in the mulching film is the same as shown by the simulation analysis results, and the simulation test results have important reference and guiding significance for the size of the actual hole opening.

Author Contributions: Conceptualization, X.Z., J.H. and J.Z.; methodology, X.Z. and Z.H.; software, Z.H. and H.Y.; validation, J.Z., X.Z. and J.H.; formal analysis, J.Z.; investigation, X.Z.; resources, J.H.; data curation, H.Y.; writing—original draft preparation, Z.H. and X.Z.; writing—review and editing, X.Z.; visualization, X.Z.; supervision, J.H.; project administration, Y.L.; funding acquisition, J.Z. All authors have read and agreed to the published version of the manuscript.

Funding: This research was funded by the National Key R&D Program (2021YFD1901104−5), Key R&D Program of Hebei (21327005D), State Key Laboratory of North China Crop Improvement and Regulation (NCCIR2024ZZ−12), and S&T Program of Hebei (23567601H). The central government guides local funds for scientific and technological development (Grant No. 236Z7202G), and the APC was funded by [2021YFD1901104−5].

Institutional Review Board Statement: Not applicable.

Data Availability Statement: The datasets used and/or analyzed during the current study are available from the corresponding author on reasonable request.

References

1. Li, L.; Wu, W.L.; Giller, P.; O'Halloran, J.; Liang, L.; Peng, P.; Zhao, G. Life cycle assessment of a highly diverse vegetable multi-cropping system in Fengqiu County, China. *Sustainability* **2018**, *10*, 983. [CrossRef]

2. Yin, D.Q.; Wang, J.Z.; Zhang, S.; Zhang, N.Y.; Zhou, M.L. Optimized design and experiments of a rotary-extensive-type flowerpot seedling transplanting mechanism. *Int. J. Agric. Biol. Eng.* **2019**, *12*, 45–50. [CrossRef]

3. Ji, J.T.; Cheng, Q.; Jin, X.; Zhang, Z.H.; Xie, X.L.; Li, M.Y. Design and test of 2ZLX-2 transplanting machine for oil peony. *Int. J. Agric. Biol. Eng.* **2020**, *13*, 61–69. [CrossRef]

4. Wen, Y.S.; Zhang, J.X.; Tian, J.Y.; Duan, D.S.; Zhang, Y.; Tan, Y.Z.; Yuan, T.; Li, X. Design of a traction double-row fully automatic transplanter for vegetable plug seedlings. *Comput. Electron. Agric.* **2021**, *182*, 106017. [CrossRef]

5. Jin, X.; Li, D.Y.; Ma, H.; Ji, J.T.; Zhao, K.X.; Pang, J. Development of single row automatic transplanting device for potted vegetable seedlings. *Int. J. Agric. Biol. Eng.* **2018**, *11*, 67–75. [CrossRef]

6. Uchiyama, H.; Wada, Y.; Hatanaka, Y.; Hirata, Y.; Taniguchi, M.; Kadota, K.; Tozuka, Y. Solubility and permeability improvement of quercetin by an interaction between α-glucosyl stevia nano-aggregates and hydrophilic polymer. *J. Pharm. Sci.* **2019**, *108*, 2033–2040. [CrossRef] [PubMed]

7. Kumar, N.; Upadhyay, G.; Choudhary, S.; Patel, B.; Naresh Chhokar, R.S.; Gill, S.C. Resource conserving mechanization technologies for dryland agriculture. In *Enhancing Resilience of Dryland Agriculture under Changing Climate: Interdisciplinary and Convergence Approaches*; Springer Nature: Singapore, 2023; pp. 657–688.

8. Hu, S.; Hu, M.; Yan, W.; Zhang, W. Design and Experiment of an Integrated Automatic Transplanting Mechanism for Picking and Planting Pepper Hole Tray Seedlings. *Agriculture* **2022**, *12*, 557. [CrossRef]

9. Quan, W.; Wu, M.; Dai, Z.; Luo, H.; Shi, F. Design and Testing of Reverse-Rotating Soil-Taking-Type Hole-Forming Device of Pot Seedling Transplanting Machine for Rapeseed. *Agriculture* **2022**, *12*, 319. [CrossRef]

10. Vlahidis, V.; Rosca, R.; Cârlescu, P.-M. Evaluation of the Functional Parameters for a Single-Row Seedling Transplanter Prototype. *Agriculture* **2024**, *14*, 388. [CrossRef]

11. Fu, J.; Cui, Z.; Chen, Y.; Guan, C.; Chen, M.; Ma, B. Simulation and Experiment of Compression Molding Behavior of Substrate Block Suitable for Mechanical Transplanting Based on Discrete Element Method (DEM). *Agriculture* **2023**, *13*, 883. [CrossRef]

12. Du, S.; Wang, W.B.; Wang, J.K. Parameters research of dibble-type transplanter for the smallest mulch-film damage. *J. Chin. Agric. Mech.* **2016**, *37*, 5–8+17. (In Chinese)

13. Feng, J.; Qin, G.; Song, W.T.; Liu, Y.J. The Kinematic analysis and design criteria of the dibble-type planting devices. *Trans. CSAM* **2002**, *33*, 48–50. (In Chinese)

14. Cui, W.; Zhao, L.; Song, J.N.; Lin, J.T. Kinematic analysis and experiment of dibble-type planting device. *Trans. CSAM* **2012**, *43* (Suppl. S1), 35–38+34. (In Chinese)
15. Li, X.Y.; Wang, Y.W.; Lu, G.C.; Zhang, B.; Zhang, H.J. Optimization design and experiment of dibble-type transplanting device. *Trans. CSAE* **2015**, *31*, 58–64. (In Chinese)
16. Liu, Y.; Mao, H.P.; Wang, T.; Li, B.; Li, Y.X. Collision optimization and experiment of tomato plug seedling in basket-type transplanter. *Trans. CSAM* **2018**, *49*, 143–151. (In Chinese)
17. Tian, Y.; Li, X.Y.; Zhang, H.J.; Chi, M.L.; Zhang, X.J. Design and the motion simulation for spatial cam of dibble-type planting device. *J. Agric. Mech. Res.* **2014**, *36*, 62–65+69. (In Chinese)
18. Zhang, X.Z.; Li, X.Y.; Tian, Y.; Chi, M.L. Three-dimensional modeling and motion simulation analysis of dibble-type planting device based on Pro/E. *J. Agric. Mech. Res.* **2014**, *36*, 87–90. (In Chinese)
19. Jin, X.; Li, S.J.; Yang, X.J.; Yan, H.; Wu, J.M.; Mao, Z.H. Motion analysis and parameter optimization for pot seedling planting mechanism based on up-film transplanting. *Trans. CSAM* **2012**, *43* (Suppl. S1), 29–34. (In Chinese)
20. Xue, X.L.; Li, L.H.; Xu, C.L.; Li, E.Q.; Wang, Y.J. Optimized design and experiment of a fully automated potted cotton seedling transplanting mechanism. *Int. J. Agric. Biol. Eng.* **2020**, *13*, 111–117. [CrossRef]
21. Li, H.; Cao, W.B.; Li, S.F.; Fu, W.; Liu, K.Q. Kinematic analysis and test on automatic pick-up mechanism for chili plug seedling. *Trans. CSAE* **2015**, *31*, 20–27. (In Chinese)
22. Zhou, M.F.; Huang, Z.J.; Shi, L.D.; Yu, G.H.; Zhao, X.; Liu, P.F. Improvement design and experimental analysis of automatic transplanting of flower plug seedlings. *J. Chin. Agric. Mech.* **2019**, *40*, 10–16. (In Chinese)
23. Zhou, M.F.; Xu, J.J.; Tong, J.H.; Yu, G.H.; Zhao Xiong Xie, J. Design and experiment of integrated automatic transplanting mechanism for taking and planting of flower plug seedling. *Trans. CSAE* **2018**, *34*, 44–51. (In Chinese)

Article

Model Development for Off-Road Traction Control: A Linear Parameter-Varying Approach

Adam Szabo [1], Daniel Karoly Doba [1], Szilard Aradi [1,*] and Peter Kiss [2]

[1] Department of Control for Transportation and Vehicle Systems, Faculty of Transportation Engineering and Vehicle Engineering, Budapest University of Technology and Economics, Műegyetem rkp. 3., H-1111 Budapest, Hungary; szabo.adam@kjk.bme.hu (A.S.); doba.daniel.karoly@kjk.bme.hu (D.K.D.)

[2] Institute of Technology, Department of Vehicle Technology, Hungarian University of Agriculture and Life Sciences, Páter Károly u. 1., H-2100 Gödöllő, Hungary; kiss.peter@uni-mate.hu

* Correspondence: aradi.szilard@kjk.bme.hu

Abstract: The number of highly automated machines in the agricultural sector has increased rapidly in recent years. To reduce their fuel consumption, and thus their emission and operational cost, the performance of such machines must be optimized. The running gear–terrain interaction heavily affects the behavior of the vehicle; therefore, off-road traction control algorithms must effectively handle this nonlinear phenomenon. This paper proposes a linear parameter-varying model that retains the generality of semiempirical models while supporting the development of real-time state observers and control algorithms. First, the model is derived from the Bekker–Wong model for the theoretical case of a single wheel; then, it is generalized to describe the behavior of vehicles with an arbitrary number of wheels. The proposed model is validated using an open-source multiphysics simulation engine and experimental measurements. According to the validated results, it performs satisfactorily overall in terms of model complexity, calculation cost, and accuracy, confirming its applicability.

Keywords: linear parameter-varying systems; traction systems; validation; wheel–terrain interaction

Citation: Szabo, A.; Doba, D.K.; Aradi, S.; Kiss, P. Model Development for Off-Road Traction Control: A Linear Parameter-Varying Approach. *Agriculture* **2024**, *14*, 499. https://doi.org/10.3390/agriculture14030499

Academic Editor: Massimiliano Varani

Received: 19 February 2024
Revised: 14 March 2024
Accepted: 18 March 2024
Published: 19 March 2024

1. Introduction

Modeling terrain–machinery interaction and terrain–vehicle interactions in real-time applications has become one of the main challenges of terramechanics in the last three decades. Related research has focused on two main topics: real-time dynamic simulation and real-time controller design. These algorithms enable the advanced control of off-road vehicles, such as through the online traction control of intelligent tractors [1]. Due to increasing energy prices and tightening emission standards, increasing the efficiency of agricultural vehicles has become crucial [2]. The higher level of automation reduces operational costs and energy consumption and increases overall productivity and safety [3].

Off-road vehicle dynamics are heavily affected by the interaction between the running gear and soft soil; hence, typical solutions developed for on-road vehicles are not applicable. Models that describe the wheel–terrain interaction range from purely analytical to purely empirical models. Empirical models are based on large amounts of experimental data and utilize different dimensionless parameters to describe the wheel–terrain interaction. Thus, their applicability is limited to the vehicle and soil type used during the measurements [4]. On the other hand, discrete element and finite element methods (DEM and FEM) are increasing in popularity in simulators due to their high accuracy [5,6] and the increase in the computational power of computers; yet, their computational cost still prevents their widespread application in control-oriented models and real-time tasks. Semiempirical models like the Bekker–Wong model [7] utilize experimental data and theoretical analysis to offer a good trade-off between accuracy and computational cost. Hence, they are attractive for real-time simulation purposes. An extensive literature review of the developed and applied techniques is presented in [8]. In the case of real-time controller design, the parameters and states of the vehicle and the terrain are divided into four categories. Some parameters, such as the moisture content in rover applications, are known beforehand. Other parameters can be measured in real time (e.g., slip angle [9]), predicted based on current states [10], or estimated from easily measurable vehicle states [11].

Still, semiempirical models are considered too complex for control-oriented model development [12]. To overcome this issue, several synthesis models have been developed. In [13,14], augmented kinematic models were proposed to incorporate the sliding phenomenon without incorporating the wheel dynamics. The results showed that including the sliding phenomenon in the observer and controller design significantly reduced the lateral deviation. Other researchers used experimental models, such as the empirical Brixius tire model. The conventional Brixius tire model includes wheel sinkage, which is often difficult to estimate; hence, in [15], a modification was proposed to include the soil reaction. In [16], an empirical model was derived from the statistical analysis of soil experiments to achieve maximum slip efficiency on stochastic terrain.

Another solution is adaping the various on-road tire models for off-road applications, such as the Pacejka tire model [17]. In [18], the slip–friction relationship was described using the Pacejka model fitted to the measurement data. Then, it was combined with the traction effectiveness to determine an optimal slip value for the traction controller. Due to the lack of available off-road tire measurement data, ref. [19] estimated the traction performance using the Bekker–Wong model and then derived the Pacejka model coefficients from these results. In [20], an adapted Burckhardt tire model was introduced for model-based control design. The coefficients of the model were derived from the Bekker–Wong model through nonlinear optimization. The model was further improved in [12] to include the antisymmetric off-road feature in a large operating range in longitudinal and lateral modes.

Contributions

Conventional tire models are widely used in controller development and can be successfully adapted to off-road circumstances. In these examples, the model parameters are mainly attained through optimization; therefore, they lose their physical meaning. Furthermore, this optimization must be performed for different terrain types and vehicle parameters.

This paper presents the development of a linear parameter-varying (LPV) model for off-road vehicles that can be utilized in model-based control design. The model for a single wheel is derived from Bekker–Wong theory. It explicitly includes the terrain and vehicle parameters, thus ensuring its generality.

Next, it is augmented for vehicles with an arbitrary number of wheels. The results of the model were verified using the Project Chrono 8.0 simulator. At last, the model was validated against experimental measurements.

The paper is organized as follows: Section 2 introduces the theory of LPV models and details the derivation of the proposed control-oriented model. Section 3 presents the verification and validation of the model. Section 4 summarizes the results, and Section 5 contains some concluding remarks.

2. Materials and Methods

2.1. Introduction to Linear Parameter-Varying Models

The LPV approach is widely used to model and control a subclass of nonlinear systems. It has many applications related to on-road vehicles, for example, controlling active [21] and semiactive [22] suspension systems, lateral vehicle control [23] and longitudinal vehicle control, such as adaptive cruise control (ACC) [24], antilock braking systems (ABS) [25], and the and the integrated control of driver assistance systems (DAS) [26,27].

Linear parameter-varying systems are time-varying state-space models written as follows:

$$\begin{bmatrix} \dot{x}(t) \\ y(t) \end{bmatrix} = \begin{bmatrix} A(\rho(t)) & B(\rho(t)) \\ C(\rho(t)) & D(\rho(t)) \end{bmatrix} \begin{bmatrix} x(t) \\ u(t) \end{bmatrix} \tag{1}$$

where $x(t)$ is the state vector, $y(t)$ is the output vector, $u(t)$ is the input vector, $\rho(t)$ is the vector of scheduling parameters, and A, B, C, and D are parameter-dependent matrices of the state-space representation.

The dynamics of the system heavily depend on a set of time-varying parameters described by $\rho(t)$, which must be well known and measurable in real-time. Parameter uncertainty is a major challenge for semiempirical approaches, as terrain parameters depend heavily on several other parameters, such as moisture content or grouser height [28]. These parameters are either exogenous variables modeling the nonstationary behavior of a system or endogenous variables representing nonlinear system dynamics, also called quasi-LPV (qLPV) systems.

Several representations are available to describe the parameter dependence of LPV systems. Linear fractional representation (LFR) models use linear fractional transformation (LFT) to separate the nonlinearities of the system from the nominal model. Such nonlinearities often include time-varying parameters and uncertainties, but LFR models can be used only for rational parameter dependence. Several equations used in semiempirical models include irrational parameter dependence, limiting the applicability of LFR models. Affine LPV models are usually described using polytopic models, where the vector of parameters evolves inside a polytope. Thus, it is written as a complex combination of the polytope's vertices. Affine LPV models deal with smooth and continuous parameter variation and are sensitive to measurement errors in the scheduling parameters. In the case of a grid-based representation, the linear dynamics can be defined at each grid point of the model. Grid-based LPV models capture the parameter dependence of the system implicitly; therefore, they can handle any parameter representation. They offer certain robustness advantages over affine LPV models in the presence of measurement errors in the scheduling parameters, which is advantageous in dealing with the uncertainty of terrain parameters. Considering the semiempirical nature of the models describing wheel–terrain interactions, grid-based LPV models are a reasonable choice for modeling the phenomena.

2.2. Model Development

The forces acting on a vehicle moving over an inclined terrain are shown in Figure 1. The longitudinal dynamics can be calculated according to the following force balance equation:

$$\dot{v} = \frac{\sum F - \sum R - F_{air} - F_{grade} - F_{drawbar}}{m} \tag{2}$$

where $\sum F$ is the sum of the tractive efforts acting on the wheels, $\sum R$ is the sum of the motion resistances acting on the wheels, F_{air} is the aerodynamic resistance, F_{grade} is the grade resistance, and $F_{drawbar}$ is the drawbar pull.

The torque balance of the wheel dynamics is written as follows:

$$\dot{\omega} = \frac{T_w - T_f - Fr}{J_w} \tag{3}$$

where ω and J_w are the angular velocity and inertia of the wheel, T_f is the friction, and T_w is the input torque of the wheel.

First, some of the forces are neglected to simplify model development. While it significantly affects the behavior of the vehicle at higher velocities, the aerodynamic resistance below 48 km/h can be neglected [29]. Hence, it is omitted from the model. The drawbar pull is the amount of horizontal force available to a vehicle at the drawbar for pulling a load. In our case, the drawbar pull is assumed to be zero, and the remaining force accelerates the vehicle. In Figure 1, the color red notes the neglected forces, while the forces taken into account are blue. Therefore, the nonlinear state-space model describing the movement of a single wheel on a flat terrain is written as follows:

$$\begin{bmatrix} \dot{v}(t) \\ \dot{\omega}(t) \end{bmatrix} = \begin{bmatrix} \frac{F-R}{m} \\ \frac{T_w - T_f - Fr}{J} \end{bmatrix} \tag{4}$$

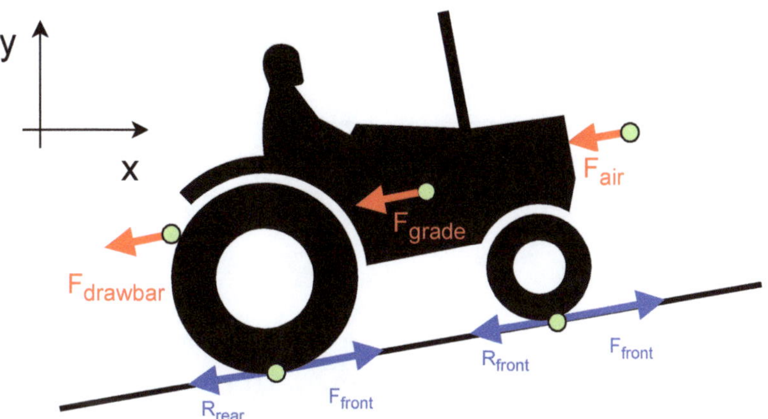

Figure 1. Forces acting on an off-road vehicle.

Based on the simplified model of the wheel–soil interaction shown in Figure 2, the equilibrium of the horizontal and vertical forces is written as follows:

$$R_c = b \int_0^{\theta_0} \sigma r \sin\theta d\theta \tag{5}$$

$$W = b \int_0^{\theta_0} \sigma r \cos\theta d\theta \tag{6}$$

where b and r are the smaller dimensions of the contact patch and radius of the wheel, and θ_0 is the contact angle of the wheel.

To predict the performance of a rigid wheel, Bekker [30] assumed that the terrain reaction on the contact patch is purely radial and is equal to the normal pressure beneath a sinkage plate:

$$\sigma r \sin\theta d\theta = pdz \tag{7}$$

$$\sigma r \cos\theta d\theta = pdx \tag{8}$$

where θ and z are the angle and sinkage shown in Figure 2, σ is the radial pressure on the wheel–terrain interface, and p is the normal pressure below a sinkage plate.

Combining (5) and (7), the motion resistance is written as follows:

$$R_c = b \int_0^z pdz \tag{9}$$

For homogeneous soils, Bernstein [31] proposed an empirical model to describe their pressure–sinkage relationship:

$$p = kz^n \tag{10}$$

where k is the sinkage modulus, and n is the sinkage exponent.

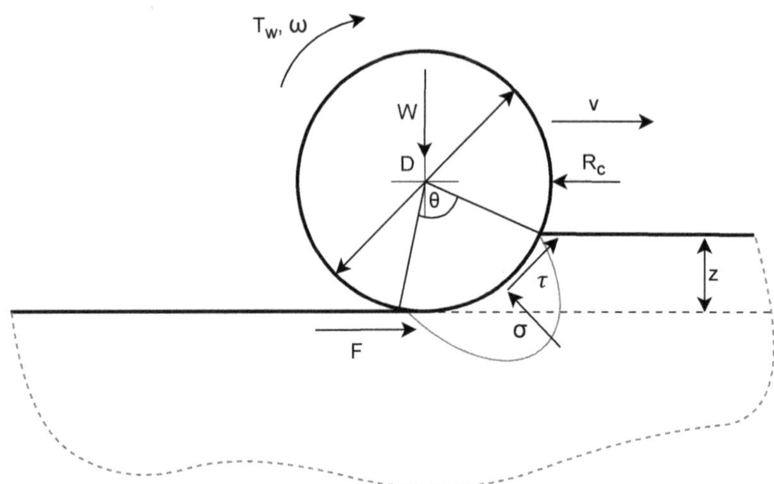

Figure 2. Rigid wheel–deformable terrain interaction.

Later, Bekker [32] separated the sinkage modulus used by Berstein into two parts to represent both the soil cohesion and the effect of the internal shearing resistance:

$$p = \left(\frac{k_c}{b} + k_\phi\right)z^n \tag{11}$$

where k_c is the cohesive modulus of deformation, k_ϕ is the frictional modulus of deformation, and b is the smaller dimension of the contact patch.

Substituting the pressure–sinkage Equation (11) into (9), the motion resistance is rewritten as:

$$R_c = b_{ti}\int_0^{z_r}(\frac{k_c}{b} + k_\phi)z^n dz \tag{12}$$

The solution of the definitive integral along the sinkage is the following:

$$R_c = b_{ti}\left[\left(\frac{k_c}{b} + k_\phi\right)\frac{z^{n+1}}{n+1}\right] \tag{13}$$

For the geometry shown in Figure 2, assuming slight sinkage and expressing the sinkage from (8), the motion resistance of wheeled vehicles can be expressed in a closed form [29]. This solution takes a highly nonlinear form; thus, it is not optimal for state-space models.

According to Bekker, the thrust–slip relationship of a tire can be determined as that of a rigid track. First, the shear stress developed at the running gear–terrain interface must be calculated to calculate the tractive effort. For plastic soils, ref. [33] proposed a modification of Bekker's equation to describe the shear stress–displacement relationship:

$$\frac{\tau}{\tau_{max}} = 1 - e^{\frac{-j}{K}} \tag{14}$$

where τ and τ_{max} are the shear stress and the maximum shear stress, respectively, j is the shear displacement, and K is the shear deformation parameter.

The maximum shear displacement that the terrain can bear is defined using the Mohr–Coulomb equation:

$$\tau_{max} = c + \sigma \tan\phi \tag{15}$$

where c is the cohesion, σ is the normal stress below the track, and ϕ is internal angle of friction of the terrain.

The shear displacement is the relative movement between the running gear and the terrain, which causes slip. The slip can be expressed as follows:

$$i = 1 - \frac{V}{r\omega} = 1 - \frac{V}{V_t} = \frac{V_t - V}{V_t} = \frac{V_j}{V_t} \tag{16}$$

where ω, V, V_t, and V_j are the rotational speed, actual forward velocity, theoretical forward speed, and slip speed of the running gear, respectively.

At a given point, which is at x distance from the front of the contact patch, the shear displacement can be written as:

$$j = V_j t = \frac{V_j x}{V_t} = ix \tag{17}$$

where t is the contact time of the point.

Then, assuming uniformly distributed normal pressure below the running gear, the equation describes the shear stress–displacement relationship:

$$F = b \int_0^l \tau_{max} \left(1 - e^{\frac{-j}{K}} \right) dl \tag{18}$$

Calculating the definitive integral along the length of the contact patch results in the following equation:

$$F = (Ac + W \tan \phi) \left[1 - \frac{k}{il} \left(1 - e^{\frac{-il}{k}} \right) \right] \tag{19}$$

where A and l are the area and length of the contact patch, while W is the normal force.

The viscous friction acting on the wheel is calculated as follows:

$$T_f = b_f \omega \tag{20}$$

where b_f is the viscous friction coefficient.

Substituting (13), (19), and (20) into (4), the nonlinear state-space model takes the following form:

$$\begin{bmatrix} \dot{v}(t) \\ \dot{\omega}(t) \end{bmatrix} = \begin{bmatrix} \dfrac{(Ac + W \tan \phi)\left[1 - \frac{k}{il}\left(1 - e^{\frac{-il}{K}}\right)\right] - b_{ti}\left[\left(\frac{k_c}{b} + k_\phi\right)\frac{z^{n+1}}{n+1}\right]}{m} \\ \dfrac{T_w - b\omega - (Ac + W \tan \phi)\left[1 - \frac{k}{il}\left(1 - e^{\frac{-il}{K}}\right)\right]r}{J} \end{bmatrix} \tag{21}$$

Next, some simplifications must be made to create an LPV representation from the nonlinear model. While the exponential term $e^{\frac{-il}{K}}$ could be handled as one of the scheduling parameters of the model, it rapidly converges to zero at higher slip values. For typical agricultural vehicles, such as tractors, it is negligible above 10% slip. Thus, it can be neglected under normal operational conditions. As a result, the tractive effort is formulated as follows:

$$F = \tau_{max} A \left(1 - \frac{k}{il} \right) \tag{22}$$

Furthermore, substituting Bekker's pressure–sinkage Equation (11) into the closed form equation of the motion resistance (13) yields:

$$R_c = b_{ti} \left[\left(\frac{k_c}{b} + k_\phi \right) \frac{p}{\frac{k_c}{b} + k_\phi} \frac{z}{n+1} \right] \tag{23}$$

As the normal pressure is the product of the normal force and the contact patch area, the equation is rewritten as:

$$R_c = b_{ti}\left[\left(\frac{k_c}{b}+k_\phi\right)\frac{mg}{\left(\frac{k_c}{b}+k_\phi\right)A_c}\frac{z}{n+1}\right] \tag{24}$$

Then, k_c and k_ϕ can be simplified from the equation of the motion resistance:

$$R_c = \frac{b_{ti}mg}{(n+1)A_c}z \tag{25}$$

Due to the applied simplification, the nonlinear state-space model takes the following form:

$$\begin{bmatrix} \dot{v}(t) \\ \dot{\omega}(t) \end{bmatrix} = \begin{bmatrix} \dfrac{\tau_{max}A_c\left(1-\frac{K}{il}\right)-\frac{b_{ti}mg}{(n+1)A_c}z}{m} \\ \dfrac{T_w-b\omega-\left[\tau_{max}A_c\left(1-\frac{K}{il}\right)\right]r}{J} \end{bmatrix} \tag{26}$$

As the aim of this model is to support traction control algorithms, the input of the model shall be the motor torque applied at the wheel:

$$u = T_w \tag{27}$$

Then, the remaining time-dependent variables in the state matrix A and the input matrix B can be collected in the scheduling parameter vector ρ. However, applying Jacobi linearization using the current state and input vector would yield inapt results. The traction force (22) and the motion resistance (25) are independent of both the state vector and the input vector. Eliminating them from the state-space model would result in an unusable model. On the other hand, the traction force is a linear function of the maximum shear stress, while the motion resistance is a linear function of the static sinkage. They must be handled either as states or inputs of the model.

Including them among the states of the system would require their differential equations. Therefore, handling them as uncontrollable system inputs is a more reasonable choice:

$$u_{uc} = \begin{bmatrix} \tau_{max} \\ z \end{bmatrix} \tag{28}$$

In agricultural applications, the terrain is usually known; hence, the maximum shear stress of the terrain is usually known beforehand. However, the shear stress of the soil also shows a relationship with its color [34], while online estimators, such as [35], have also been developed. Hence, it can also be determined using vision-based methods. Such methods are also applicable for determining e wheel sinkage [36].

At last, the state equation of the LPV model for a single wheel can be written as follows:

$$\begin{bmatrix} \dot{v}(t) \\ \dot{\omega}(t) \end{bmatrix} = \begin{bmatrix} 0 & 0 \\ 0 & \frac{-b}{J} \end{bmatrix}\begin{bmatrix} v(t) \\ \omega(t) \end{bmatrix} + \begin{bmatrix} 0 \\ \frac{1}{J} \end{bmatrix}T_w + \begin{bmatrix} \dfrac{A_c-\frac{Kb}{i}}{m} & \dfrac{-g}{(n+1)A_c} \\ \dfrac{\frac{Kbr}{i}-A_c}{J} & 0 \end{bmatrix}\begin{bmatrix} \tau_{max} \\ z \end{bmatrix} \tag{29}$$

In the case of an LTI model, the state and input matrices are constants. In (29), several parameters of the state and input matrices vary over time or change based on the inputs or states of the model. These parameters are collected in the vector of scheduling parameters:

$$\rho = \begin{bmatrix} k & n & A_c & \frac{1}{i} \end{bmatrix} \tag{30}$$

Next, the state-space model must be augmented to describe the behavior of a vehicle with an arbitrary number of wheels. First, $v(t)$ shall be the longitudinal speed of the vehicle, and each added wheel shall have its rotational speed as a new state of the model. Then, the traction forces and motion resistances must be summarized for all wheels to determine the longitudinal velocity of the vehicle:

$$
\begin{bmatrix} \dot{v}(t) \\ \dot{\omega}_1(t) \\ \dot{\omega}_2(t) \\ \vdots \\ \dot{\omega}_n(t) \end{bmatrix} = \begin{bmatrix} 0 & 0 & 0 & \cdots & 0 \\ 0 & \frac{-b_1}{J_1} & 0 & \cdots & 0 \\ 0 & 0 & \frac{-b_2}{J_2} & \cdots & 0 \\ \vdots & \vdots & & \ddots & \vdots \\ 0 & 0 & 0 & \cdots & \frac{-b_n}{J_n} \end{bmatrix} \begin{bmatrix} v(t) \\ \omega_1(t) \\ \omega_2(t) \\ \vdots \\ \omega_n(t) \end{bmatrix} + \begin{bmatrix} 0 & 0 & 0 & \cdots & 0 \\ \frac{1}{J_1} & 0 & 0 & \cdots & 0 \\ 0 & \frac{1}{J_2} & 0 & \cdots & 0 \\ \vdots & & \ddots & & \vdots \\ 0 & 0 & 0 & \cdots & \frac{1}{J_n} \end{bmatrix} \begin{bmatrix} T_{w1} \\ T_{w2} \\ \vdots \\ T_{wn} \end{bmatrix} +
$$

$$
\begin{bmatrix} \frac{A_{c1}-\frac{k_1 b_1}{i_1}}{m_1} & \frac{-g\cos\alpha_1}{(n_1+1)A_{c1}} & \frac{A_{c2}-\frac{k_2 b_2}{i_2}}{m_2} & \frac{-g\cos\alpha_2}{(n_2+1)A_{c2}} & \cdots & \frac{A_{cn}-\frac{k_n b_n}{i_n}}{m_n} & \frac{-g\cos\alpha_n}{(n_n+1)A_{cn}} \\ \frac{\frac{k_1 b_1 r_1}{i_1}-A_{c1}}{J_1} & 0 & 0 & 0 & \cdots & 0 & 0 \\ 0 & 0 & \frac{\frac{k_2 b_2 r_2}{i_2}-A_{c2}}{J_2} & 0 & \cdots & 0 & 0 \\ \vdots & & & & \ddots & & \vdots \\ 0 & 0 & 0 & 0 & \cdots & \frac{\frac{k_n b_n r_n}{i_n}-A_{cn}}{J_n} & 0 \end{bmatrix} \begin{bmatrix} \tau_{1,max} \\ z_1 \\ \tau_{2,max} \\ z_2 \\ \vdots \\ \tau_{n,max} \\ z_n \end{bmatrix}
\tag{31}
$$

where the indices 1, 2, and n refer to the states of the corresponding wheel.

2.3. Implementation Details

The developed model was implemented using a Matlab R2023b function. There were five groups of input arguments for the function:

- The actual state vector of the model;
- The model parameters (such as the wheel inertia, assumed to be constant);
- The inputs of the model over time;
- The actual scheduling parameter vector;
- The simulation time.

The function determines the actual inputs of the model based on the simulation time. Thus, variable step solvers can be used for simulation. Using the derived LPV model, the function determines and returns the gradient of the state vector. During simulation, a function handle was created for the function of the state-space model, whose arguments were the simulation time and the state vector. Then, the function handle was passed to the selected ordinary differential equation (ODE) solver. The results presented in this paper qwew attained using a variable-step Runge–Kutta (ODE45) method.

3. Results

3.1. Simulation Results

First, the developed model was verified using Project Chrono [37], an open-source multiphysics simulation engine. It has various applications, such as vehicle dynamics, terramechanics, granular flows, and seismic engineering. Built upon the Bekker–Wong soil model, it is capable of modeling the operation of wheeled and tracked vehicles on deformable terrain. Custom parameters defining soil types can be assigned to every point of the managed map. The simulator features dynamic models of various wheeled and tracked vehicles with numerous suspension, drivetrain, and steering system configurations. Furthermore, custom vehicle models can also be implemented.

Project Chrono supports multi-ore, distributed computing modes and can harness GPU power for particle motion modeling. It can be faster than real time when using rigid surfaces, although its performance significantly degrades with increasing deformable terrain size. The environment created in Project Chrono is presented in Figure 3. It consisted of flat terrain created using the Soil Contact Model (SCM) deformable terrain class [38], built upon the Bekker–Wong theory, but also taking into account contact kinematics and the pressure distribution in the footprint. The terrain was modeled using a mesh, which could be deformed only in the vertical direction.

Figure 3. Project Chrono simulation environment.

On the other hand, the LPV model was a lumped model, which also neglected the velocity of the impact on the sinkage. Hence, for the same soil type, the contact area remained constant. In Project Chrono, the wheel is created slightly above the ground to avoid unintended collision detection and numerical instabilities at the start of the simulation.

After the wheel touches the ground, a constant torque of ca. 3140 Nm is applied to the wheel, which is also used as the input of the state-space model. The comparison consisted of test cases with different terrains. For example, the parameters used to simulate a cohesionless, purely frictional terrain (LETE Sand) are summarized in Table 1.

Table 1. Terrain and vehicle parameters—Project Chrono simulation.

Terrain parameters	Sinkage coefficient [-]	1.1
	Cohesion [Pa]	0
	Shear deformation modulus [m]	0.01
	Friction angle [°]	30
	Contact area [m^2]	0.224
Vehicle parameters	Wheel inertia [kgm^2]	20
	Vehicle mass [kg]	500
	Wheel width [m]	0.4
	Wheel radius [m]	0.5

First, the test case was simulated in Chrono, and the simulated states were exported to a text file. Then, ithey were imported to Matlab to feed the relevant inputs of the simulation to the LPV model and to perform the comparison.

The comparison of the models is presented in Figure 4. The first diagram shows the longitudinal velocity, the second diagram presents the angular velocity, and the third diagram shows the wheel slip. The slip of the LPV model envelops that of the simulator, and the quantitative results are nearly identical; the maximum difference between the calculated slips is below 2%. An increasing oscillation could also be observed in the outputs of the simulator, which was caused by the ideal motor driving the wheel and the spring-like modeling of the contact.

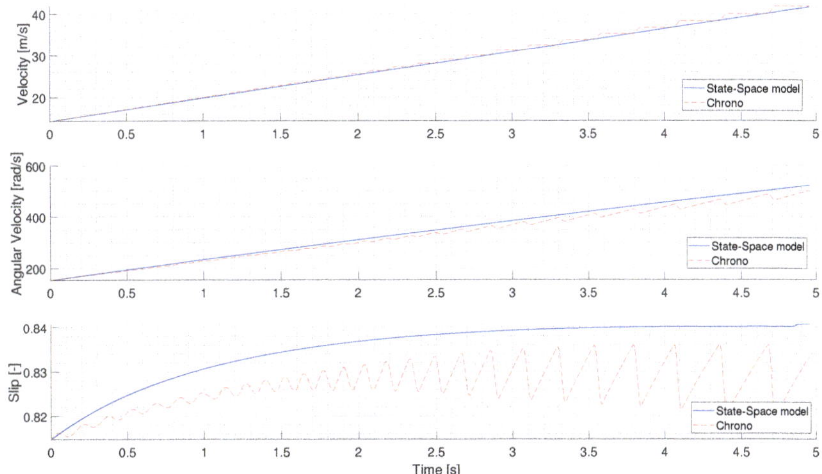

Figure 4. Verification of the proposed LPV model using Project Chrono—high slip.

The tests were carried out using different wheel torques. Another example is presented in Figure 5, where the input torque is ca. 1570 Nm. As can be seen, the difference between the two models increases in the case of lower slip values. The LPV model overestimates the angular and longitudinal velocities compared to the Chrono simulation. While there is a difference between the exact values, the main characteristics of the two models are similar. Furthermore, overestimating the values errs on the side of caution.

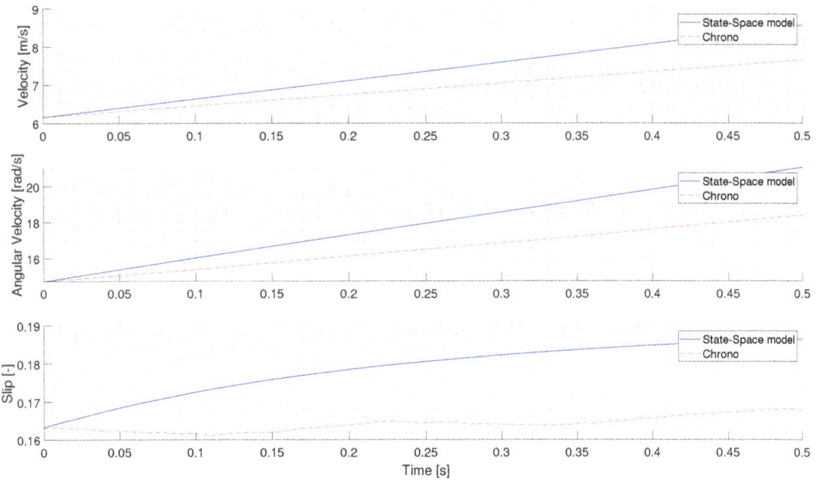

Figure 5. Verification of the proposed LPV model using Project Chrono—low slip.

3.2. Experimental Measurements

The model was also validated against experimental measurements. The measurements used to validate the developed model were obtained previously to analyze the energetic balance of tractors. Several studies have presented results regarding the energy balance aspects, such as [39,40].

The measurements were conducted using a John Deere 6600 tractor and a Dyna-Cart dynamometer vehicle shown in Figure 6 on a flat, sandy-clay terrain. The measurements aimed to determine the energy absorption of the terrain; hence, the traction force was nearly constant during the measurements. The original drive shafts were replaced with shafts equipped with strain gauge stamps to determine the torque of the rear wheels. The measurements were conducted on straight sections, with the rear axle differential locked. Therefore, only one measurement point was sufficient to determine the rear wheel speeds, to which an electronic tachometer was used. The longitudinal velocity of the vehicle was measured using a radar.

Figure 6. John Deere 6600 tractor and Dyna-Cart dynamometer vehicle.

Each measurement was conducted on virgin, undisturbed track sections to determine vertical soil deformation. The vertical terrain profile was measured in front of the tractor, in the expected wheel path, and behind the tractor, in the remaining wheel path, to determine the vertical soil deformation under the tractor wheels. Soil samples were taken from various parts of the field, and their composition, density, volume weight, water content, and porosity were determined in a laboratory to determine the soil properties of the experimental area. However, the Bekker–Wong terrain parameters of the soil were not determined during the measurement presented in [29,34] for similar types of soils with the same moisture content. The same applies to the inertia of the wheel: the vehicle mass is, in this case, half of the mass on the rear axle of the tractor, which was measured using a weight-bridge. The parameters used for validation are summarized in Table 2.

Figure 7 presents the validation results. As in the case of the verification results, the first diagram shows the longitudinal velocity, the second diagram presents the angular velocity, and the third diagram shows the wheel slip. During the measurements, the total fraction force was nearly constant, although there was a slight oscillation in the torque measured at the wheels, which was used as the input of the model. The oscillation of the torque resulted in a similar behavior in the simulated angular velocity, although it was nearly constant during the measurements, which indicated that the reduced inertia of the wheel and the transmission system was underestimated.

Table 2. Terrain and vehicle parameters—measurement.

Terrain parameters	Sinkage coefficient [-]	0.85
	Cohesion [Pa]	2200
	Shear deformation modulus [m]	0.08
	Friction angle [°]	30
	Contact area [m^2]	0.224
Vehicle parameters	Wheel inertia [kgm^2]	20
	Vehicle mass [kg]	1662.5
	Wheel width [m]	0.46
	Wheel radius [m]	0.9

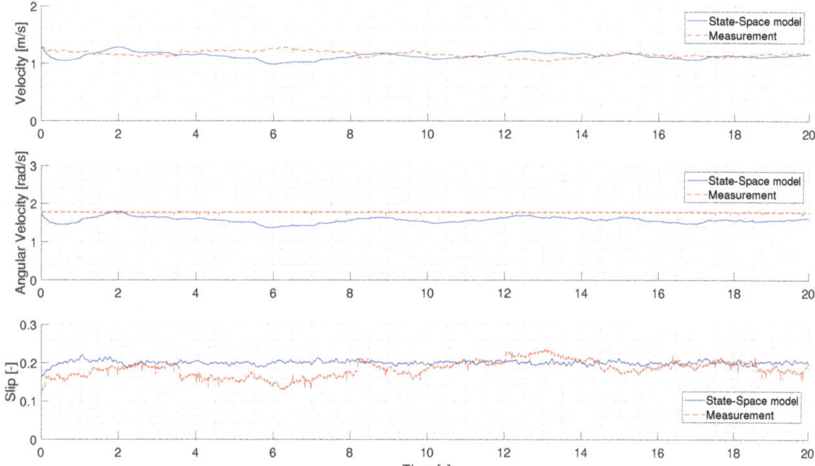

Figure 7. Validation of the proposed LPV model.

The absolute differences between the simulated and measured velocities are small. The maximum absolute error of the longitudinal velocity is below 0.3 m/s and 0.5 rad/s in the case of the angular velocity. On the other hand, the relative errors are significant due to the low reference values, which is also reflected in the slip, where the absolute error of the model reaches 0.06. While the difference could be caused by both parameter uncertainty and simplifications of the model, it is interesting to note that the errors significantly decrease after 8 seconds.

4. Discussion

The deviation of the LPV model compared to both the Chrono simulators and the experimental measurements are summarized in Table 3. Quantitatively, the most significant deviance can be seen in the case of the angular velocity error during the high-slip simulation, which comes from the sawtooth-like form of the angular velocity simulated in Chrono. The absolute differences are low in all other cases.

According on the results, the proposed model can describe the behavior of off-road vehicles with an accuracy acceptable for control design. Furthermore, it can run in real-time: the simulation of 10 s of driving required approximately 0.0365 s. Yet, the proposed model has some drawbacks, which must be discussed. Due to the applied simplifications, the operating domain of the model was narrowed. As Figure 5 shows, the accuracy of the model decreases in the case of lower slip values, which is caused by neglecting the exponential term in (19). Similarly, its performance degrades in the presence of a significant slope or if the velocity of the vehicle increases.

Table 3. Performance metrics of the LPV model.

		High-Slip Simulation	Low-Slip Simulation	Measurement
Velocity [m/s]	Max	1.406	0.908	0.283
	Mean	0.314	0.456	0.075
Angular velocity [rad/s]	Max	40.631	2.650	0.423
	Mean	14.706	1.416	0.202
Slip [-]	Max	0.02	0.019	0.076
	Mean	0.008	0.015	0.022

On the contrary, grid-based LPV models often suffer from the curse of dimensionality. Narrowing the operation domain of the model reduces the number of grid points, simplifying control synthesis. Depending on the requirements of the control design, it is possible to include the neglected forces in the LPV model.

Some model parameters, such as the area of the contact patch or the terrain parameters, are difficult to measure or estimate in real time. Even if the terrain is traversed in advance, the accuracy of these parameters is uncertain, or they could have changed since the measurement. While this is a common drawback of semiempirical models, robust control techniques, such as the LPV $\mathcal{H}$-∞ algorithm, can handle parameter uncertainty.

On the other hand, the number of terrain parameters was reduced compared to the original equations, and grid-based LPV models can provide some robustness against measurement errors in the scheduling parameters. Furthermore, contrary to several of the adapted on-road tire models, the proposed model takes into account the characteristics of the vehicle; hence, the same model can be used for different vehicles without any intermediate steps, contrary to the adapted on-road models.

5. Conclusions

An LPV model was developed for off-road vehicles, which aims to model terrain–vehicle interactions in real time with acceptable accuracy and support the development of model-based control algorithms. First, a nonlinear state-space model was derived from the Bekker–Wong equations; then, it was simplified and reformulated using the LPV approach.

Next, the model was verified using Project Chrono, a multiphysics simulator, and validated against experimental measurements. Based on the results, the accuracy of the developed model is suitable for control design purposes. Furthermore, off-road traction control algorithms have to estimate the terrain parameters, which can be computationally expensive and inaccurate. The proposed model requires only three terrain parameters: the shear deformation coefficient, the sinkage exponent, and the maximum shear stress. Thus, the uncertainty caused by the accuracy of the estimated parameters can be reduced.

The proposed model can be used to design model-based estimator algorithms, such as different variants of the Kalman filter algorithm. However, in this case, the uncontrollable and controllable inputs can be combined into a single input vector to simplify the design process.

Traction control algorithms usually aim to control either the slip or the longitudinal velocity of the vehicle. However, in the presented form of the model, the latter is an uncontrollable state as it is independent of the controllable input and the controllable states of the model. It depends only on the uncontrollable inputs of the system. A possible solution is using the slip or its reciprocal as a system state instead of a scheduling parameter. This way, both the slip and the longitudinal velocity become controllable states of the system. The nonlinearities caused by the inclusion of the slip among the state variables can also be handled by the LPV representation. LPV-based slip ratio controllers are already presented in the literature for on-road vehicles, such as the method presented in [41].

Further work will focus on the generalization of the presented model and the improvement in the model's accuracy. The possibility of considering the effects of steep slopes, different pressure–sinkage values, shear stress–displacement models, or even organic terrain must be analyzed.

Author Contributions: Conceptualization, A.S. and P.K.; methodology, A.S. and P.K.; software, D.K.D.; validation, P.K. and D.K.D.; writing—original draft preparation, A.S. and D.K.D.; writing—review and editing, S.A.; visualization, S.A.; supervision, S.A. All authors have read and agreed to the published version of the manuscript.

Funding: This research was supported by the Ministry of Culture and Innovation and the National Research, Development and Innovation Office within the Cooperative Technologies National Laboratory of Hungary (grant No. 2022-2.1.1-NL-2022-00012).

Data Availability Statement: The data presented in this study are available upon request from the corresponding author.

Conflicts of Interest: The authors declare no conflicts of interest.

Abbreviations

The following abbreviations are used in this manuscript:

ABS	Antilock Braking System
ACC	Adaptive Cruise Control
DAS	Driver Assistance System
DEM	Discrete Element Method
FEM	Finite Element Method
LFR	Linear Fractional Representation
LFT	Linear Fractional Transformation
LPV	Linear Parameter-Varying
qLPV	Quasi-Linear Parameter-Varying
SCM	Soil Contact Model

References

1. Sunusi, I.I.; Zhou, J.; Wang, Z.Z.; Sun, C.; Ibrahim, I.E.; Opiyo, S.; Soomro, S.A.; Sale, N.A.; Olanrewaju, T.; et al. Intelligent tractors: Review of online traction control process. *Comput. Electron. Agric.* **2020**, *170*, 105176. [CrossRef]
2. Luo, Z.; Wang, J.; Wu, J.; Zhang, S.; Chen, Z.; Xie, B. Research on a Hydraulic Cylinder Pressure Control Method for Efficient Traction Operation in Electro-Hydraulic Hitch System of Electric Tractors. *Agriculture* **2023**, *13*, 1555. [CrossRef]
3. Stentz, A.; Dima, C.; Wellington, C.; Herman, H.; Stager, D. A system for semi-autonomous tractor operations. *Auton. Robot.* **2002**, *13*, 87–104. :1015634322857 [CrossRef]
4. Upadhyaya, S.; Wulfsohn, D. Traction prediction using soil parameters obtained with an instrumented analog device. *J. Terramech.* **1993**, *30*, 85–100. [CrossRef]
5. Rubinstein, D.; Shmulevich, I.; Frenckel, N. Use of explicit finite-element formulation to predict the rolling radius and slip of an agricultural tire during travel over loose soil. *J. Terramech.* **2018**, *80*, 1–9. [CrossRef]
6. Jiang, M.; Dai, Y.; Cui, L.; Xi, B. Experimental and DEM analyses on wheel-soil interaction. *J. Terramech.* **2018**, *76*, 15–28. [CrossRef]
7. Wong, J.Y.; Reece, A. Prediction of rigid wheel performance based on the analysis of soil-wheel stresses part I. Performance of driven rigid wheels. *J. Terramech.* **1967**, *4*, 81–98. [CrossRef]
8. He, R.; Sandu, C.; Khan, A.K.; Guthrie, A.G.; Els, P.S.; Hamersma, H.A. Review of terramechanics models and their applicability to real-time applications. *J. Terramech.* **2019**, *81*, 3–22. [CrossRef]
9. Johnson, D.K.; Botha, T.R.; Els, P.S. Real-time side-slip angle measurements using digital image correlation. *J. Terramech.* **2019**, *81*, 35–42. [CrossRef]
10. Linström, B.V.; Els, P.S.; Botha, T. A real-time non-linear vehicle preview model. *Int. J. Heavy Veh. Syst.* **2018**, *25*, 1–22. [CrossRef]
11. Sandu, C.; Kolansky, J.; Botha, T.; Els, S. 6.3. Multibody Dynamics Techniques for Real-Time Parameter Estimation. *Adv. Auton. Veh. Des. Sev. Environ.* **2015**, *44*, 221.

12. Vieira, D.; Orjuela, R.; Spisser, M.; Basset, M. An adapted Burckhardt tire model for off-road vehicle applications. *J. Terramech.* **2022**, *104*, 15–24. [CrossRef]
13. Lenain, R.; Thuilot, B.; Cariou, C.; Martinet, P. High accuracy path tracking for vehicles in presence of sliding: Application to farm vehicle automatic guidance for agricultural tasks. *Auton. Robot.* **2006**, *21*, 79–97. [CrossRef]
14. Cariou, C.; Lenain, R.; Thuilot, B.; Berducat, M. Automatic guidance of a four-wheel-steering mobile robot for accurate field operations. *J. Field Robot.* **2009**, *26*, 504–518. [CrossRef]
15. Kim, J.; Lee, J. Traction-energy balancing adaptive control with slip optimization for wheeled robots on rough terrain. *Cogn. Syst. Res.* **2018**, *49*, 142–156. [CrossRef]
16. Salama, M.A.; Vantsevich, V.V.; Way, T.R.; Gorsich, D.J. UGV with a distributed electric driveline: Controlling for maximum slip energy efficiency on stochastic terrain. *J. Terramech.* **2018**, *79*, 41–57. [CrossRef]
17. Pacejka, H.B.; Besselink, I.J.M. Magic Formula Tyre Model with Transient Properties. *Veh. Syst. Dyn.* **1997**, *27*, 234–249. [CrossRef]
18. Alexander, A.; Sciancalepore, A.; Vacca, A. Online Controller Setpoint Optimization for Traction Control Systems Applied to Construction Machinery. In Proceedings of the BATH/ASME 2018 Symposium on Fluid Power and Motion Control, Bath, UK, 12–14 September 2018; p. V001T01A068. [CrossRef]
19. Saunders, A.; White, D.; Compere, M. Estimating Pacejka (PAC2002) Tire Coefficients for Pneumatic Tires on Soft Soils With Application to BAJA SAE Vehicles. In Proceedings of the ASME 2019 International Mechanical Engineering Congress and Exposition, Salt Lake City, UT, USA, 11–14 November 2019; Volume 59414, p. V004T05A081.
20. Vieira, D.; Orjuela, R.; Spisser, M.; Basset, M. Longitudinal Vehicle Control based on Off-road Tire Model for Soft Soil Applications. *IFAC-PapersOnLine* **2021**, *54*, 304–309. [CrossRef]
21. Gaspar, P.; Szaszi, I.; Bokor, J. Active Suspension Design using LPV Control. *IFAC Proc. Vol.* **2004**, *37*, 565–570. [CrossRef]
22. Poussot-Vassal, C.; Sename, O.; Dugard, L.; Gáspár, P.; Szabó, Z.; Bokor, J. A new semi-active suspension control strategy through LPV technique. *Control Eng. Pract.* **2008**, *16*, 1519–1534. [CrossRef]
23. Németh, B. Robust LPV design with neural network for the steering control of autonomous vehicles. In Proceedings of the 2019 18th European Control Conference (ECC), Naples, Italy, 25–28 June 2019; pp. 4134–4139.
24. Németh, B.; Gáspár, P.; Orjuela, R.; Basset, M. LPV-based Control Design of an Adaptive Cruise Control System for Road Vehicles. *IFAC-PapersOnLine* **2015**, *48*, 62–67. [CrossRef]
25. Vu, V.T.; Sename, O.; Dugard, L.; Gaspar, P. The Design of an H∞/LPV Active Braking Control to Improve Vehicle Roll Stability. *IFAC-PapersOnLine* **2019**, *52*, 54–59. [CrossRef]
26. Gáspár, P.; Németh, B.; Bokor, J. Design of an LPV-based integrated control for driver assistance systems. *IFAC Proc. Vol.* **2012**, *45*, 511–516. [CrossRef]
27. Gáspár, P.; Szabó, Z.; Bokor, J. LPV design of adaptive integrated control for road vehicles. *IFAC Proc. Vol.* **2011**, *44*, 662–667. [CrossRef]
28. Li, J.; Sun, S.; Sun, C.; Liu, C.; Tang, W.; Wang, H. Analysis of Effect of Grouser Height on Tractive Performance of Tracked Vehicle under Different Moisture Contents in Paddy Soil. *Agriculture* **2022**, *12*, 1581. [CrossRef]
29. Wong, J.Y. *Terramechanics and Off-Road Vehicle Engineering: Terrain Behaviour, Off-Road Vehicle Performance and Design*; Butterworth-Heinemann: Oxford, UK, 2009.
30. Bekker, M.G. *Theory of Land Locomotion*; University of Michigan Press: Ann Arbor, MI, USA, 1956.
31. Bernstein, R. Probleme zur experimentellen Motorpflugmechanik. *Der Mot.* **1913**, *16*, 199–206.
32. Bekker, M.G. *Introduction to Terrain-Vehicle Systems. Part I: The Terrain. Part II: The Vehicle*; University of Michigan Press: Ann Arbor, MI, USA 1969.
33. Janosi, Z. The Analytical Determination of Drawbar Pull as A Function of Slip for Tracked Vehicles in Deformable Soils. In Proceedings of the International Society for Terrain-Vehicle Systems, The 1st International Conference, Turin, Italy, 1961; Volume 707.
34. El Hariri, A.; Ahmed, A.E.E.; Kiss, P. Sandy Loam Soil Shear Strength Parameters and Its Colour. *Appl. Sci.* **2023**, *13*, 3847. [CrossRef]
35. Iagnemma, K.; Shibly, H.; Dubowsky, S. On-line terrain parameter estimation for planetary rovers. In Proceedings of the 2002 IEEE International Conference on Robotics and Automation (Cat. No. 02CH37292), Washington, DC, USA, 11–15 May 2002; Volume 3, pp. 3142–3147.
36. Iagnemma, K.; Kang, S.; Brooks, C.; Dubowsky, S. Multi-sensor terrain estimation for planetary rovers. In Proceedings of the 8th International Symposium on Artificial Intelligence, Robotics, and Automation in Space, Nara, Japan, 19–23 May 2003.
37. Tasora, A.; Serban, R.; Mazhar, H.; Pazouki, A.; Melanz, D.; Fleischmann, J.; Taylor, M.; Sugiyama, H.; Negrut, D. *Chrono: An Open Source Multi-Physics Dynamics Engine*; Springer: Berlin/Heidelberg, Germany, 2016; pp. 19–49.
38. Krenn, R.; Hirzinger, G. SCM—A soil contact model for multi-body system simulations. In Proceedings of the 11th European Regional Conference of the International Society for Terrain-Vehicle Systems, Bremen, Germany, 5–8 October, 2009.
39. Kiss, P.; Laib, L. Tractor energy balance under instationary conditions: 6th Mini Conference on Vehicle System Dynamics, Identification and Anomalies (VSDIA'98). In Proceedings of the 6th Mini Conference on Vehicle System Dynamics, Identification and Anomalies, VSDIA'98, Budapest, Hungary, 9–11 November 1998; pp. 407–414.

40. Kiss, P. Effect of Soil Deformation on the Energy Balance of Tractors. *Hung. Agric. Eng.* **1999**, *12*, 35–39.
41. Jia, F.; Liu, Z. A LPV traction control approach for independent in-wheel electric motor vehicle. In Proceedings of the 11th World Congress on Intelligent Control and Automation, Shenyang, China, 29 June–4 July 2014; pp. 1992–1997. [CrossRef]

 *agriculture*

Article

Study the Flow Capacity of Cylindrical Pellets in Hopper with Unloading Paddle Using DEM

Huinan Huang [1,2], Yan Zhang [1], Defu Wang [1,2], Zijiang Fu [1], Hui Tian [1], Junjuan Shang [1], Mahmoud Helal [3,4] and Zhijun Lv [1,*]

[1] College of Mechanical and Electrical Engineering, Henan Agricultural University, Zhengzhou 450002, China; huanghuinan@henau.edu.cn (H.H.); zhangyan801@stu.henau.edu.cn (Y.Z.); wangdefu@henau.edu.cn (D.W.); fzj@stu.henau.edu.cn (Z.F.); tianhui@henau.edu.cn (H.T.); shangjunjuan@henau.edu.cn (J.S.)
[2] Key Laboratory of Swine Facilities Engineering, Ministry of Agriculture, Harbin 150030, China
[3] Department of Mechanical Engineering, Faculty of Engineering, Taif University, P.O. 11099, Taif 21944, Saudi Arabia; mo.helal@tu.edu.sa
[4] Department of Production and Mechanical Design, Faculty of Engineering, Mansoura University, Mansoura 35516, Egypt
* Correspondence: lvzhijun@henau.edu.cn

Abstract: The hopper is an important piece of basic equipment used for storing and transporting materials in the agricultural, grain, chemical engineering, coal mine and pharmaceutical industries. The discharging performance of hoppers is mainly affected by material properties and hopper structure. In this work, the flow capacity of cylindrical pellets in the hopper with the unloading paddle is studied. A series of numerical simulation analyses with the aid of the discrete element method (DEM) platform are carried out. Then, the discharging process is illustrated, and the flow capacity of pellets in the hopper is analyzed by the mass flow index (MFI), the dynamic discharging angle (DDA) formed in the discharging process and porosity among pellets. Furthermore, the effect of parameters such as hopper half angle, rotation speed of the unloading paddle and outlet diameter of the hopper is investigated. The results show that MFI increases with an increase in hopper half angle or outlet diameter and a decrease in rotation speed. Meanwhile, DDA and porosity decrease with the increase in the hopper half angle or outlet diameter and the decrease in the rotation speed. Finally, the MFI ~0.24 is identified as the criterion to distinguish the mass flow from the funnel flow for the hopper with an unloading paddle, and the optimization results are decided as follows: hopper half angle greater than 60°, outlet diameter greater than 60 mm and rotation speed between 45 rpm and 60 rpm. These results should be useful for providing a theoretical reference for the optimization design of feeding devices for swine feeders.

Keywords: discrete element method; cylindrical pellet; hopper; unloading paddle; flow capacity

Citation: Huang, H.; Zhang, Y.; Wang, D.; Fu, Z.; Tian, H.; Shang, J.; Helal, M.; Lv, Z. Study the Flow Capacity of Cylindrical Pellets in Hopper with Unloading Paddle Using DEM. *Agriculture* **2024**, *14*, 523. https://doi.org/10.3390/agriculture14040523

Academic Editor: Massimiliano Varani

Received: 28 February 2024
Revised: 20 March 2024
Accepted: 21 March 2024
Published: 25 March 2024

1. Introduction

Cylindrical pellet feed is a bulk solid feed used widely in the swine industry [1–5]. A feeder typically consists of a hopper, an unloading unit (optional design according to feed requirements) and a feed trough. The cylindrical pellets fall into the feed trough from the hopper by the unloading unit or gravity. However, it is also indicated that cylindrical pellets are blocked once arching behavior occurs in the discharging process, which seriously affects the feeding program due to the suspension of the feeding process of the swine. Meanwhile, if the discharging process cannot ensure cylindrical pellets "first in first out" from the hopper in a continuous feeding process, feed quality will be affected, thus affecting the health of swine. Therefore, the flow capacity of cylindrical pellets in the hopper during the discharging process is important to the design, optimization and application of feeders [6,7].

Recently, scholars have focused their research on discharge characteristics in the hopper [8–14]. For instance, Mahajan et al. [15] modified the design to improve the feeding accuracy of the belt weigh feeder system, which ensured uniform feeding of material by adjusting the gate opening of the feeding hopper and adjusting the vibrating motors attached to the feeding hopper. Chandravanshi et al. [16] carried out the theory analysis for the vibratory feeder. Furthermore, the movement of particles being delivered was analyzed at different frequencies in the experiment. Zhang et al. [17] revealed that the transition from mass flow to funnel flow existed during the discharging process for a conical silo with a flapper at the outlet of the silo. Specifically, the height of the flow pattern transition could be determined by DEM simulations, and the impact of the outlet size of the silo on the critical height was investigated. Fernandez et al. [18] evaluated the effect of the relative performance of different screw options, including a variable screw pitch, variable screw flight outside diameters and variable core diameters, on spherical particles in a horizontal screw feeder so as to evaluate the relative performance of different screw options. To date, all studies tend to optimize the parameters of the mechanism in accordance with different groups of experiments. Many measures have been adopted in the design of feeders, including adding vibrating motors attached to hoppers, adding a flapper at the outlet of the silo and setting up the screw device in the elevation direction with the longitudinal section of hoppers [19,20]. Although many kinds of feeders have been designed and applied in different kinds of industries, the flow capacity of pellets discharged from the hopper with an unloading paddle is seldom studied.

The discrete element method (DEM) is used extensively for the simulation of granules and in powder technology [21–24]. Ma et al. [25] proposed an intelligent calibration method for microscopic parameters based on the measured landslide accumulation morphology and the discrete element numerical simulation method. DEM is often used to investigate the particle flow in a feeder and obtain some useful findings [26]. The main advantage of DEM is that it can describe some dynamic information that is difficult to obtain by experiment [27]. In this work, DEM is adopted to investigate the flow capacity of cylindrical pellets in a hopper with an unloading paddle.

The goal of the work is to investigate the effect of the main parameters on the discharging process of cylindrical pellets in the hopper with an unloading paddle and to determine these parameters using the mass flow theory. Using DEM and physical experiments, we achieved the following objectives: to investigate the effect of main parameters, such as hopper half angle, rotation speed of the unloading paddle and outlet diameter of the hopper, on the flow capacity of cylindrical pellets; to evaluate the flow capacity by introducing the mass flow index (MFI), the dynamic discharging angle (DDA) and porosity; to decide the criterion for distinguishing between mass flow and funnel flow; and to determine the optimized parameters for the design of the hoppers with an unloading paddle.

2. Technical Solution and DEM Simulations

2.1. Research Technology Route

For the purpose of exploring the flow capacity of pellets in the hopper with an unloading paddle, numerical simulation and physical experiments are applied in this work. The research technology route is presented in Figure 1, which is determined according to the research characteristics. The work is carried out step by step according to the progress sequence: kinematics analysis of an unloading paddle, analyzing the influence law of the unloading paddle on discharging, determining the key parameters affecting the discharging of the hopper, studying the virtual discharge process, determining the MFI value used to define the flow patterns and optimizing parameters for the experimental research.

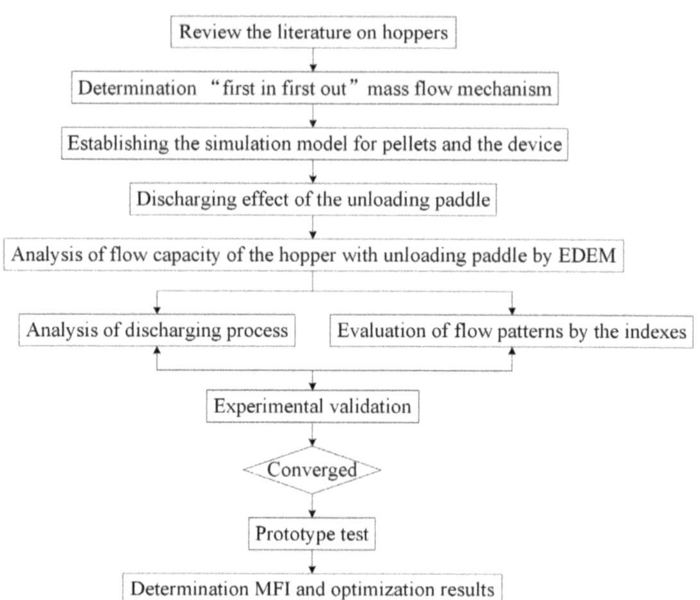

Figure 1. Research procedures.

2.2. DEM Model Description

2.2.1. Mechanical Contact Model

A three-dimensional DEM based on the soft-sphere model is used to simulate the flow process of cylindrical pellets (simplified as pellets in the following) in a hopper with an unloading paddle. The translation and rotation of each particle in a system are described by Newton's law of motion. Therefore, the equations (translational and rotational motion) of a particle i interacting with another particle j can be given by:

$$\begin{cases} m_i \frac{dv_i}{dt} = m_i g + \sum\limits_{j=1}^{n_i} (F_n^k + F_n^c + F_t^k + F_t^c) \\ I_i \frac{d\omega_i}{dt} = \sum\limits_{j=1}^{n_i} (M_t + M_f) \end{cases} \tag{1}$$

where v_i and ω_i are the translational and angular velocities of particle i respectively, with the gravity $m_i g$ and the moment of inertia I_i. n_i is the number of particle j in collision with particle i. F_n^k and F_t^k are the elastic contact force in both normal and tangential components at the contact point, respectively. F_n^c and F_t^c are the viscous damping force in both normal and tangential components at the contact point, respectively. M_t and M_f are the torque generated by tangential force and rolling friction torque respectively.

Considering the numerous studies, the Hertz–Mindlin is the most widespread contact model that can be applied in DEM simulations [17,28]. Thus, the no-slip Hertz–Mindlin contact model is adopted for the calculation of the above forces and torques of pellets; this model combines Hertz's theory in the normal direction and Mindlin's no-slip model in the tangential direction [29–31]. All forces and torques in the contact model, according to the above equations, are summarized in Table 1 [32–34]. The simulation of the pellet flow process is performed by the professional EDEM 2.7 (DEM Solution, Edinburgh, UK) software, which was installed on an Intel Xeon CPU E5-2690 v4 with 24 GB of RAM and a 64-bit Windows 7 Professional operating system. With the existing configuration, it takes approximately six CPU hours to simulate 1 s of real-time. When the number of cores in the simulator engine can be adjusted to six, the time step is 2.5×10^{-5} s for all simulations, ensuring the real situation and increasing calculation speed.

Table 1. Components of forces and torque acting on pellet i.

Forces and Torques	Symbols	Equations
Normal elastic force/N	F_n^k	$-4/3E^*\sqrt{R^*}\delta_n^{3/2}n$
Normal damping force/N	F_n^c	$-c_n\left(8m_{ij}E^*\sqrt{R^*\delta_n}\right)^{1/2}v_{n,ij}$
Tangential elastic force/N	F_t^k	$-\mu_s\left|F_n^k\right|\left[1-(1-\delta_t/\delta_{t,\max})^{3/2}\right]\hat{\delta}_t,(\delta_t<\delta_{t,\max})$
Tangential damping force/N	F_t^c	$-c_t\left(6\mu_s m_{ij}\left|F_n^k\right|\sqrt{1-\left|v_t\right|/\delta_{t,\max}}/\delta_{t,\max}\right)^{1/2}v_{t,ij},(\delta_t<\delta_{t,\max})$
Torque by tangential forces/N	M_t	$R_{ij}\times\left(F_n^k+F_t^c\right)$
Rolling friction torque/N·m	M_f	$\mu_{r,ij}\left|F_{n,ij}\right|\hat{\omega}_{t,ij}^n$
Coulomb friction force/N	f_t	$-\mu_s\left|F_n^k\right|\hat{\delta}_t,(\delta_t\geq\delta_{t,\max})$

where $\frac{1}{R^*}=\frac{1}{|R_i|}+\frac{1}{|R_j|}$, $E^*=\frac{E}{2(1-v^2)}$, $\hat{\omega}_{t,ij}=\frac{\omega_{t,ij}}{\left|\omega_{t,ij}\right|}$, $\hat{\delta}_t=\frac{\delta_t}{\left|\delta_t\right|}$, $\delta_{t,\max}=\mu_s\frac{2-v}{2(1-v)}$, $v_{ij}=v_j-v_i+\omega_j\times R_j-\omega_i\times R_i$, $v_{n,ij}=(v_{ij}\times n)\times n$, $v_{t,ij}=(v_{ij}\times n)\times n$. Note that tangential forces $(F_t^k+F_t^c)$ should be replaced by f_t when $\delta_t\geq\delta_{t,\max}$.

2.2.2. Simulation Condition

In this work, the prototype of a simulated pellet is pellet feed used in the swine industry, and the appearance is similar to that of a cylinder. The geometrical dimensions of 100 pellets are randomly measured by the Vernier Calipers; they can get a pellet with an average length L of 7.98 mm and an average radius R of 2.5 mm. At the same time, if the density measurement of a pellet is 1538 kg·m^{-3}, then the discrete element modeling of a pellet is established according to the above data [16,35–41]. In order to truly restore the state of the pellets and consider the computation of the simulation test, it is usually necessary to adjust the number and size of subspheres. Finally, the virtual model of the pellet is composed of 45 overlapping spheres of the same radius as EDEM. The theoretical radius of the nine spheres in the top section is shown in Figure 2a; the coordinate values of the nine spheres are determined based on the actual pellet diameter, as shown in Table 2, and the virtual model of the pellet is shown in Figure 2b.

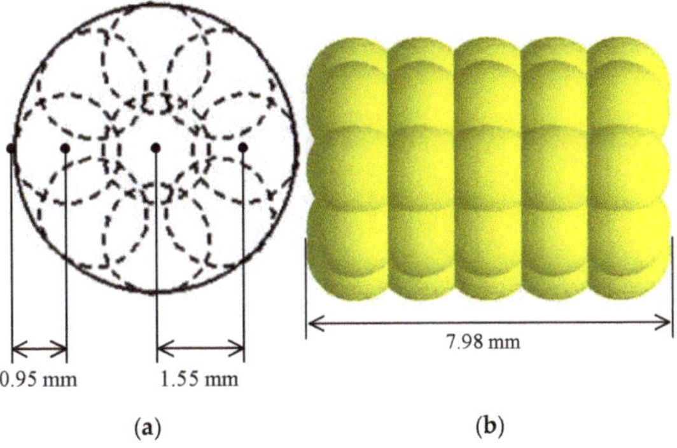

Figure 2. Model of a pellet in DEM simulation: (**a**) 2D model of a pellet and (**b**) 3D model of a pellet.

The team independently designed a swine feeder; its structure diagram is shown in Figure 3. It shows that the device contains a hopper, a rotating shaft, an unloading unit and a feeding trough. As the key part of the feeding device, the unloading unit consists of an unloading paddle and a fixed disk. The unloading paddle is arranged 9 mm below the discharge outlet of the hopper based on the requirements of pellets' unloading. The fixed disk mentioned above is designed as a stainless-steel disk with thickness h_F = 3 mm and is by Equation (2). The structural parameters of the unloading paddle are thickness h_P = 3 mm, rotation diameter $D_r = (D_F + 6)$ mm, maximum width B_1 = 20 mm and radius of curvature r = 70 mm, respectively, as shown in Figure 4. Furthermore, the back wall of the unloading paddle is a line shape, and the front wall is designed as an arc curve for pushing pellets on the fixed disk.

$$D_F = 2 \times \left(\frac{h_1}{\tan \beta_r} + \frac{D_2}{2} \right) + h_1 \qquad (2)$$

where D_F and D_2 are the diameters of the fixed disk and the discharge outlet, respectively (mm). β_r is the angle of repose of pellets (°). h_1 = 9 mm (the distance between the outlet and the unloading paddle mentioned above).

Table 2. Parameter setting of cylindrical pellet modeling.

Spherical Unit Number	X/mm	Y/mm	Z/mm
1	0.000945	−0.000321	0.001012
2	1.36691	−0.00098	0.733574
3	−1.36502	0.000337	−0.73155
4	−0.731617	−0.000114	1.36697
5	0.733507	−0.000529	−1.36495
6	0.450455	−0.000641	1.49029
7	1.49022	−0.000936	−0.448498
8	−1.48833	0.000293	0.450522
9	−0.448565	-2×10^{-6}	−1.48826

X, Y and Z in the Table 2 represent the spatial coordinate values of the unit sphere in the EDEM 2.7 software.

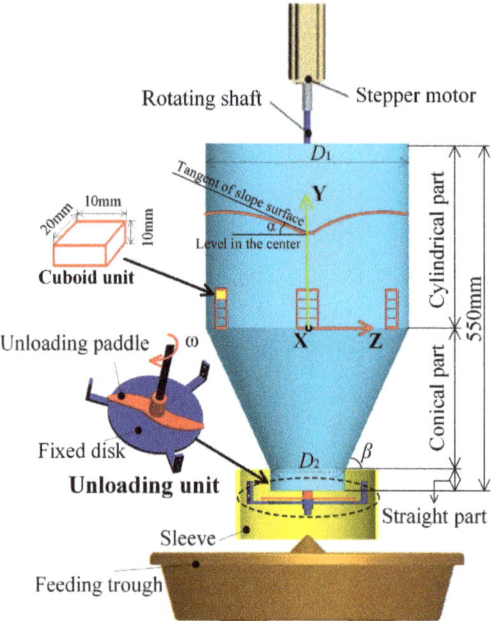

Figure 3. The schematic diagram of the feeder used in DEM simulation.

In the simulations, the coordinate axis is placed in the center of the intersection between the cylindrical part and conical part in the hopper, the positive direction of the X axis is pointed to the inside of the hopper, and the global variable parameters used between the pellet and the feeding system are determined based on the physical test results of the basic characteristic parameters and contact parameters of pellets, feeding system, pellet-pellet and pellet-feeding system, as shown in Table 3. The values of the characteristic parameters used in the simulation tests are the same as in our previous work [42].

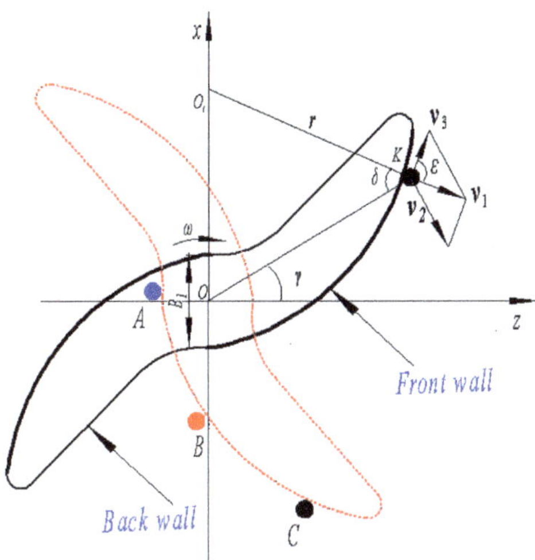

Figure 4. The motion analysis of the pellets on the unloading unit. Noting that A, B, and C represent three pellets randomly selected near the front wall of the unloading paddle and corresponding to Figure 6.

Table 3. Global variable parameters and their values in simulations.

Type	Parameters	Values
Pellet feed	Density, ρ_f (kg·m^{-3})	1538
	Poisson ratio, ν_f	0.26
	Shear modulus, G_f (Pa)	7.1×10^7
hopper	Hopper half angle, β (°)	45–75
	Outlet diameter, D_2 (mm)	60–140
	Hopper diameter, D_1 (mm)	290
	Density, ρ_h (kg·m^{-3})	2000
	Poisson ratio, ν_h	0.3
	Shear modulus, G_h (Pa)	1.1×10^9
Rotating shaft	Density, ρ_R (kg·m^{-3})	7800
	Poisson ratio, ν_R	0.3
	Shear modulus, G_R (Pa)	7×10^{10}
feeding trough	Density, ρ_t (kg·m^{-3})	1340
	Poisson ratio, ν_t	0.32
	Shear modulus, G_t (Pa)	1.2×10^9
Feed-feed	Restitution coefficient, e_{ff}	0.65
	Coefficient of static friction, $\mu_{s,ff}$	0.4
	Coefficient of rolling friction, $\mu_{r,ff}$	0.01
Feed-hopper	Restitution coefficient, e_{fh}	0.48
	Coefficient of static friction, $\mu_{s,ff}$	0.4
	Coefficient of rolling friction, $\mu_{r,fh}$	0.01
Feed-rotating shaft	Restitution coefficient, e_{fr}	0.62
	Coefficient of static friction, $\mu_{s,fr}$	0.27
	Coefficient of rolling friction, $\mu_{r,fr}$	0.01
Feed-feeding trough	Restitution coefficient, e_{ft}	0.5
	Coefficient of static friction, $\mu_{s,ft}$	0.38
	Coefficient of rolling friction, $\mu_{r,ft}$	0.01
Simulation	Time step, Δt (s)	2.5×10^{-5}

Pellet feed in Table 3 refers to the columnar pellet feed (processed by the pelleting machine) for nursery.

In order to improve the accuracy of the simulation calculation, the simulation started with the random generation of pellets from the top of the hopper in each simulation test, and all pellets in an unlimited style fell onto the bottom of the hopper under gravity until more than 100 mm high in the cylindrical part was reached, which ensured the flow capacity of pellets in the hopper. When all the pellets are in a steady state, the unloading paddle is driven by the stepper motor and pellets in the hopper fall into the feeding trough. During the simulation test, the position and velocity data for each pellet are automatically saved every 0.01 s.

2.3. Evaluation Index

2.3.1. MFI Theory

MFI [14,17,43] is used to define the flow patterns in the hopper with an unloading paddle accurately. The MFI value is the ratio of the average vertical velocity of the pellet near the wall to the average vertical velocity of the pellet in the center, it is calculated as follows:

$$v_W = \frac{\sum\limits_{i=1}^{t} v_i}{t} \tag{3}$$

Then

$$\text{MFI} = \frac{v_W}{v_C} \tag{4}$$

where v_c and v_w are the average vertical velocity of pellets in the center and near the wall, respectively (m/s). v_i is the average vertical velocity of pellets in each cuboid unit at time step i (m/s). t is the total time step when the top surface at the center of the hopper flows downward from Y = 100 mm to the position of Y = 45 mm (s).

To observe the flow pattern and analyze the important data conveniently, the v_i mentioned above can be obtained in a series of cuboid units (in the height range of 0 mm $\leq$ Y $\leq$ 40 mm) as shown in Figure 3. The left (right) cuboid unit is 10 mm away from the hopper wall.

2.3.2. DDA and Porosity Theory

The dynamic discharging angle (DDA) and porosity are used to describe the flow characteristics of pellets in the hopper during the discharging process. The DDA is defined as the maximum angle between the tangent of the top slope surface and the level in the center when the top surface flows downward from 100 mm to the position of Y = 45 mm, as shown in Figure 3. The tangent of the top slope surface above is formed by the tangent between the lowest point of the level in the center and the slope surface. The mean value of the DDA at the different layer heights is calculated by applying Equation (5). Meanwhile, the porosity in each cuboid unit with a range of 0 mm $\leq$ Y $\leq$ 40 mm is calculated in Equation (6).

$$\alpha_i = \arctan\left(\frac{y_r - y_l}{z_r - z_l}\right) \tag{5}$$

where α_i is the DDA formed in the discharging process (°). (y_l, z_l) are the coordinate average in the intersection of the tangent of the top slope surface and the level in the center at the time step i (mm, mm). (y_r, z_r) are the coordinate average at the intersection of the top slope surface and the tangent at the time step i (mm, mm).

$$P = \frac{V - \overline{N} \cdot V_p}{V} \tag{6}$$

where P and $\overline{N}$ are the porosity and the average number of pellets in each cuboid unit, respectively, when the top surface flows downward from 100 mm to the position of Y = 45 mm (/,/). V is the volume of each cuboid unit (mm^3). V_p is the volume of each pellet (mm^3).

2.4. Simulation Tests Arrangement

The hopper half angle, outlet diameter of the hopper and rotation speed of the unloading paddle are important parameters in the design of the hopper with the unloading unit [44,45]. For the purpose of studying the effect of the parameters on the flow capacity of pellets inside the hopper, experimental factors and levels are determined by pretest and the Beverloo Equation [46], as well as the prediction of mass discharge rate suggested by Zheng [47]. The hopper half angle varies from 45 to 75°, the rotation speed transitions from 30 rpm to 60 rpm, and the outlet diameter changes from 60 mm to 140 mm, as shown in Table 4.

Table 4. List of simulation cases.

Test	Hopper Half Angle β (°)	Rotation Speed n (rpm)	Outlet Diameter D_2 (mm)	MFI (/)
1	45	45	100	0.10
2	50	45	100	0.13
3	55	45	100	0.18
4	60	45	100	0.24
5	65	45	100	0.60
6	70	45	100	0.67
7	75	45	100	0.75
8	65	30	100	0.64
9	65	60	100	0.45
10	65	45	60	0.30
11	65	45	80	0.54
12	65	45	120	0.64
13	65	45	140	0.68

3. Analysis of the Discharging Effect of the Unloading Paddle

3.1. Effect of the Unloading Paddle on Pellets on the Fixed Disk

With the rotation of the unloading paddle, such as from the initial solid line position to the dotted line position, as shown in Figure 4, pellets on the fixed disk are pushed and discharged. In Figure 4, A pellet K that is about to leave the edge of the fixed disk is randomly selected as the research object for theoretical analysis to reflect the movement law of pellets on the fixed disk. Firstly, the relative formula in $\Delta O_1 OK$ is shown in Equations (7)–(9).

$$\cos(\frac{\pi}{2} - \gamma) = \frac{l_{OO_1}{}^2 + l_{OK}{}^2 - r^2}{2 l_{OO_1} l_{OK}} \tag{7}$$

$$\frac{\sin(\frac{\pi}{2} - \gamma)}{r} = \frac{\sin \delta}{l_{OO_1}} (0 < \gamma < \frac{\pi}{2}) \tag{8}$$

$$l_{OO_1} = r - \frac{B_1}{2} \tag{9}$$

The relation between the absolute velocity and the following velocity can be obtained through the kinematic analysis of the pellet K, as shown in Equation (7).

$$\frac{\sin \varepsilon}{v_2} = \frac{\sin\left[\frac{\pi}{2} + (\varepsilon - \delta)\right]}{v_1} \tag{10}$$

$$v_2 = \omega \cdot l_{OK} \tag{11}$$

where v_1 is the absolute velocity (m/s). v_2 is the following velocity (m/s).

From Figure 4 and Equation (10), v_1 gradually increases from the inside to the outside along the curved unloading paddle, meanwhile the direction of v_1 is anisotropic, which is to the realization of the uniform discharge. In Figure 4, angle ε is the angle between the relative velocity v_3 and the absolute velocity v_1 (°). Based on the calculation, angle ε is

$97.3° \pm 1°$ when the rotation speed changes from 30 to 60 rpm, which indicates pellets left the fixed disk approximately along the arc normal direction. Then, the direction of pellet absolute velocity is anisotropic along the curved unloading paddle, which is beneficial to uniform discharge.

Then, the DEM 2.7 software is used to conduct a virtual simulation of the movement process of pellets by the unloading paddle, and the discharging process is also tested according to the rotation speed of 45 rpm and the outlet diameter of 100 mm. After simulation, the vector display method is used to visually process the pellets on the fixed disk, the arrow direction is specified by the velocity magnitude. Meanwhile, the pellets are given different colors to reflect the velocity magnitude; red, green and blue are the min color, mid color and max color, respectively. The simulation snapshots in Figure 5a revealed that pellets showed an anisotropy distribution. As the unloading paddle rotates counterclockwise, it can be seen that the inside pellets were pushed gradually to the outside of the fixed disk by the unloading paddle and finally, the outside pellets left the fixed disk at a rapid velocity approximately along the arc normal.

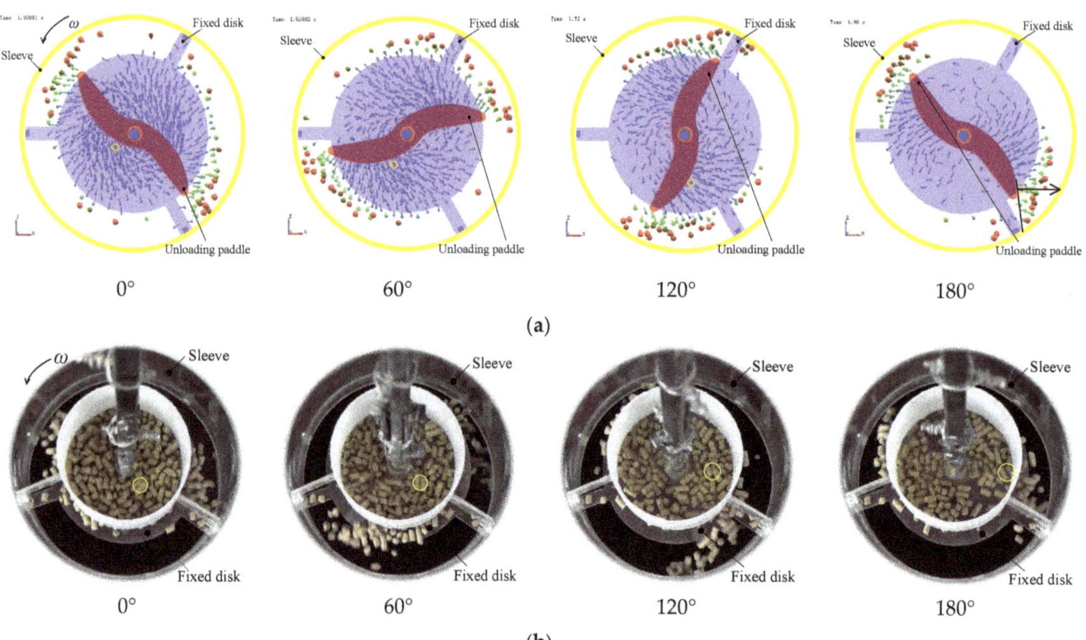

Figure 5. Qualitative comparison between simulation (**left**) and experiment (**right**) (n = 45 rpm, D2 = 100 mm). (**a**) The simulation test in the straight part; (**b**) the discharging experiment in the straight part.

Additionally, the discharge experiment in the straight part of the hopper is carried out under the same parameter conditions. Then, a pellet is randomly selected in the central area of the fixed disk and circled with yellow marks to analyze the details. The results show that the pellets discharged from the fixed disk are basically evenly distributed, as shown in Figure 5b, and the movement rule of pellets at the yellow mark is approximately consistent with the simulation, which proves the correctness of the virtual simulation.

Further, the trajectory of three pellets near the inside, middle and outside of the paddle on the fixed disk is tracked by the simulation so as to clarify the discharging process. As shown in Figure 6, the trajectory of the C pellet is a straight line, and it can be seen that the direction of the C pellet is approximately along the arc normal combining with position 1 of the unloading paddle. While the trajectory of the B pellet consists of an arc and a straight line, which is due to the B pellet first being pushed towards the edge of the fixed disk by the unloading paddle, the B pellet leaves the fixed disk approximately along the arc normal of the unloading paddle, similar to the C pellet combining with position 2 of the unloading paddle. Finally, the trajectory of the A pellet is similar to that of the B pellet; the motion direction of the A pellet is approximately along the arc normal, combining with position 3 of the unloading paddle. Nevertheless, the curvature of the arc in the A pellet trajectory is slightly larger than that of the B pellet; this may be because the velocity of the A pellet increases gradually when the A pellet moves outside.

In other words, the above analysis and experiment verify that the inside pellets gradually move outside as the unloading paddle rotates and eventually move away from the fixed disk approximately along the normal arc of the unloading paddle.

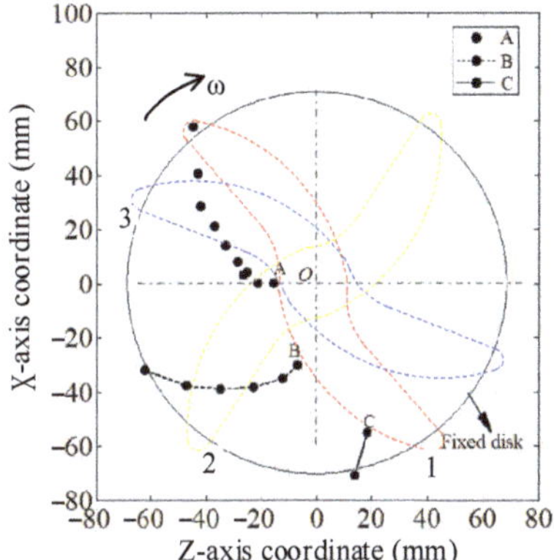

Figure 6. The movement trajectory of three pellets on the fixed disk. A is the pellet near the inside, B is the pellet near the middle and C is the pellet near the outside of the unloading paddle. Noting that red, yellow and blue represent the three positions in the clockwise rotation of the unloading paddle on the fixed disk.

3.2. Effect of the Unloading Paddle on Pellets in the Hopper

According to Figure 6, the unloading paddle pushes pellets on the fixed disk away. In order to discuss the effect of the unloading paddle on pellets in the hopper during the discharging process, the movement trajectory of the three pellets is tracked at a distance of 20 mm, 36 mm and 60 mm from the hopper center, respectively. When the hopper half angle is 75°, the rotation speed is 45 rpm and the outlet diameter is 100 mm. As shown in Figure 7, the pellets A, B and C move down a vertical line in the cylindrical part. Then, the pellets A, B and C have a motion trend towards the hopper center when falling into the lower conical part of the hopper, which is because the shrinkage of the hopper wall impels pellets near the hopper wall to move toward the hopper center.

Near the straight part of the outlet, the pellets A and B begin to have both a downward displacement and a reciprocating motion away from the hopper center, while the pellet C basically moves downward. The pellet A spreads towards the outside of the outlet more obviously than the pellets B and C after entering the straight part of the outlet, which may be because when the unloading paddle rotates, the center pellets gradually move to the outer edge of the fixed disk with the rotation, while a small number of particles on the edge of the straight part fall to the edge of the fixed disk and are directly sent to the feeding trough by the unloading paddle. Therefore, the pellets in the center are most frequently and actively sent by the unloading paddle, resulting in the radial displacement of the pellets in the center during the falling process. As a result, the pellets in the center of the upper layer will continue to be carried out along with the pellets on the fixed disk, but the energy may be lost during the transfer process, resulting in the movement trajectory (flow characteristics) of the pellets above the straight part not being greatly affected. Finally, pellets A, B and C are discharged through the gap between the fixed disk and the outlet of the hopper. Therefore, pellets are distributed into the feeding trough layer by layer by the unloading paddle, which is beneficial to evenly discharge pellets from the hopper [48].

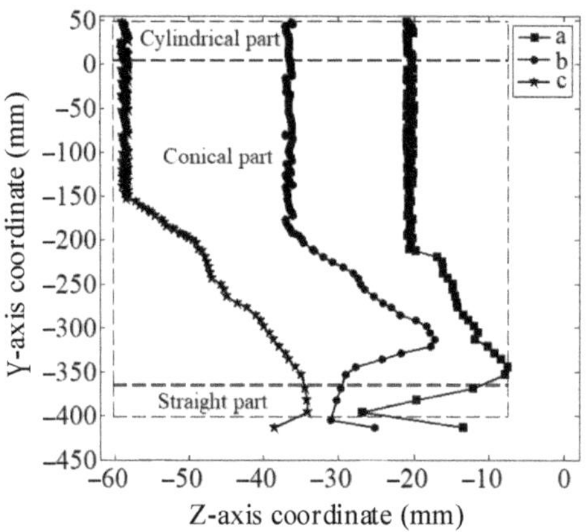

Figure 7. The movement trajectory of pellets in the hopper during the discharging process ($\beta = 75°$, $n = 45$ rpm and $D_2 = 100$ mm).

The above analyses show that the motion of the unloading paddle and the shrinkage of the hopper wall have certain effects on the discharging process of pellets. Thus, it is necessary to study the effect of the hopper half angle, the outlet diameter and the rotation speed on the flow capacity of pellets in the hopper with the unloading paddle.

4. Effect Analysis of Main Parameters on Flow Capacity and Bench Test

4.1. The Effect of Hopper Half Angle on Flow Capacity

Numerous studies have shown that the hopper half angle is one of the key structural parameters in the design of hoppers, meanwhile, which has a significant influence on flow patterns [17,47]. The hopper half angle increases from 45 to 75° with a step of 5° when the rotation speed is 45 rpm and the outlet diameter is 100 mm. Simulation results from test 1 to test 7 are analyzed.

4.1.1. Analysis of Discharging Process

It is worthy to study the complex pellet-flowing phenomena in the hopper during the discharging process [49]. Figure 8a,b show the discharging process under the hopper half angle of 45° and 75°, respectively, when the rotation speed is 45 rpm and the outlet diameter is 100 mm. Different pellet-coloring layer classification methods are used to conveniently distinguish different flow patterns of pellets in the hopper.

In Figure 8a, at 6 s, pellets in the center of layer A have been discharged from the hopper, and pellets in the center of layer B~F are in turn filled toward the hopper center. A narrow flow zone is formed in the center of each layer of pellets, and a symmetrical dip is formed on the top surface, which is consistent with that presented by Drescher et al. [50]. However, pellets near the wall of layer B~F remained relatively in their original posture, which forms the stagnant zone proposed in [21]. Then, from 6 s to 12 s, pellet layers in the flow zone continue to flow to the outlet, and the symmetrical dip on the top surface gradually increases with the discharging process. As the number of pellets in layer F are filled into the flow zone, this stagnant zone slowly decreases. From 12 s to 24 s, pellets in layer F are poured into the flow zone continuously and pellets in the stagnant zone of layer E begin to fill up the flow zone because most pellets in layer F have been discharged from the hopper. The symmetrical dip is even more pronounced at this moment, and the stagnant zone decreases obviously with the discharging process. In short, the above phenomena demonstrate the characteristics of "fast in the center, slow near the wall". Nevertheless, in livestock and poultry industry, continuous or regular filling of pellets into the hopper of the feeder is necessary to ensure feed supply, which will cause pellets near the wall to further extend retention time in the hopper, and the flow pattern may affect the feeding quality in the long run.

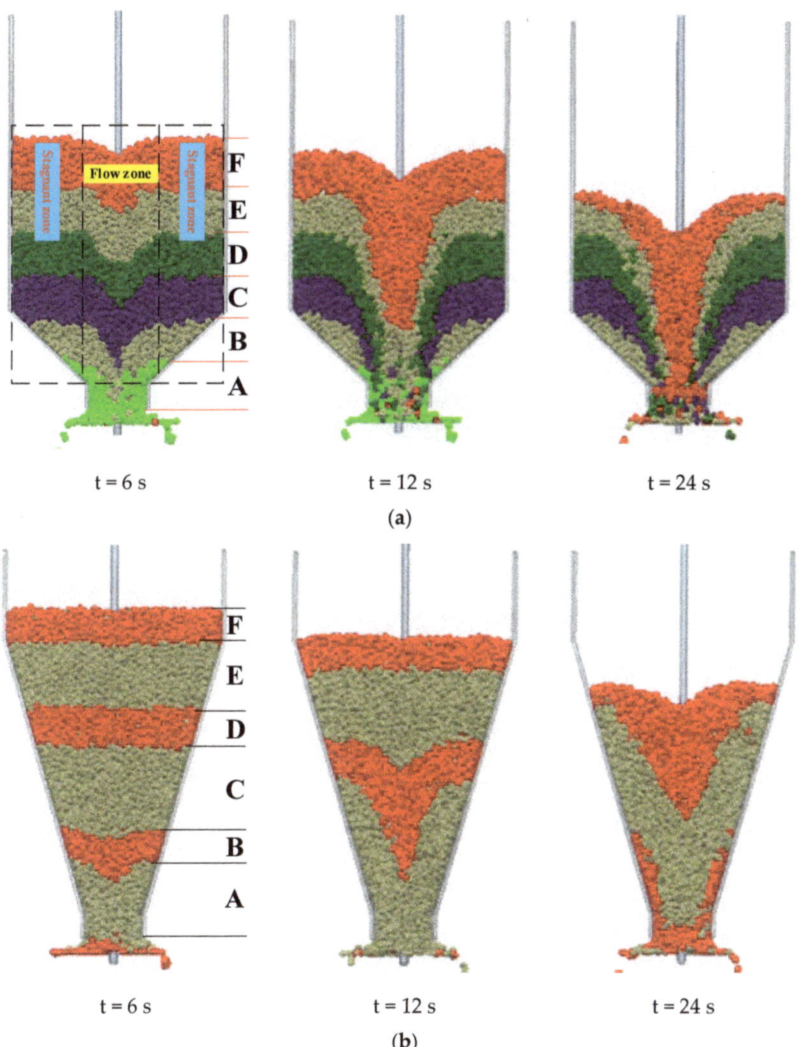

Figure 8. Different snapshots of the discharging process. (**a**) When the hopper half angle is 45°, the rotation speed is 45 rpm and the outlet diameter is 100 mm; (**b**) when the hopper half angle is 75°, the rotation speed is 45 rpm and the outlet diameter is 100 mm.

Figure 8b demonstrates that pellets in layer A are discharging, pellets in layer B form a small, inverted cone and pellets in other layers keep falling as a whole at 6 s. Then, from 6 s to 12 s, pellets in layer A have been discharged from the hopper and pellets in layer B have basically been discharged. Meanwhile, pellets in layer C are flowing to the outlet of the hopper, layer D and layer E have formed an inverted cone. While layer F in the cylindrical part has been downward as a whole, no inverted cone has appeared until pellets completely enter the conical part. From 12 s to 24 s, pellets in layer D have basically been discharged and pellets in layer E have basically reached the outlet, whereas layer F forms an inverted cone at this moment, and the symmetrical dip formed by the top surface begins to appear. It can be seen that the pellet layer in the cylindrical part falls overall during the discharging process, while the pellet layer in the conical part forms an inverted cone. This can be explained for the following reasons: the hopper wall in the conical part gradually shrinks and friction exists between the hopper wall and pellets. This reflects the characteristics of mass flow in the conical part according to the principle of volume conservation, which is basically consistent with Fullard's explanation [51]. In short, the discharge rule of "pellets in the lower layer are firstly discharged from the hopper, and pellets on the upper layer are subsequently discharged" is shown; therefore, the flow zone is filled all over the hopper during the discharging process. The discharging process can be described as "first in, first out".

4.1.2. Evaluation of Flow Capacity

This section investigated the effect of the hopper half angle on flow capacity, MFI, DDA and porosity were adopted to explain the above discharging process.

Figure 9a shows the effect of the hopper half angle on the MFI and the calculation of the MFI according to Equation (4). Apparently, the value of MFI increases when the hopper half angle increases from 45° to 75° in a step of 5°. For the hoppers at 70° and 75°, the value of MFI is basically constant, approximately 0.7 and 0.85, respectively, which means little difference in the vertical velocity of pellets in the center and near the wall during the discharging process. For the hopper at 65°, the value of MFI is stable firstly at approximately 0.65 and then decreases and stabilizes at approximately 0.56, which may be because the top pellet layer begins to form a small inverted cone when the layer is near the boundary between the cylindrical part and conical part during the discharging process, resulting in an increase in the difference in vertical velocity in the hopper center and near the hopper wall. Especially when the hopper half angle is less than 60°, the value of MFI decreases from 0.39 to 0.18 during the discharging process. The average value of MFI < 0.24 is clearly identified by the data processing, which reflects the difference in vertical velocity. Evidently, the value of MFI has a significant downward trend when the hopper half angle decreases from 65° to 60°. This may be because the component force of pellet gravity downward along the hopper wall starts to obviously decrease in the conical part. Meanwhile, the friction between the pellets and the hopper wall keeps increasing, which results in a small velocity of pellets near the hopper wall and a faster passage of pellets in the center of the cylindrical part.

In Figure 10a, it is observed that the value of DDA (as Equation (5)) increases with the decrease in the hopper half angle. Additionally, the variation of DDA for hopper half angle between 55° and 65° is large, which has a value between 7° and 25°, while the variation of DDA for hopper half angle between 65° and 75° and 45° and 55° tends to be gentle, which has a value between 2° and 7° and 25° and 30°, respectively. The smaller the value of DDA, the smaller the difference in vertical velocity of pellets in the center and near the wall during the discharging process. In this work, the hopper half angle is 63° when the value of DDA is less than 10°.

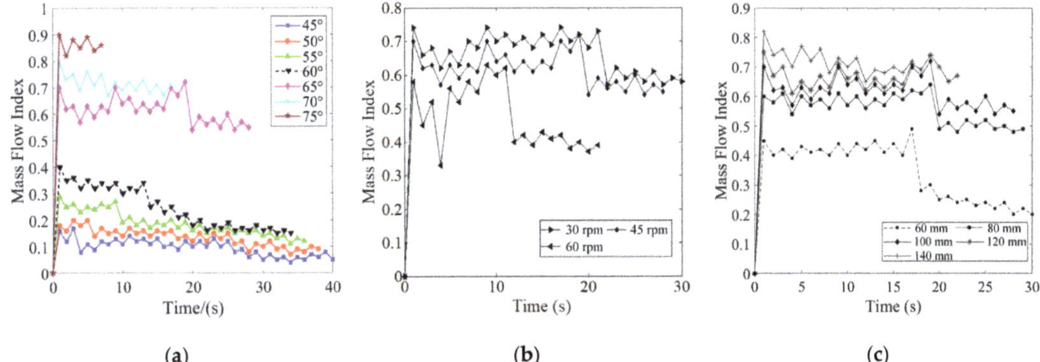

Figure 9. Effect of the main parameters on the MFI. (**a**) The hopper half angle (n = 45 rpm and D_2 = 100 mm); (**b**) the rotation speed (D_2 = 100 mm and β = 65°); (**c**) the outlet diameter (n = 45 rpm and β = 65°).

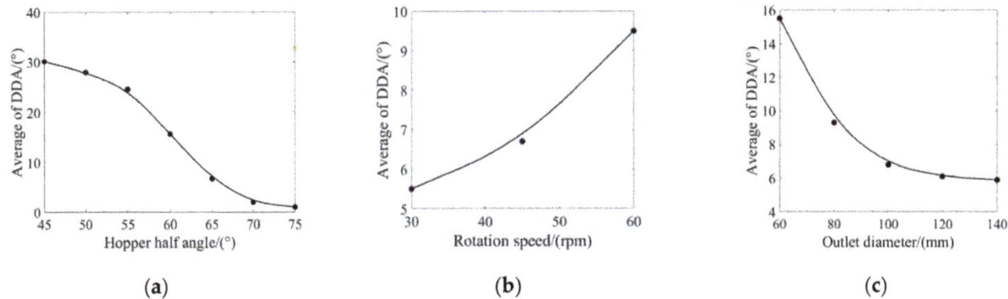

Figure 10. Effect of main parameters on the DDA. (**a**) The hopper half angle; (**b**) The rotation speed; (**c**) The outlet diameter.

Figure 11a shows that the porosity of cuboid units (see Equation (6)) in the center obviously increases and that near the wall decreases slightly with the decrease in the hopper half angle, which illustrates that the flow characteristics of pellets in the center have changed more significantly. When the hopper half angle is between 45° and 50°, the porosity among pellets in the center and near the wall is approximately 0.5 and 0.25, respectively, and the pellets in the center are in a loose state because of faster vertical velocity, resulting in a decrease in the contact friction force among the pellets in the center field. The movement of the pellets at the center of the cylindrical part is more intense, which explains the discharge phenomenon in Figure 8a. Additionally, the porosity among pellets in the center and near the wall is approximately 0.3 and 0.29, respectively. When the hopper half angle is between 70° and 75°, pellets in the center and near the wall for the cylindrical part fall almost simultaneously, which verifies the discharging process of the cylindrical part in Figure 8b. In other words, these investigations reveal that the decrease in the hopper half angle accelerates the appearance of the uneven flow. Especially, a rapid transition zone occurs at the hopper half angle from 55° to 65°, which is basically in agreement with the above analysis.

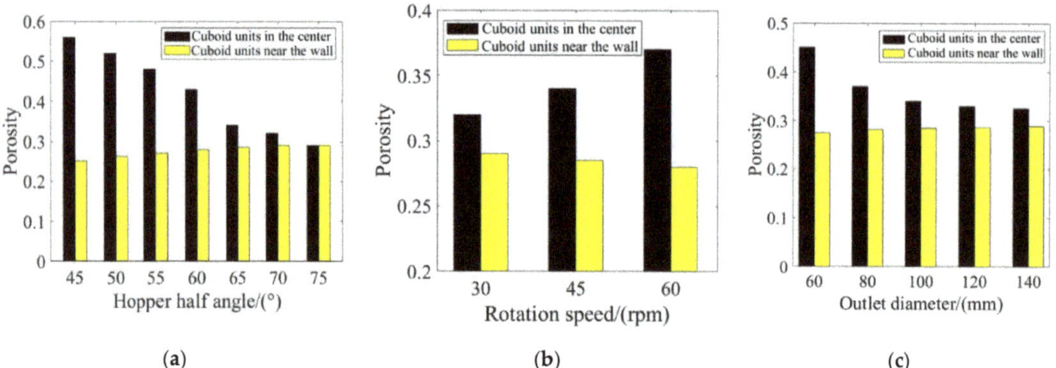

Figure 11. Effect of main parameters on porosity. (**a**) The hopper half angle; (**b**) the rotation speed; (**c**) the outlet diameter.

4.2. The Effect of Rotation Speed on Flow Capacity

In order to investigate the effect of rotation speed on the flow capacity of pellets in the hopper, the simulation results of test 5, test 8 and test 9 are analyzed. The rotation speed increases from 30 rpm to 60 rpm in step of 15 rpm when the hopper half angle is 65° and the outlet diameter is 100 mm, respectively.

4.2.1. Analysis of the Discharging Process

As shown in Figure 12a (the rotation speed is 30 rpm), at 12 s, pellets in layer A and layer B have been discharged from the hopper. Meanwhile, layer C in the conical part forms a large-inverted cone and layer D begins to form a small, inverted cone just when entering the conical part. While layer E and layer F keep falling as a whole. Then, from 12 s to 24 s, pellets in layer C have basically been discharged. Meanwhile, layer D has formed a large, inverted cone. While a small, inverted cone begins to appear when layer F is close to the boundary between the cylindrical part and the conical part, this is affected by the flow characteristics of pellets in the conical part. In other words, the pellets in the cylindrical part keep falling as a whole. The bottom pellet layer in the conical part is discharged from the hopper first, and the upper pellet layer is discharged subsequently. The inverted cone in the conical part is getting bigger and bigger as pellet layers lower, according to the principle of volume conservation of pellet layers.

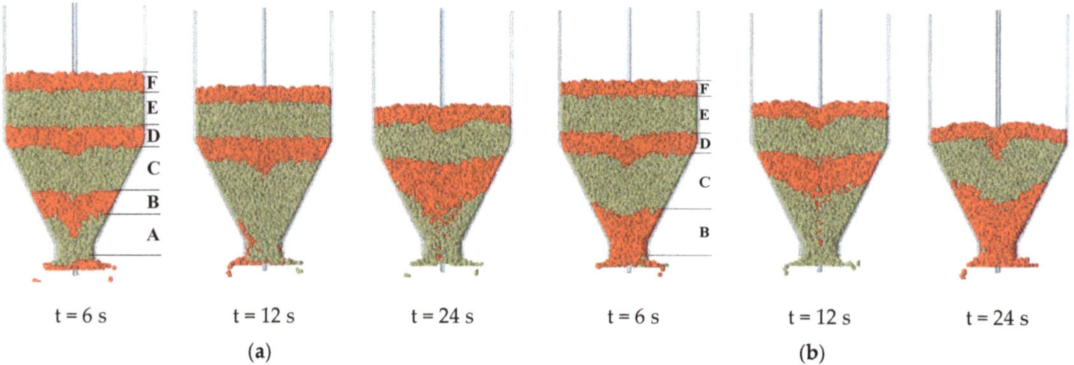

Figure 12. Comparison of flow characteristics for different rotation speeds. (**a**) When the rotation speed is 30 rpm, the hopper half angle is 65° and the outlet diameter is 100 mm; (**b**) when the rotation speed is 60 rpm, the hopper half angle is 65° and the outlet diameter is 100 mm.

Figure 12b shows the discharging process when the rotation speed is 60 rpm. From 0 s to 12 s, layers A and B have been discharged, and layer C in the conical part is flowing to the outlet of the hopper, and a large, inverted cone is formed in layer D. Furthermore, layer F just arises as a small, inverted cone when it is close to the boundary between the cylindrical part and the conical part. The discharging characteristics at this moment are approximately the same as those at 24 s in Figure 11a. From 12 s to 24 s, layer C has been discharged and layer D is being discharged; layer F begins to enter the conical part and the inverted cone is more obvious. Similarly, the reason for the inverted cone formation in the discharging process is basically consistent with the analysis in Figure 12a.

Combining Figure 12a,b, it can be seen that the discharging characteristics of "first in, first out" do not change with the increase in the rotation speed; only the speed of forming an inverted cone is accelerated, which means the discharging process is accelerated.

4.2.2. Evaluation of Flow Capacity

In order to investigate the effect of rotation speed on flow capacity, MFI, DDA and porosity as a function of rotation speed are illustrated in Figures 9b, 10b and 11b.

Figure 9b shows the effect of the rotation speed on the MFI. Apparently, the value of MFI decreases with an increase in the rotation speed. The value of MFI is stable initially at approximately 0.7 and then decreases and stabilizes at approximately 0.6 when the rotation speed is 30 rpm, which means there is little difference in the vertical velocity of pellets in the center and near the wall in the cylindrical. Furthermore, the value of the MFI decreases obviously during the rotation speed increase from 45 rpm to 60 rpm, which is between 0.38 and 0.55. Nevertheless, the average values of MFI are more than 0.3 when the rotation speed changes from 30 rpm to 60 rpm, the flow pattern of pellets is still confirmed by the mass flow. On the whole, the rotation speed has little effect on the flow patterns of pellets in the hopper, which is less than that of the hopper half angle.

Figure 10b shows that the average of DDA increases from 5.5° to 9.5° when the rotation speed increases from 30 rpm to 60 rpm, which indicates that the increase in the rotation speed does not cause a great change in DDA.

Figure 11b shows the porosity of cuboid units (as Equation (6)) as the center increases and that near the wall decreases with the increase in the rotation speed. The porosity in the center is 0.32 at 30 rpm and 0.38 at 60 rpm, and that near the wall is approximately 0.29 at 30 rpm and 0.28 at 60 rpm, which also indicates that the increase in the rotation speed does not cause a great change in porosity in the center or near the wall. Based on the above analysis, when the rotation speed is 30 rpm, pellets in the center and near the wall of the cylindrical part fall almost simultaneously, conforming to the discharging process of the cylindrical part in Figure 12a. Only when the rotation speed is 60 rpm, the difference in porosity in the center and near the wall is big, which illustrates that the vertical velocity of the pellets in the center is bigger, explaining the discharge phenomenon in Figure 12b. In other words, because the variation of porosity with the rotation speed is small, it is a revelation that the rotation speed has no significant effect on the flow characteristics of pellets.

4.3. The Effect of Outlet Diameter on Flow Capacity

In order to determine the effect of the outlet diameter of the hopper on the flow capacity of pellets, the simulation results of test 5, test 10 and test 13 are analyzed. The outlet diameter increases from 60 mm to 140 mm in a step of 20 mm when the hopper half angle is 65° and the rotation speed is 45 rpm.

4.3.1. Analysis of Discharging Process

As shown in Figure 13a (the outlet diameter is 100 mm), from 0 s to 12 s, layers A and B have been discharged, layer C in the conical part basically has reached the outlet of the hopper, layer D in the cylindrical part just enters the conical part and an inverted cone forms in layer D. While layers E and F in the cylindrical part have been downward as a whole. From 12 s to 24 s, the inverted cone formed by layer D is more obvious; furthermore, a small inverted cone begins to appear when layer F is near the boundary between the cylindrical part and the conical part. In conclusion, the pellet layers in the cylindrical part basically fall down as a whole, while the pellet layers in the conical part form the inverted cone and fall down layer by layer as a whole.

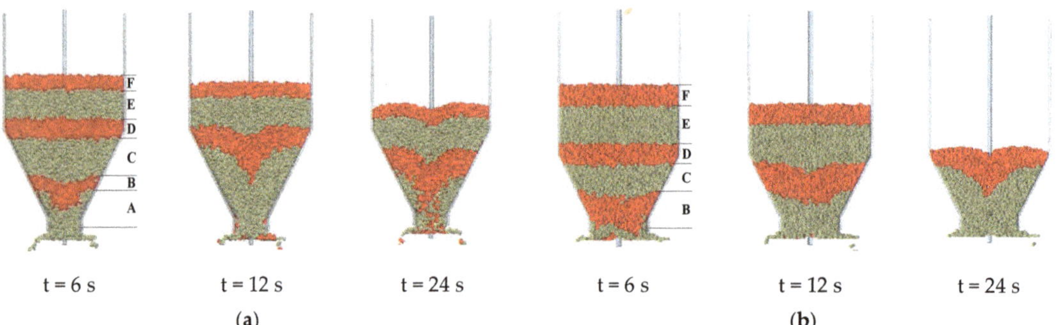

Figure 13. Comparison of flow characteristics for different outlet diameters of the hopper. (**a**) When the outlet diameter is 100 mm, the rotation speed is 45 rpm and the hopper half angle is 65°; (**b**) when the outlet diameter is 140 mm, the rotation speed is 45 rpm and the hopper half angle is 65°.

As shown in Figure 13b (the outlet diameter is 140 mm), from 0 s to 12 s, layer D in the cylindrical part has entered the conical part, and a large, inverted cone has formed. While layers E and F in the cylindrical part have been downward as a whole. From 12 s to 24 s, layer F in the cylindrical part enters the conical part completely and forms an inverted cone. Pellets in layer E are discharging. In general, the pellet layers in the cylindrical part fall overall during the discharging process; no inverted cone appears until pellets completely enter the conical part. The pellet layers in the conical part form the inverted cone, and the inverted cone increases with the shrinkage of the hopper wall, which can be explained by the principle of volume conservation of pellet layers, and pellets in the lower layer are discharged first from the hopper and pellets in the upper layer are subsequently discharged.

A comprehensive analysis of Figure 13a,b shows that the flow characteristics are "first in, first out". Only the speed of forming an inverted cone and the discharging process are accelerated when the outlet diameter is increased.

4.3.2. Evaluation of Flow Capacity

In order to investigate the effect of outlet diameter on flow capacity, MFI, DDA and porosity as a function of outlet diameter are illustrated in Figures 9c, 10c and 11c.

The variation of MFI is plotted over the outlet diameter in Figure 9c according to Equation (4), which reveals that the increase in outlet diameter causes an increase in MFI. Furthermore, the value of MFI is basically stable at approximately 0.75 for the outlet diameter of 140 mm, which means a little difference in the vertical velocity of pellets in the center and near the wall during the discharging process. The value of MFI is basically stable at approximately 0.7 when the outlet diameter is 120 mm. The value of MFI is stable initially at approximately 0.6~0.65 and then decreases and stabilizes at approximately 0.5~0.56 when the outlet diameter is between 100 mm and 80 mm. In particular, the value of the MFI shows a sharp decrease when the outlet diameter changes from 80 mm to 60 mm, which is similar to the phenomenon of a linear decrease in the DDA when the outlet diameter increases from 60 mm to 80 mm in Figure 10c. This may be because the discharge rate decreases when the outlet diameter changes from 80 mm to 60 mm and the pellets in the hopper are in a state of compaction. Additionally, the value of MFI is less than 0.2 after 25 s when the outlet diameter is 60 mm, which illustrates the large difference in the vertical velocity of pellets in the center and near the wall. The flow pattern should be in transition between the mass flow and the funnel flow at this moment and the outlet diameter has a great influence on the flow pattern when the outlet diameter is less than 60 mm.

On the whole, when the outlet diameter is more than 60 mm, the values of MFI are more than 0.24, whatever the outlet diameter is. Flow patterns are almost unaffected and the flow pattern of pellets is still confirmed by the mass flow, which illustrates that the effect of the outlet diameter on flow patterns is less than that of the hopper half angle in the previous section. In conclusion, combined with the pretest results and the feed discharge requirements of swine, the outlet diameter should be 100–120 mm when the hopper is designed.

Figure 10c shows a slight variation in the value of DDA when the outlet diameter is from 100 mm to 140 mm, the value is approximately 6°, which means the vertical velocity of pellets in the center is basically the same as that of the pellets near the wall. However, the DDA is almost linearly increasing as the outlet diameter decreases from 100 mm to 60 mm, and the value of DDA is approximately 9° for an outlet diameter of 80 mm, but the value is approximately 16° for an outlet diameter of 60 mm, which means the vertical velocity of pellets in the center is obviously bigger than that of the pellets near the wall.

Figure 11c shows the porosity of cuboid units in the center increases and that near the wall decreases with the decrease in the outlet diameter, and the variation trend of porosity with the increase in the outlet diameter is nearly the same as shown in Figure 10c. The porosity among pellets in the center and near the wall is approximately 0.32 and 0.29, respectively, when the outlet diameter is between 120 mm and 140 mm, which indicates pellets in the center and near the wall for the cylindrical part fall almost simultaneously. Only the porosity among pellets in the center increases slightly when the outlet diameter is 100 mm. It can be observed that the movement of the pellets at the center of the cylindrical part is more intense compared to the outlet diameter of 140 mm. Concurrently, the analysis verifies the discharge phenomenon in Figure 13a,b. Additionally, the porosity among pellets in the center and near the wall is approximately 0.37 and 0.28, respectively. When the outlet diameter is 80 mm, which indicates the vertical velocity of pellets in the center increases slightly, the velocity difference between the center and hopper wall begins to get bigger. While the porosity among pellets in the center and near the wall is approximately 0.45 and 0.27, respectively, when the outlet diameter is 60 mm, it can be observed that the vertical velocity of the pellets at the center of the cylindrical part increases rapidly. In other words, these investigations reveal that the appearance of the uneven flow is accelerated when the outlet diameter is less than 60 mm, and the flow characteristics are almost unaffected when the outlet diameter is more than 60 mm.

In order to formulate an explicit decision criterion for flow patterns according to the MFI, referring to the relevant literature, Johanson and Jenike [52] insisted that a velocity ratio of MFI ~0.3 could be used as the boundary line between the funnel flow and mass flow in hoppers, which meant the funnel flow is defined under MFI < 0.3, reversely, if MFI > 0.3, the flow pattern is the mass flow. Based on the results of this study for the hopper with the unloading paddle, the evaluation method is also feasible, but if the boundary value of MFI is determined to be 0.24, then the flow pattern of pellets in the hopper can be confirmed to be the funnel flow when the value of MFI is less than 0.24, the mass flow is determined when the value of MFI is more than 0.24.

In conclusion, the above studies indicate that the flow pattern of pellets can be identified as the mass flow when the hopper half angle exceeds 60°. The characteristics of "first in, first out" in the mass flow could ensure the quality of feed. Otherwise, it is confirmed by the funnel flow. Therefore, according to the pre-test and the design requirements of the existing feeder structure, the hopper half angle should be in the range of 65° to 75° when the hopper is designed.

4.4. Experimental Validation

To clarify the DEM model, physical experiments were carried out with experimental equipment in the Laboratory of Animal Husbandry Mechanization at Northeast Agricultural University. To ensure that the experiment conditions were consistent with the simulation, the component materials of the experiment equipment were the same as those of the model components in the simulation process. The same structural and operating parameters are selected between the experiment and simulation test, noting that the hopper half angle is 45° and 75°, respectively, when the rotation speed and the outlet diameter are 45 rpm and 100 mm. In order to more clearly represent the problems found in the pre-test, the hopper is composed of clear plastic, and pellets are filled into the area by two parallel clear sheets to be observed visually. The clear sheet is 25 mm away from the hopper center to reduce the influence on the flow behavior of pellets in the hopper. The colored and original pellets crossed in the hopper at the initial time, which made it easy to visually reflect the flow and pattern of pellets.

Snapshots of different kinds of posture during the discharging process at a hopper half angle of 45° are shown in Figure 14. All layers of pellets in the center initially formed a small flow zone and a small DDA at 6 s; however, the positions of pellets near the wall hardly changed at this time. From 6 s to 12 s, little pellets on the top surface in the cylindrical part are poured into the flow zone first due to the pellet repose angle characteristics and the DDA increases, so the stagnant zone formed by pellets near the wall decreases slightly. Afterwards, a lot of pellets on the top surface in the cylindrical part are poured into the flow zone with the pellet layers down from 12 s to 21 s, the DDA increases constantly, and the stagnant zone decreases obviously. Finally, the last photo revealed that the pellet layers had entered the conical part at approximately 39 s. The top two pellet layers in the cylindrical part have entered the feeding trough completely, whereas the few pellets in the conical part still stay near the wall. The stagnant zone decreases obviously at this moment. The above phenomena were in line with Figure 8a in the simulation test. In a word, the average value of DDA in the cylindrical part is 29.6° during discharging, which conforms to the characteristics of "fast in the center, slow near the wall" in the funnel flow.

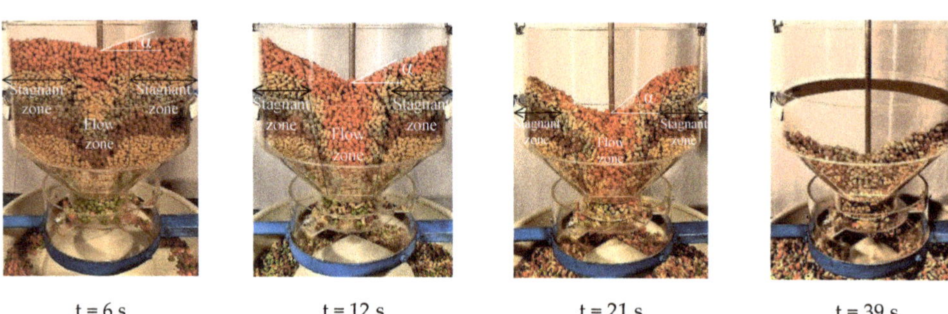

t = 6 s   t = 12 s   t = 21 s   t = 39 s

Figure 14. Snapshots for flow patterns of pellets at different times in the hopper ($\beta = 45°$, $n = 45$ rpm and $D_2 = 100$ mm).

In order to investigate the flow pattern of pellets in the hopper, the pellet-tracking tests were conducted, and the experimental equipment is presented in Figure 15. Firstly, the red pellet is placed in the center and near the wall on the top surface, respectively, in the experiment. The position of the red pellet is obtained by the downward displacement of the black marker on the fishing line with a scale of 10 mm, and it is recorded by the high-speed camera (PCC 2.8 new version Phantom V9.1 camera, Vision Research Inc. Wayne, NJ, USA, shooting frequency is 1000 f/s).

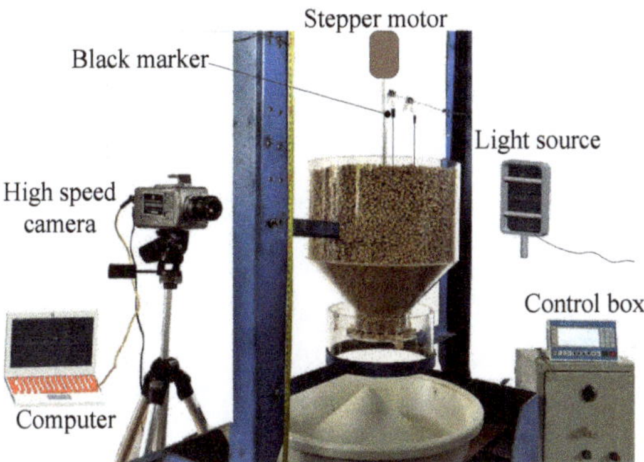

Figure 15. The experimental equipment.

Figure 16a shows the discharging process of the pellets in the hopper under the hopper half angle of 45°. The position of the black marker in the center is in a fast-straight descent line, while the black marker near the wall hardly moves from 0 s to 16.78 s. In contrast, Figure 16b shows the black marker in the center and near the wall, which basically dropped the same height in the same discharging time.

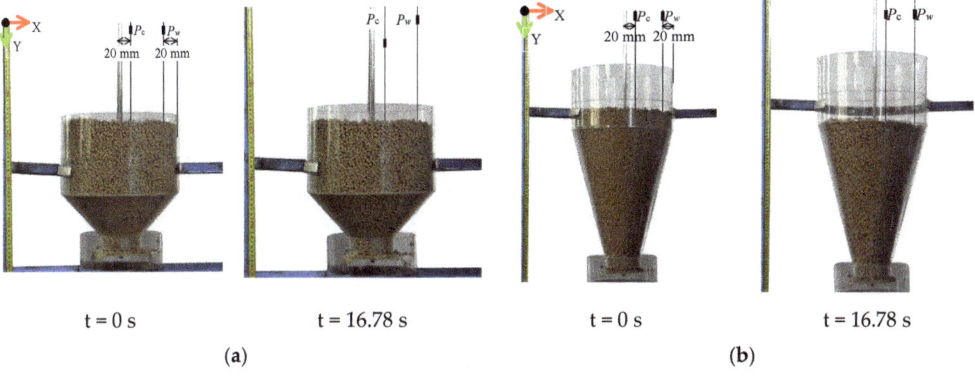

Figure 16. High-speed photograph diagrams of different pellet positions on the front faces. (**a**) When the hopper half angle is 45°, the rotation speed is 45 rpm and the outlet diameter is 100 mm; (**b**) when the hopper half angle is 75°, the rotation speed is 45 rpm and the outlet diameter is 100 mm.

The function of MFI and discharging time is established by collecting the coordinates of the black marker to quantitatively analyze the accuracy of the simulation test, as shown in Figure 17. As expected, when the rotation speed and the outlet diameter are 45 rpm and 100 mm, the average value of MFI is 0.79 under the hopper half angle of 75°, however, the average value of MFI is 0.17 under the hopper half angle of 45°, which is in agreement with the results in the simulation test. Notwithstanding, the experimental results are slightly bigger than the simulation test results. The error could be explained in two aspects. Firstly, the size of the pellet model established by DEM 2.7 software is the average of 100 pellets, which is different from the distribution of pellets in the experiment. Secondly, it may come from a system error caused by the operator in the experiment.

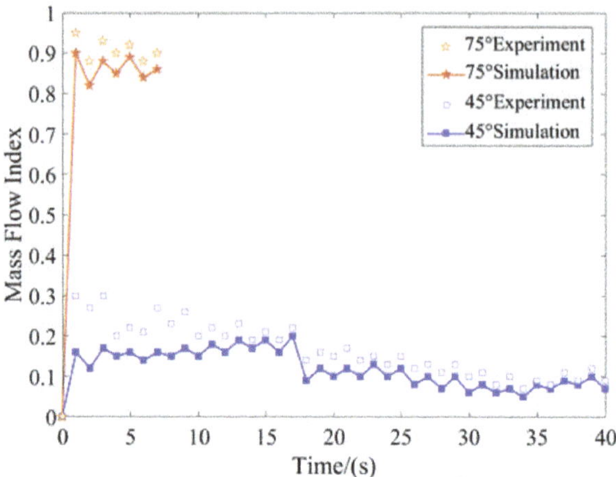

Figure 17. MFI of pellets discharged from the hopper.

5. Discussion

Combined with the discussion above, it is confirmed that the hopper half angle determines the flow pattern mainly, which proves Lu et al.'s [14] point. On the other hand, the work analyzes the effects of structural and motion parameters of the hopper with an unloading paddle on the flow capacity of pellets. The deficiencies in the flow capacity of pellets were investigated:

1.　Environmental impact:

The pellets will be deliquesced by environmental factors during the feeding process, so that their own moisture content changes in real-time, the pellet group in the hopper is reunited, and the vertical velocity of the pellet in the center decreases more significantly than near the wall, resulting in the vertical velocity difference decreasing in the pellets in the center and near the wall. Finally, the error in measurement results is caused.

2.　Test bench construction error:

In order to track the velocity and trajectory change of pellets, a pellet was randomly selected in the center and near the wall and tied with a fishing line, while the other end of the fishing line was used to adjust the counterweight (straw). In the process of the downward movement of pellets, the quality of the two counterweights might be inconsistent and the sliding friction between the fishing line and the frame was caused. The difference in vertical velocity of pellets in the center and near the wall changes, resulting in an error in measurement results.

3.　Pellets are broken during discharge:

A small amount of broken pellets will be produced continuously in the fixed disk during the unloading paddle rotation. The friction among the pellets on the fixed disk causes the falling of the pellets in the straight part to lag, and the vertical velocity of the pellets in the flow zone of the hopper decreases by layer-by-layer upward conduction, resulting in a decrease in the vertical velocity difference of the pellets in the center and near the wall. Finally, the error in measurement results is caused.

6. Conclusions

In this work, DEM was applied to simulate the discharging of pellets in the hopper of the feeder with an unloading paddle. The impacts of the hopper half angle, rotation speed and outlet diameter on the flow capacity of pellets were investigated. The results in the experiment and simulation test were in good agreement, which verifies the confidence of the results. The detailed conclusions can provide important support for the design of the hopper with an unloading paddle.

1. The Mass Flow Index (MFI) can define the flow patterns of pellets in the hopper with an unloading paddle, and the MFI~0.24 is identified to distinguish the mass flow and the funnel flow by the DEM simulation tests.
2. The MFI decreases with an increase in the rotation speed from 30 rpm to 60 rpm, the outlet diameter from 60 mm to 140 mm and the hopper half angle from 45° to 75°. Additionally, the flow pattern is mainly determined by the hopper half angle; the rotation speed and the outlet diameter have little influence on the flow patterns of pellets in the hopper, but the outlet diameter has a greater influence on the flow pattern when the outlet diameter is less than 60 mm.
3. For the design of the hopper with an unloading paddle, the optimum parameters are as follows: the hopper half angle is between 65° and 75°, the outlet diameter is between 100 mm and 120 mm and the rotation speed is between 45 rpm and 60 rpm.

Author Contributions: Conceptualization, H.H. and Y.Z.; methodology, Z.L. and D.W.; software, H.H. and Z.F.; resources, H.T. and J.S.; data curation, H.H., D.W. and Y.Z.; writing—original draft preparation, H.H., Y.Z. and M.H. All authors have read and agreed to the published version of the manuscript.

Funding: This research was funded by the Henan Province Science and Technology Research Project (222102110351) and the Chinese Natural Science Foundation (51405076).

Institutional Review Board Statement: Not applicable.

Informed Consent Statement: Not applicable.

Data Availability Statement: The data used to support the findings of this study are available from the corresponding author upon request.

Acknowledgments: The authors would like to thank the Laboratory of Animal Husbandry Mechanization; also, the authors are grateful to Taif University (KSA) and Mansoura University (Mansoura, Egypt) for their support and gratefully appreciate the reviewers who provided helpful suggestions for this manuscript.

Conflicts of Interest: The authors declare no conflicts of interest.

References

1. Vukmirovic, D.; Colovic, R.; Rakita, S.; Briek, T.; Duragic, O.; Sola-Oriol, D. Importance of feed structure (particle size) and feed form (mash vs. pellets) in pig nutrition—A review. *Anim. Feed Sci. Technol.* **2017**, *233*, 133–144. [CrossRef]
2. Lv, M.; Yan, L.; Wang, Z.; An, S.; Wu, M.; Lv, Z. Effects of feed form and feed particle size on growth performance, carcass characteristics and digestive tract development of broilers. *Anim. Nutr.* **2015**, *1*, 252–256. [CrossRef] [PubMed]
3. Pampuro, N.; Busato, P.; Cavallo, E. Gaseous Emissions after Soil Application of Pellet Made from Composted Pig Slurry Solid Fraction: Effect of Application Method and Pellet Diameter. *Agriculture* **2018**, *8*, 119. [CrossRef]
4. Pampuro, N.; Bisaglia, C.; Romano, E.; Brambilla, M.; Pedretti, E.F.; Cavallo, E. Phytotoxicity and Chemical Characterization of Compost Derived from Pig Slurry Solid Fraction for Organic Pellet Production. *Agriculture* **2017**, *7*, 94. [CrossRef]
5. Bao, Z.; Li, Y.; Zhang, J.; Li, L.; Zhang, P.; Huang, F.R. Effect of particle size of wheat on nutrient digestibility, growth performance, and gut microbiota in growing pigs. *Livest. Sci.* **2016**, *183*, 33–39. [CrossRef]
6. Tao, H.; Jin, B.; Zhong, W.; Wang, X.; Ren, B.; Zhang, Y.; Xiao, R. Discrete element method modeling of non-spherical granular flow in rectangular hopper. *Chem. Eng. Process. Process Intensif.* **2010**, *49*, 51–158. [CrossRef]
7. Nguyen, V.D.; Cogne, C.; Guessasma, M.; Bellenger, E.; Fortin, J. Discrete modeling of granular flow with thermal transfer: Application to the discharge of silos. *Appl. Thermal Eng.* **2009**, *29*, 1846–1853. [CrossRef]
8. Zhao, Y.; Yang, S.L.; Zhang, L.Q.; Chew, J.W. Understanding the varying discharge rates of lognormal particle size distributions from a hopper using the Discrete Element Method. *Powder Technol.* **2019**, *342*, 356–370. [CrossRef]
9. Xiao, H.Y.; Fan, Y.; Jacob, K.V.; Umbanhowar, P.B.; Kodam, M.; Koch, J.F.; Lueptow, R.M. Continuum modeling of granular segregation during hopper discharge. *Chem. Eng. Sci.* **2019**, *193*, 188–204. [CrossRef]
10. Engisch, W.E.; Muzzio, F.J. Method for characterization of loss-in-weigh feeder equipment. *Powder Technol.* **2012**, *228*, 395–403. [CrossRef]
11. Engisch, W.E.; Muzzio, F.J. Feedrate deviations caused by hopper refill of loss-in-weigh feeders. *Powder Technol.* **2015**, *283*, 389–400. [CrossRef]

12. Carcel, C.R.; Starr, A.; Nsugbe, E. Estimation of powder mass flow rate in a screw feeder using acoustic emissions. *Powder Technol.* **2018**, *336*, 122–130. [CrossRef]

13. Chen, S.; Baumes, L.A.; Gel, A.; Adepu, M.; Emady, H.; Jiao, Y. Classification of particle height in a hopper bin from limited discharge data using convolutional neural network models. *Powder Technol.* **2018**, *339*, 615–624. [CrossRef]

14. Lu, H.F.; Guo, X.L.; Jin, Y.; Gong, X.; Zhao, W.; Barletta, D.; Poletto, M. Powder discharge from a hopper-standpipe system modelled with CPFD. *Adv. Powder Technol.* **2017**, *28*, 481–490. [CrossRef]

15. Mahajan, N.P.; Deshpande, S.B.; Kadwane, S.G. Design and implementation of advanced controller in plant distributed control system for improving control of non-linear belt weigh feeder. *J. Proc. Control.* **2018**, *62*, 55–65. [CrossRef]

16. Chandravanshi, M.L.; Mukhopadhyay, A.K. Dynamic analysis of vibratory feeder and their effect on feed particle speed on conveying surface. *Measurement* **2017**, *101*, 145–156. [CrossRef]

17. Zhang, Y.X.; Jia, F.G.; Zeng, Y.; Han, Y.L.; Xiao, Y.W. DEM study in the critical height of flow mechanism transition in a conical silo. *Powder Technol.* **2018**, *331*, 98–106. [CrossRef]

18. Fernandez, J.W.; Cleary, P.W.; Mcbride, W. Effect of screw design on hopper drawdown of spherical particles in a horizontal screw feeder. *Chem. Eng. Sci.* **2011**, *66*, 5585–5601. [CrossRef]

19. Wang, Y.F.; Li, T.Y.; Muzzio, F.J.; Glasser, B.J. Predicting feeder performance based on material flow properties. *Powder Technol.* **2017**, *308*, 135–148. [CrossRef]

20. Mellmann, J.; Hoffmann, T.; Fürll, C. Mass flow during unloading of agricultural bulk materials from silos depending on particle form, flow properties and geometry of the discharge opening. *Powder Technol.* **2014**, *253*, 46–52. [CrossRef]

21. Weinhart, T.; Labra, C.; Luding, S.; Ooi, J.Y. Influence of coarse-graining parameters on the analysis of DEM simulations of silo flow. *Powder Technol.* **2016**, *293*, 138–148. [CrossRef]

22. Emden, H.K.; Kacianauskas, R. Discrete element analysis of experiments on mixing and bulk transport of wood pellets on a forward acting grate in discontinuous operation. *Chem. Eng. Sci.* **2013**, *92*, 105–117. [CrossRef]

23. Moysey, P.A.; Baird, M.H.I. Size segregation of spherical nickel pellets in the surface flow of a packed bed: Experiments and Discrete Element Method simulations. *Powder Technol.* **2009**, *196*, 298–308. [CrossRef]

24. Yuan, H.; Cai, Y.; Liang, S.F.; Ku, J.S.; Qin, Y. Numerical Simulation and Analysis of Feeding Uniformity of Viscous Miscellaneous Fish Bait Based on EDEM Software. *Agriculture* **2023**, *13*, 356. [CrossRef]

25. Ma, C.H.; Chen, L.; Yang, K.; Yang, J.; Tu, Y.; Cheng, L. Intelligent calibration method for microscopic parameters of soil–rock mixtures based on measured landslide accumulation morphology. *Comput. Methods Appl. Mech. Eng.* **2024**, *422*, 116835. [CrossRef]

26. Hou, Q.F.; Dong, K.J.; Yu, A.B. DEM study of the flow of cohesive particles in a screw feeder. *Powder Technol.* **2014**, *256*, 529–539. [CrossRef]

27. Tao, H.; Zhong, W.Q.; Jin, B.S. Comparison of construction method for DEM simulation of ellipsoidal particles. *Chin. J. Chem. Eng.* **2013**, *21*, 800–807. [CrossRef]

28. Saeed, M.K.; Siraj, M.S. Mixing study of non-spherical particles using DEM. *Powder Technol.* **2019**, *344*, 617–627. [CrossRef]

29. Jiang, Y.J.; Fan, X.Y.; Li, T.H.; Xiao, S.Y. Influence of particle-size segregation on the impact of dry granular flow. *Powder Technol.* **2018**, *340*, 39–51. [CrossRef]

30. Han, Y.L.; Jia, F.G.; Meng, X.Y.; Cao, B.; Zeng, Y.; Xiao, Y.W. Numerical analysis of similarities of particle flow behavior in stirred chambers. *Powder Technol.* **2019**, *344*, 286–301. [CrossRef]

31. Zeng, Y.; Jia, F.G.; Meng, X.Y.; Han, Y.L.; Xiao, Y.W. The effects of friction characteristic of particle on milling process in a horizontal rice mill. *Adv. Powder Technol.* **2018**, *29*, 1280–1291. [CrossRef]

32. Zhou, Z.Y.; Zhu, H.P.; Yu, A.B.; Wright, B.; Zulli, P. Discrete particle simulation of gas-solid flow in a blast furnace. *Comput. Chem. Eng.* **2008**, *32*, 1760–1772. [CrossRef]

33. Yang, W.J.; Zhou, Z.Y.; Yu, A.B. Discrete particle simulation of solid flow in a three-dimensional blast furnace sector model. *Chem. Eng. J.* **2015**, *278*, 339–352. [CrossRef]

34. Zhang, T.F.; Gan, J.Q.; Pinson, D.; Zhou, Z.Y. Size-induced segregation of granular materials during filling a conical hopper. *Powder Technol.* **2018**, *340*, 331–343. [CrossRef]

35. Toson, P.; Khinast, J.G. Impulse-based dynamics for studying quasi-static granular flows: Application to hopper emptying of non-spherical particles. *Powder Technol.* **2017**, *313*, 353–360. [CrossRef]

36. Scherer, V.; Wirtz, S.; Krause, B.; Wissing, F. Simulation of reacting moving granular material in furnaces and boilers an overview on the capabilities of the discrete element method. *Energy Procedia* **2017**, *120*, 41–61. [CrossRef]

37. You, Y.; Zhao, Y.Z. Discrete element modelling of ellipsoidal particles using super-ellipsoids and multi-spheres: A comparative study. *Powder Technol.* **2018**, *331*, 179–191. [CrossRef]

38. Zhong, W.Q.; Yu, A.B.; Liu, X.J.; Tong, Z.B.; Zhang, H. DEM/CFD-DEM Modelling of non-spherical particulate systems: Theoretical developments and applications. *Powder Technol.* **2016**, *302*, 108–152. [CrossRef]

39. Cabiscol, R.; Finke, J.H.; Kwade, A. Calibration and interpretation of DEM parameters for simulations of cylindrical tablets with multi-sphere approach. *Powder Technol.* **2018**, *327*, 232–245. [CrossRef]

40. Qian, Q.; An, X.Z.; Zhao, H.Y.; Dong, K.J.; Yang, X.H. Numerical investigations on random close packings of cylindrical particles with different aspect ratios. *Powder Technol.* **2019**, *343*, 79–86. [CrossRef]

41. Coetzee, C.J. Review: Calibration of the discrete element method. *Powder Technol.* **2017**, *310*, 104–142. [CrossRef]

42. Huang, H.N.; Wang, D.F.; Li, B.Q.; Zhang, J.F.; Li, J.Q. Design and Experiment of Discharging Performance of Feeder for Nursery. *Trans. Chin. Soc. Agric. Mach.* **2018**, *49*, 161–169.
43. Kumar, R.; Patel, C.M.; Jana, A.K.; Gopireddy, S.S. Prediction of hopper discharge rate using combined discrete element method and artificial neural network. *Adv. Powder Technol.* **2018**, *29*, 2822–2834. [CrossRef]
44. Faqih, A.N.; Alexander, A.W.; Muzzio, F.J.; Tomassone, M.S. A method for predicting hopper flow characteristics of pharmaceutic powders. *Chem. Eng. Sci.* **2007**, *62*, 1536–1542. [CrossRef]
45. Höhner, D.; Wirtz, S.; Scherer, V. A numerical study on the influence of particle shape on hopper discharge within the polyhedral and multi-sphere discrete element method. *Powder Technol.* **2012**, *226*, 16–28. [CrossRef]
46. Beverloo, W.A.; Leniger, H.A.; Velde, J.V.D. The flow of granular solids through orifices. *Chem. Eng. Sci.* **1961**, *15*, 260–269. [CrossRef]
47. Zheng, Q.J.; Xia, B.S.; Pan, R.H.; Yu, A.B. Prediction of mass discharge rate in conical hoppers using elastoplastic model. *Powder Technol.* **2017**, *307*, 63–72. [CrossRef]
48. Xu, K.M.; Wang, J.; Wei, Z.B.; Deng, F.F.; Wang, Y.W.; Cheng, S.M. An optimization of the MOS electronic nose sensor array for the detection of Chinese pecan quality. *J. Food Eng.* **2017**, *203*, 25–31. [CrossRef]
49. Song, J.; Yang, H.; Li, R.; Chen, Q.; Zhang, Y.J.; Wang, Y.J.; Kong, P. Improved PTV measurement based on Voronoi matching used in hopper flow. *Powder Technol.* **2019**, *355*, 172–182. [CrossRef]
50. Drescher, A.; Ferjani, M. Revised model for plug/funnel flow in bins. *Powder Technol.* **2004**, *141*, 44–54. [CrossRef]
51. Fullard, L.A.; Davies, C.E.; Wake, G.C. Modelling powder mixing in mass flow discharge: A kinematic approach. *Adv. Powder Technol.* **2013**, *24*, 499–506. [CrossRef]
52. Johanson, J.R. Stress and velocity fields in the gravity flow of bulk solids. *J. Appl. Mech.* **1964**, *31*, 499–506. [CrossRef]

 *agriculture*

Article

Optimized Design for Vibration Reduction in a Residual Film Recovery Machine Frame Based on Modal Analysis

Xinzhong Wang [1,*], Tianyu Hong [1], Weiquan Fang [1] and Xingye Chen [1]

School of Agricultural Engineering, Jiangsu University, Zhenjiang 212013, China;
2222116046@stmail.ujs.edu.cn (T.H.); 2112116003@stmail.ujs.edu.cn (W.F.); 2212116030@stmail.ujs.edu.cn (X.C.)
* Correspondence: xzwang@ujs.edu.cn

Abstract: The technology of plastic film mulching is widely applied in Xinjiang, but it also brings about serious issues of residual film pollution. Currently, the 1MSF-2.0 residual film recovery machine can effectively address the problem. However, it faces challenges such as high overall machine weight and noticeable frame vibrations, which affect the stability of the entire machine operation. The frame, as the installation foundation, needs to bear loads and impact. Therefore, the reliability of the frame is crucial for the stability of the entire machine. Improving the frame's vibration is of great importance. In response to the significant vibration issues during the operation of the 1MSF-2.0 residual film recovery machine, this paper utilized Workbench 2020 R2 to establish a finite element model of the machine frame and conducted static analysis to obtain strength information, thereby initially understanding the optimization space of the frame. Building upon this, Mechanical was employed to solve the first 14 natural frequencies and mode shapes of the frame, and the accuracy of the theoretical analysis was verified through modal testing. After analyzing the frequency characteristics of external excitation forces, it was found that the fourth-order natural frequency of the frame fell within the frequency range of the excitation force of the shaft of the straw grinder, causing resonance in the frame and necessitating structural optimization. The optimal results indicated that the optimized frame increased in mass by 4.41%, reduced the maximum stress value by 2.56 MPa, and increased the fourth-order natural frequency to 22.7 Hz, avoiding the frequency range of the excitation force of the shaft of the straw grinder, thus improving the resonance issue. This paper provides a reference for optimizing the design of the frame of the residual film recovery machine.

Keywords: residual film recovery machine; static analysis; modal analysis; modal testing; dimensional optimization

Citation: Wang, X.; Hong, T.; Fang, W.; Chen, X. Optimized Design for Vibration Reduction in a Residual Film Recovery Machine Frame Based on Modal Analysis. *Agriculture* **2024**, *14*, 543. https://doi.org/10.3390/agriculture14040543

Academic Editor: Massimiliano Varani

Received: 7 March 2024
Revised: 23 March 2024
Accepted: 27 March 2024
Published: 29 March 2024

1. Introduction

The technology of plastic film mulching can effectively increase the yield of cotton in Xinjiang and is widely applied there [1–3]. However, the problem of plastic film pollution in Xinjiang has become increasingly severe due to the long-term use of plastic film. This pollution issue seriously hinders the sustainable development of agriculture. Research has shown that residual plastic film can cause a yield reduction of approximately 10% in cotton and corn crops [4]. According to statistics, China's annual plastic film usage has exceeded 1.4 million tons, with a recycling rate of approximately 80%. This indicates that approximately 300,000 tons of plastic film remain in the soil every year. The control and management of plastic film pollution has become an urgent task. Currently, the primary method for managing residual film pollution is to use residual film recovery machines. The focus of this paper is on the 1MSF-2.0 residual film recovery machine. Previous research and experiments have shown that this residual film recovery machine can effectively recover residual film from Xinjiang's cotton fields, significantly mitigating the problem of residual film pollution. However, there are still some key issues that need to be optimized. Due to the complexity of the residual film recycling process, the 1MSF-2.0

residual film recovery machine has numerous moving components, which results in a significant amount of vibrational force. As a result, the machine experiences noticeable vibrations, significantly impacting its reliability and stability during operation [5]. The frame, serving as the installation foundation for the residual film recovery machine, plays a crucial role in bearing and resisting impacts. Improving the vibration issues of the frame and preventing resonance under vibrational forces are of vital importance in enhancing the reliability of the entire machine [6].

Modal analysis is used to calculate the natural frequencies and mode shapes of structures. It is the most fundamental dynamic analysis and an important area of study in vibration reduction design [7]. When the natural frequency of a structure is close to the frequency of external excitation forces, the structure can experience resonance. In such cases, it is necessary to optimize the structure to shift its natural frequencies away from the frequency range of the external excitation forces, thereby improving the structure's vibration characteristics. In the work by Yaoming Li et al. [8], they solved for the modal frequencies and mode shapes of the chassis frame of a rice-wheat combined harvester. By optimizing the structure of the frame through changes in beam cross-sectional dimensions, they were able to shift the modal frequencies away from the excitation frequencies, thus enhancing the overall reliability of the machine. In the research conducted by Jianjun Gao et al. [9], they performed theoretical modal analysis on the chassis frame of a combined harvester. Then modal testing was used to validate the accuracy of their theoretical analysis. This allowed them to optimize the structure of the frame effectively, thereby preventing frame resonance and improving the stability and reliability of the machine. In the research conducted by Jinming Zhang et al. [10], simulation was used to determine the vibration characteristics of a wheeled tractor under different operating conditions. They obtained the natural frequencies of vertical vibrations for the front axle, cabin, driver's seat, and implement as 2.09 Hz, 2.84 Hz, 2.84 Hz, and 2.09 Hz, respectively. Subsequently, they performed multi-objective optimization on the stiffness and damping parameters of the front axle, obtaining the optimal stiffness and damping settings to improve the tractor's vibration performance and ride comfort. A modal analysis on the chassis frame of the 4JZ-1700 crawler-type pepper harvester using ANSYS was conducted by Xinzhong Wang et al. [11]. The natural frequencies of the frame were found to be between 23 and 76 Hz, providing a theoretical basis for the anti-vibration design of the frame. Yingcan Liu et al. [12] addressed the issue of strong vibrations in the residual film recovery machine. They performed static and modal analysis on the machine's frame using ANSYS WorkBench 19.0 software. Through an orthogonal experimental design method, they optimized the structure, resulting in a reduction of the maximum stress and the maximum amplitude, effectively avoiding resonance phenomena. Junxian Guo et al. [13] focused on the intense vibrations of the cotton stalk return and residual film recycling integrated machine. The Lanczos Method algorithm in ANSYS Workbench was used to solve for the modal frequencies and modes of the frame. After optimization, the first two natural frequencies of the frame deviated from the external excitation frequency, resulting in a significant reduction in vibration amplitudes and enhanced vibration damping. Changjiang Liang et al. [14] conducted modal analysis on the frame of the residual film recovery machine using finite element analysis software. The results showed a significant difference between the natural frequencies of the frame and the external excitation frequencies, effectively avoiding resonance occurrences. Shuangshuang Zhong et al. [15] conducted modal analysis on the roller of the mulch shredder using ANSYS finite element software, obtaining the first eight natural frequencies and modes to verify the design requirements of the frame. Panfeng Zhang et al. [16] used ANSYS software and finite element simulation analysis to prevent the occurrence of resonance in the film-lifting component of the rotary tillage nail tooth plastic film recycling machine. Modal analysis and transient dynamic analysis were performed on the nail tooth component. The data confirmed that no resonance occurred during the operation of the nail tooth, thus ensuring the reliability of the entire machine. Zhenhua Yu et al. [17] conducted finite element stress analysis and modal analysis on the disc-type film-gathering device of the residual film recovery machine. The stress distribution and dynamic characteristics of the film-gathering device under working loads were obtained, providing a reference for the design improvement of the equipment.

This paper addressed the issue of noticeable vibrations during the operation of the 1MSF-2.0 residual film recovery machine. Through methods such as static analysis and modal analysis, the strength information and vibration characteristics of the frame were solved for. Combining the characteristics of external excitation forces, the frame of the residual film recovery machine was optimized in size to enhance its reliability and stability during operation, thereby the effectiveness of residual film recovery was further improved.

2. Materials and Methods

2.1. Establishment of Finite Element Model for the Frame and Static Analysis

2.1.1. Establishment of Finite Element Model for the Frame

The frame of the residual film recovery machine is primarily composed of rectangular steel tubes and steel plates welded together. The overall dimensions of the frame are 2950 mm in length, 2560 mm in width, and 1559 mm in height, with a weight of 354.25 kg. At the front end of the frame, it is welded to the straw grinder. The frame is equipped with picking rollers, film stripping and pressure relief devices, baling machines, and some transmission components, serving a load-bearing function. Additionally, during operation, the frame must withstand external impacts and vibrations. Therefore, the frame needs to meet certain strength and stiffness requirements. The three-dimensional model of the frame, established using Autodesk Inventor Professional 2020, is shown in Figure 1.

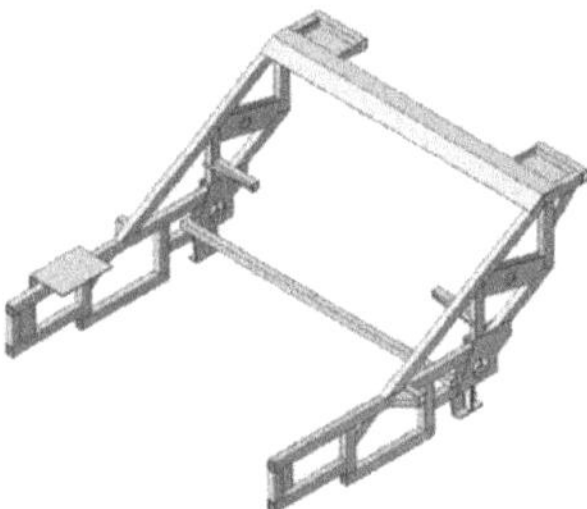

Figure 1. The three-dimensional model of the frame.

Meshing is an important step in creating a finite element model, and the quality of the meshing directly affects the mathematical representation of the physical model and the computational workload. Therefore, it is necessary to set an appropriate mesh size when establishing a finite element model. Typically, the element quality after meshing should not be lower than 0.7. In this paper, multiple mesh sizes were considered, and it was determined that a mesh size of 7 mm resulted in an element quality of 0.780. When the mesh size exceeded 8 mm, the element quality dropped below 0.700. Therefore, selecting a mesh size of 7 mm not only ensures the quality of the meshing but also reduces the computational workload.

The three-dimensional model of the frame was imported into Workbench 2020 R2 to establish the finite element model. The material of the frame is Q235 structural steel, with a density of 7850 kg/m^3, an elastic modulus of 210 GPa, and a Poisson's ratio of 0.3. The material parameters in Workbench 2020 R2 were set, with a mesh size of 7 mm, and the mesh was divided to obtain the finite element model of the frame, as shown in Figure 2. It consists of 813,440 nodes and 413,333 elements. Figure 2 also displays the meshing details of the upper crossbeam of the frame.

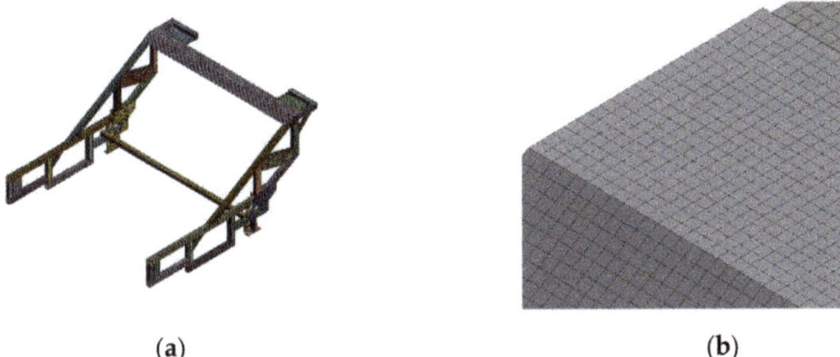

(a)   (b)

Figure 2. Finite element model of the frame: (**a**) The overall meshing of the frame. (**b**) The meshing of the upper crossbeam of the frame.

2.1.2. Static Analysis of the Frame

During the process of optimizing the frame design, it is essential to ensure that the strength of the frame meets the required standards. Static analysis can provide information about the strength of the frame, serving as a reference for optimizing its design. Therefore, conducting static analysis on the frame is necessary.

The forces acting on the frame mainly originate from the tractor's tractive force and ground resistance, as well as the gravitational forces of the picking rollers, film stripping and pressure relief devices, baling machines, and some transmission components mounted on the frame.

The magnitude of tractive force and resistance is measured through experiments. During the experiment, the torque at the picking roller was measured using the TQ201 wireless torque sensor node (manufactured by Beijing Beetech Inc., Beijing, China). The experimental setup was as shown in Figure 3. The resistance during the operation of the residual film recovery machine mainly comes from the resistance of the picking roller against the soil; it is assumed that the tractive force is solely used to overcome this resistance. Therefore, based on the measured torque magnitude and the relationship between force and torque, the magnitudes of tractive force and resistance can be calculated. In this paper, the calculated magnitudes of tractive force and resistance are 2500 N.

Figure 3. Experimental setup for torque measurement: 1. TQ201 wireless torque sensor node and 2. battery.

The forces exerted by the picking rollers, film stripping and pressure relief devices, baling machines, and some transmission components on the frame are primarily due to gravity. The weights of these components were determined by measuring their volumes using Autodesk Inventor Professional 2020, combined with their densities.

The constraints on the frame mainly come from the depth limitation roller and the rear suspension of the tractor. The depth limitation roller restricts three translational degrees of freedom and two rotational degrees of freedom of the frame, allowing only the freedom of rotation around the axis of the depth limitation roller. Additionally, the rear suspension of the tractor is connected to the frame, limiting the freedom of rotation around the axis of the depth limitation roller. Therefore, when applying constraints during static analysis, the frame is constrained according to the above conditions.

During static analysis, the positions, magnitudes, directions, and loading modes of the loads are as indicated in Table 1. The schematic diagram of load application positions is as shown in Figure 4. These positions on the frame all have symmetry.

Table 1. The positions, magnitudes, directions, and loading modes of the loads.

Positions	Magnitudes/N	Directions	Loading Modes
Picking support side plate	7298	Vertical downward	Uniformly distributed load
Picking support side plate	2500	Horizontal backward	Uniformly distributed load
Peel-off roller bearing seat	3310	Vertical downward	Uniformly distributed load
Baling machine bearing seat	8388	Vertical downward	Uniformly distributed load
Transport roller bearing seat	2992	Vertical downward	Uniformly distributed load
Straw grinder connection hole	2500	Horizontal forward	Uniformly distributed load

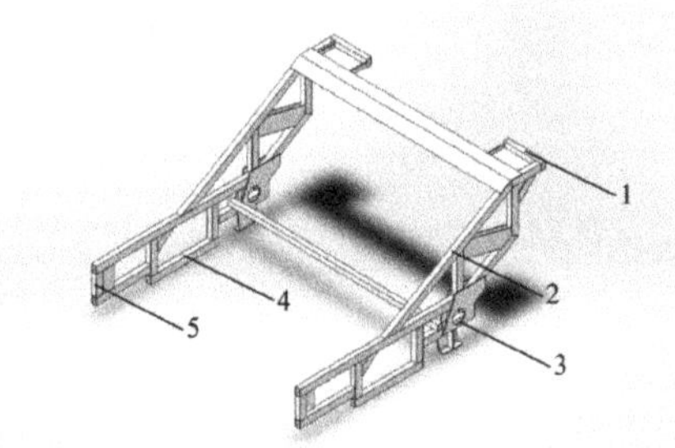

Figure 4. The schematic diagram of load application positions: 1. Baling machine bearing seat, 2. peel-off roller bearing seat, 3. transport roller bearing seat, 4. picking support side plate, and 5. straw grinder connection hole.

After applying constraints and loads, the solution was computed using Workbench 2020 R2. Through post-processing of the results, stress analysis contour plots, strain analysis contour plots, and total deformation analysis contour plots of the frame, a finite element model was obtained.

2.2. Modal Analysis of the Frame

2.2.1. Calculated Modal Analysis of the Frame

Modal analysis is used to determine the vibrational characteristics of a structure. Calculated modal analysis can be divided into free modal analysis and constrained modal analysis. Constrained modal analysis requires simulating the actual working conditions of the structure [18]. When calculating this, appropriate boundary conditions, such as loads and constraints, need to be applied to the model. Due to the complexity of the actual working conditions of the residual film recovery machine frame in this paper, using constrained modal analysis will result in significant differences between simulated and actual working conditions, making it difficult to ensure the accuracy and reliability of the analysis results. Free modal analysis assumes that the structure is in a free state and does not consider the actual working conditions. The analysis is relatively simple and can reflect the vibrational characteristics of the structure. Therefore, free modal analysis was adopted for the modal analysis of the residual film recovery machine frame in this paper. Since the frame was in a free state, the first six modes obtained from modal analysis were rigid modes with natural frequencies of 0 [19].

Based on the static analysis of the frame, modal analysis was performed using the finite element model of the frame in Mechanical, without applying loads and constraints, to obtain the first 14 non-zero mode shapes and natural frequencies.

2.2.2. Test Modal Analysis of the Frame

To validate the accuracy of the finite element model of the frame and the calculated modal analysis, it is necessary to conduct modal testing on the frame [20–22]. The principle of modal testing is as follows: excite the structure with a vibrational force input, measure the signals at various points on the structure, and use modal parameter identification methods to obtain the modal parameters of the structure. By comparing the test modal parameters with the calculated modal parameters, the accuracy of the results can be verified.

In this paper, the force measurement method was used for the modal testing of the frame, employing a single-point excitation approach. The force measurement method requires simultaneous measurement of the excitation force and response during the test. By using this method, modal parameters can be estimated. Therefore, during modal testing, a force hammer equipped with a force sensor must be used to apply measurable excitation forces. The force measurement method can estimate all modal parameters, including natural frequencies, mode shapes, damping or damping ratios, modal mass, and modal stiffness, with high accuracy.

When selecting measurement points, it is important to choose points that adequately represent the shape of the frame. Too few measurement points will result in an incomplete representation of the frame's shape, while too many points will lead to excessive computational burden, prolonged calculation time, or even computational infeasibility. As shown in Figure 5, this paper selected 18 important response points of the frame structure as the measurement points, and chose the midpoint position of the lower beam as the excitation point (Point A). The reliability of the results was ensured through two measurements and analyses [8].

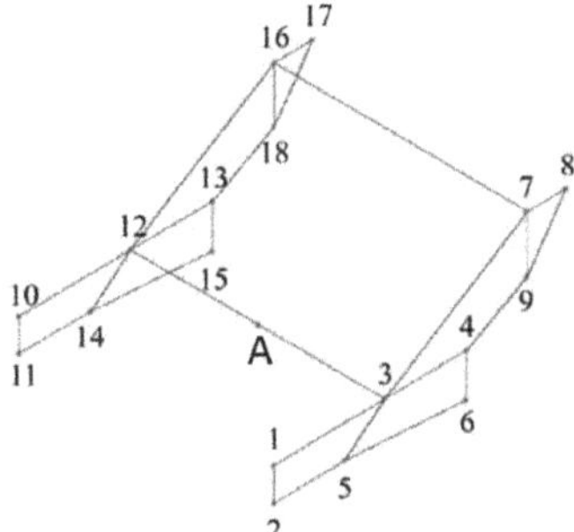

Figure 5. The positions of 18 measurement points and 1 excitation point in modal testing.

During modal testing, it is essential to maintain consistency between the testing conditions and the calculated modal analysis. Since the calculated modal analysis conducted in this paper was based on free modal analysis, during the testing, the frame was suspended using a crane with a flexible connection between the crane and the frame to approximate a free-state condition. The experimental instruments are illustrated in Figure 6.

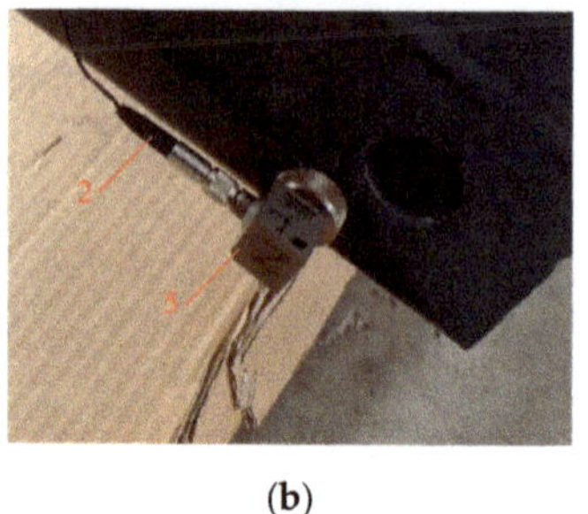

(a)   (b)

Figure 6. The experimental instruments: (**a**) Data acquisition system section: 1. force hammer, 2. accelerometer bus, 3. DH5902N rugged data acquisition system, and 4. laptop. (**b**) Sensor section: 2. accelerometer bus and 5. accelerometer.

The experimental instruments mainly included a force hammer, triaxial accelerometer, DH5902N rugged data acquisition system, and DHDAS V2.0 software (all manufactured by Donghua Testing Technology Co., Ltd., Taizhou, China). The experimental process was as follows:

1. Instrument setup: The instruments were connected. The force sensor on the force hammer was connected to the first channel of the data acquisition system, while the accelerometer was connected to the second, third, and fourth channels. The instruments and DHDAS software were started, and then the Analysis/Frequency Response Analysis function was selected.

2. Parameter setting: The sensor sensitivity was input into the sensitivity setting column of the corresponding channel. Each measurement point was struck with the force hammer and the waveform was observed. The instrument connections, wire connections, sensor, and instrument functionality were checked until the waveform was correct. The range was appropriately adjusted by striking the points with suitable force until the force and response waveforms were neither overloaded nor too small.

3. Excitation and data collection: The frame was struck with the force hammer to induce forced vibration. The accelerometer installed on the frame collected the vibration acceleration–time domain signals. The DH5902N rugged data acquisition system and DHDAS software were used to collect, save, and analyze the signals. According to the Nyquist sampling theorem, the sampling frequency must be at least twice the highest frequency present in the original signal in order to accurately reconstruct the digital signal from the original signal [23]. In order to ensure the accuracy of the sampled signal, sampling frequency was set to 2 kHz, with 800 spectral lines, and the acceleration–time domain signals collected at each measurement point were measured sequentially. For example, when measuring point 1, the acceleration–time domain signals collected by the triaxial accelerometer installed on the frame were as shown in Figure 7.

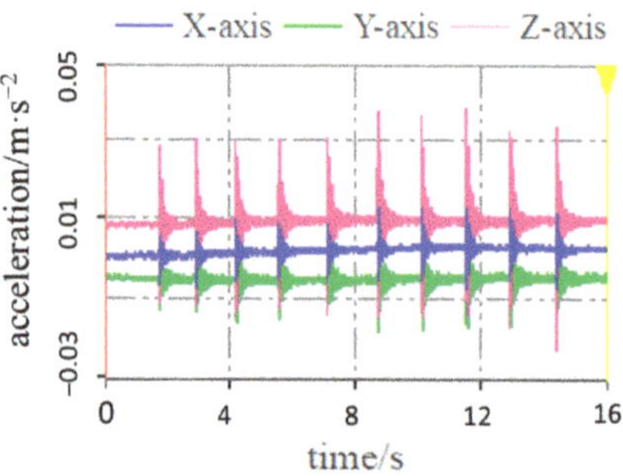

Figure 7. The acceleration–time domain signals collected by the triaxial accelerometer sensor.

The DHDAS software was utilized for the spectrum analysis and frequency response analysis of the acceleration–time domain signals. In the spectrum analysis settings, we chose amplitude output, selected the average spectrum for calculation parameters, set the number of spectral lines to 800, set the frequency spacing to 0.1 Hz, set the number of averages to 10, and selected linear averaging. In the frequency response analysis, we added input and output measurement point information. The results of the spectrum analysis and frequency response analysis are shown in Figure 8.

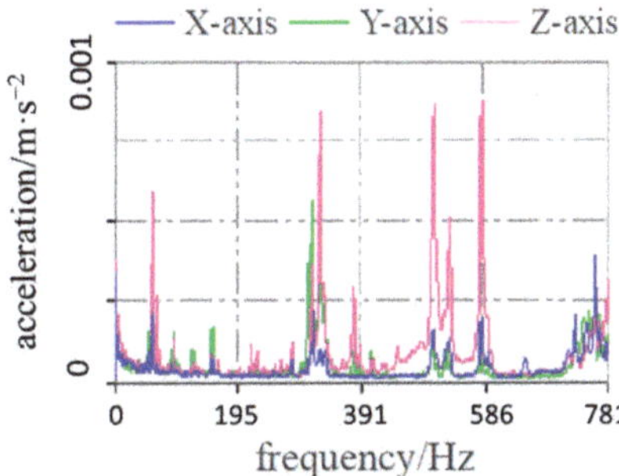

Figure 8. The results of the spectrum analysis and frequency response analysis.

We established the corresponding model in the DHDAS software based on the 18 selected measurement points and 1 excitation point in this article. The model should adequately depict the shape of the framework, as shown in Figure 5. The measurement point numbers on the model were edited; these numbers should correspond to the modal testing. We imported the transfer functions of each measurement point into the data interface, performed modal parameter identification, and solved for the first 14 modal frequencies, damping ratios, and mode shapes.

2.3. Optimization of the Framework Dimensions

Material selection, dimension optimization, shape optimization, and topology optimization are important methods for frame optimization design [24]. Currently, commonly used materials for frames are Q235 steel and Q345 steel. Compared to Q235 steel, Q345 steel has poor welding performance and is prone to weld detachment issues. Furthermore, improving the performance of the frame usually requires more than just changing the material, structural optimization is also necessary. Dimension optimization establishes an accurate mathematical model and ensures that design variables meet specific design requirements by adding constraints. It is widely used in frame optimization design due to its precision and versatility. Shape optimization involves describing variables using partial differential equations [25]. It is more complex and less commonly applied than dimension optimization, and it is usually combined with dimension optimization in practical applications. Topology optimization offers a larger design space and greater degrees of freedom, making it suitable for lightweight design. In this study, depending on the specific circumstances, the dimension optimization method was to optimize the frame design. The residual film recovery machine consists of numerous key components and transmission parts. To ensure the reliability of the machine operation, the installation positions and alignments of each component on the frame have been thoroughly considered and calculated. Furthermore, the installation positions and alignments of each component are interrelated. During the optimization design of the frame, it is essential to avoid significant changes in the overall size and shape of the frame structure. Therefore, in this paper's optimization design of the rectangular steel tubes in the frame, the wall thickness of the rectangular steel tubes was chosen as the design variable. This approach ensured that the overall shape of the frame would not undergo significant changes. There are a total of 16 rectangular steel tubes that can be optimized in terms of dimensions. The positions of the rectangular steel tubes are shown in Figure 9. Due to the symmetry of the frame, the identical rectangular steel tubes on both sides of the frame were considered as one design variable.

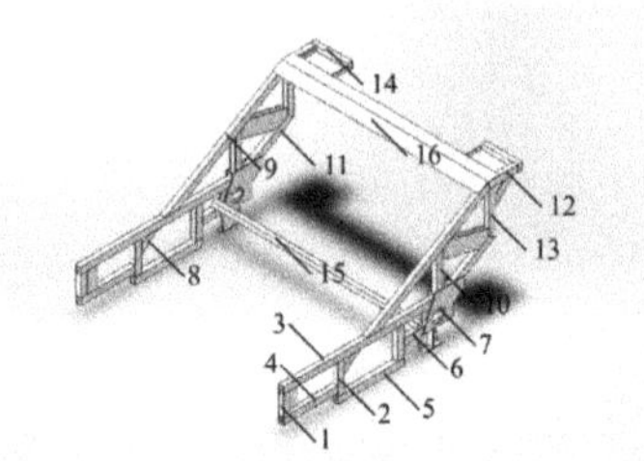

Figure 9. The positions of the 16 design variables.

This paper aimed to adjust the natural frequency of the residual film recovery machine frame to avoid the excitation frequency range of the straw grinder's stirring cage shaft. To achieve this goal, a dimensional optimization of the rectangular steel tubes constituting the frame was conducted.

Changing the dimensions of the frame to adjust the natural frequency may lead to a significant increase or decrease in the mass of the frame. The results of static analysis indicate that the frame has a certain margin of static strength but it is not sufficient because during actual operation, the frame may experience impacts where instantaneous stress could significantly increase. If the optimized frame significantly reduces in mass, although it can adjust the natural frequency, it cannot guarantee the strength of the frame. Therefore, when applying constraint conditions, it is necessary to impose constraints on the mass of the frame. Therefore, this paper imposed a constraint on the frame's mass reduction, limiting it to no more than 5%. A total of 16 locations of the rectangular steel tubes' wall thickness were chosen as design variables, forming the basis for the optimization design model. The selection of the rectangular steel tube wall thickness adhered to national standards.

The wall thickness of the 16 rectangular steel tubes is denoted in Formula (1).

$$X = [x_1, x_2, \ldots, x_{16}] \tag{1}$$

The constraints are as follows: the fourth-order natural frequency ω of the frame is greater than the maximum excitation frequency f_{max} of the straw grinder's stirring cage shaft; the mass $W(x)$ of the residual film recovery machine frame is not less than the specified minimum value W_{min}; and the maximum stress σ in the frame does not exceed the specified maximum value σ_{max}. The mathematical model for optimizing the frame design was obtained as in Formula (2).

$$\begin{cases} s.t. \begin{cases} \omega \geq f_{max} \\ W(x) \geq W_{min} \\ \sigma \leq \sigma_{max} \end{cases} \\ X = [x_1, x_2, \ldots, x_{16}] \end{cases} \tag{2}$$

Based on the static and modal analyses, Workbench 2020 R2 offers two optimization methods: direct optimization and response surface optimization [26,27]. Direct optimization involves selecting the optimal solution from sample points obtained from experimental designs. The more sample points there are, the more accurate the optimization results, but this also leads to longer computation times. Response surface optimization, on the other hand, utilizes fitted response surfaces for optimization. The accuracy of the optimization results depends on how well the response surface is fitted to the data. In this paper, there were a total of 16 discrete design variables. Since the number of sample points was relatively small, and to obtain accurate computational results, this paper intended to employ the direct optimization method to optimize the dimensions of the rectangular steel tubes in the frame. The Direct Optimization module was added to Workbench 2020 R2, aiming to calculate the optimal solution for the 16 design variables.

In Workbench 2020 R2, according to national standards, the range of values for the wall thickness of 16 rectangular steel tubes was determined and inputted into the parameter setting table. Since each parameter can only control the wall thickness of two opposite sides of the rectangular steel tube, the parameter relationship table needed to be updated to include the relationship between the wall thickness of adjacent sides. Objectives and constraints were set according to Formula (2). The optimization method chosen was Screening. After completing the setup, the calculation was performed, resulting in three sets of optimal sample points. By comparing these three sets, one set of optimal sample points was selected as the optimized design scheme for the frame.

The optimized frame underwent both static analysis and modal analysis to evaluate the feasibility of the optimization solution. By comparing the results of these analyses to those of the original frame, the effectiveness of the optimization approach could be examined.

3. Results

3.1. Results of Static Analysis

After post-processing the results of the static analysis, the stress analysis contour plot, strain analysis contour plot, and total deformation analysis contour plot of the frame finite element model were obtained as shown in Figure 10.

The main material used for the frame is Q235 structural steel, with a maximum yield strength of 235 MPa. The calculation formula for allowable stress is shown in Formula (3).

$$[\sigma] = \frac{\sigma_s}{n} \tag{3}$$

In Formula (3), $[\sigma]$ represents the allowable stress, σ_s represents the maximum yield strength, and n represents the safety factor. Taking $n = 1.5$ [10], the allowable stress $[\sigma]$ for Q235 structural steel can be calculated as 156.67 MPa.

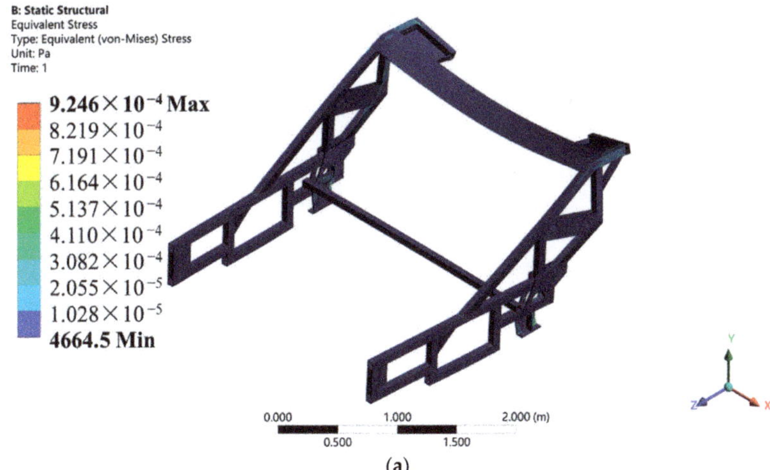

(a)

Figure 10. *Cont.*

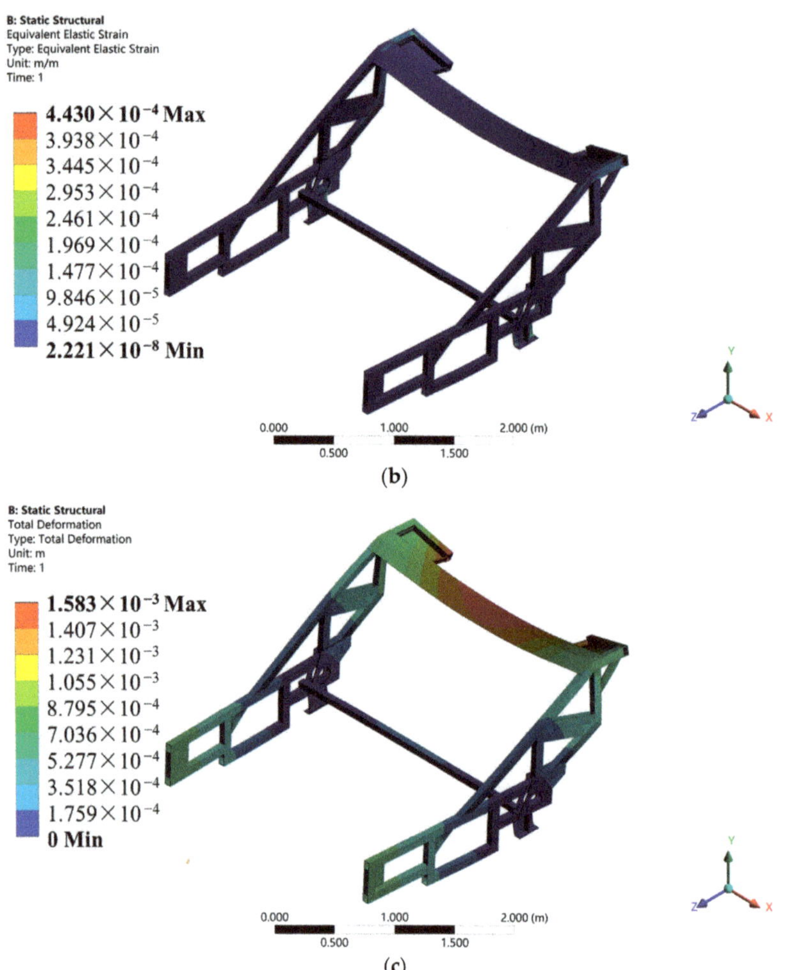

Figure 10. The results of static analysis of the frame: (**a**) stress analysis contour plots; (**b**) strain analysis contour plots; and (**c**) total deformation analysis contour plots.

The results of static analysis indicate that the maximum stress in the frame is approximately 92.46 MPa, with an average value of 4.03 MPa. These values are below the allowable stress for Q235 structural steel, demonstrating that the frame meets the strength requirements under static loading conditions.

Analyzing the stress contour plot, strain contour plot, and total deformation contour plot of the frame reveals that, under static loading, the overall stress, strain, and total deformation of the frame are relatively small. There is noticeable variation only in the upper beam, this is mainly because the upper beam is connected to the baling machine, which has a relatively large weight. The baling machine exerts its load on the upper beam through bearings, causing a significant force on the upper beam.

3.2. Results of Modal Analysis

Building upon the static analysis of the frame, modal analysis of the finite element model of the frame was conducted using Mechanical. The first 14 natural frequencies of the frame obtained from this analysis are presented in Table 2, while the mode shapes for the first four modes are depicted in Figure 11.

Table 2. Comparison between calculated modal analysis and test modal analysis.

Order	Calculated Modal Analysis		Test Modal Analysis		Relative Error/%
	Calculated Frequency	Mode Shapes	Test Frequency	Mode Shapes	
1	6.306	Torsion	5.859	Torsion	7.093
2	9.192	Bending	7.813	Bending	5.004
3	11.495	Bending	11.603	Bending	−0.940
4	21.314	Bending	21.908	Bending	−2.787
5	35.302	Bending	35.157	Bending	0.411
6	42.909	Bending	41.415	Bending	3.482
7	50.949	Bending	51.055	Bending	−0.208
8	53.713	Bending	54.751	Bending	−1.932
9	61.642	Bending	61.908	Bending	−0.432
10	69.063	Torsion, Bending	67.563	Bending	2.172
11	76.664	Bending	76.730	Bending	−0.086
12	83.454	Torsion	82.028	Torsion	1.709
13	92.378	Bending	92.631	Bending	−0.274
14	95.845	Torsion, Bending	95.334	Bending	0.533

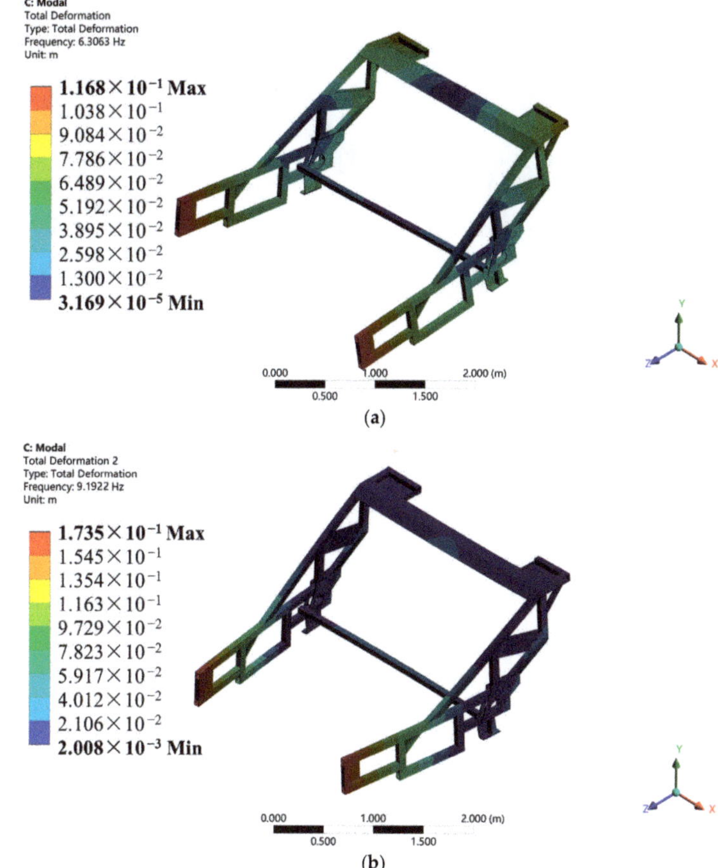

Figure 11. *Cont.*

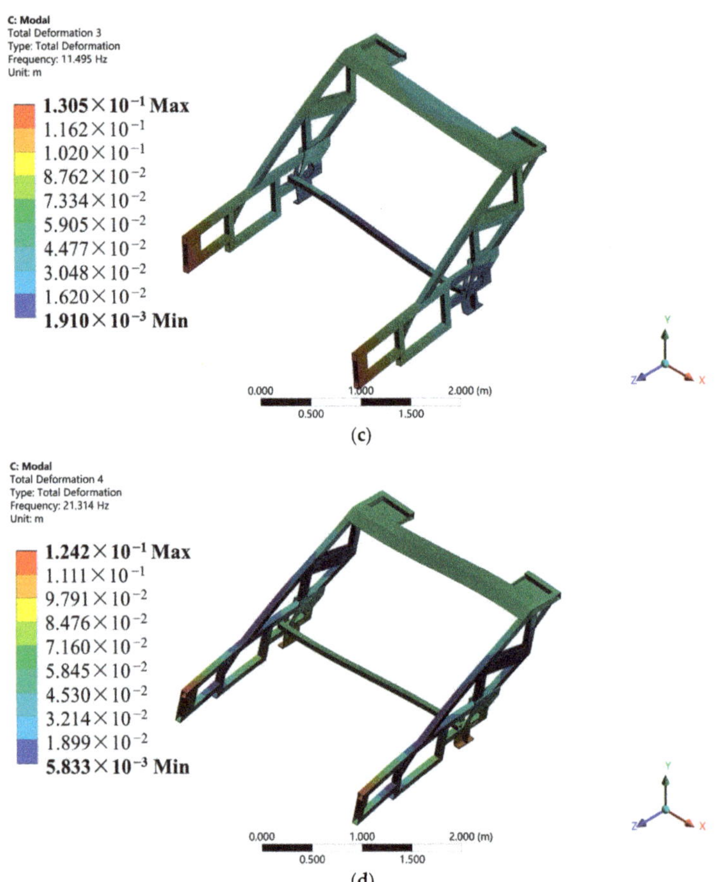

Figure 11. The first four modes of the frame in calculated modal analysis: (**a**) mode 1, (**b**) mode 2, (**c**) mode 3, and (**d**) mode 4.

The three directions of the frame in the three-dimensional coordinate system are shown in Figure 11. Analyzing the first four non-zero mode shapes of the frame, the following observations can be made.

Mode 1: The primary deformation involves left–right torsion, with the maximum total deformation located at the front end of the frame. The torsion of the welded structures on both sides of the frame mainly occurs in the Y direction, with relatively minor deformations in the X and Z directions.

Mode 2: The primary deformation is characterized by forward–backward bending, with the maximum total deformation located at the front end of the frame. The bending of the welded structures on both sides of the frame primarily occurs in the X direction, with minor deformations in the Y and Z directions.

Mode 3: Similar to Mode 2, the primary deformation involves forward–backward bending, with the maximum total deformation located at the front end of the frame. The bending of the welded structures on both sides of the frame primarily occurs in the X direction, with minor deformations in the Y and Z directions.

Mode 4: The primary deformation is characterized by left–right bending, with the maximum total deformation located at the front end of the frame. Although there are no significant shape changes in the welded structures on both sides of the frame, noticeable displacement occurs due to a significant deformation of the two beams.

The natural frequencies of the frame obtained from modal testing are presented in Table 2, while the first four mode shapes are depicted in Figure 12. The red dots in Figure 12 represent the corresponding 18 measurement points and 1 excitation point in Figure 5.

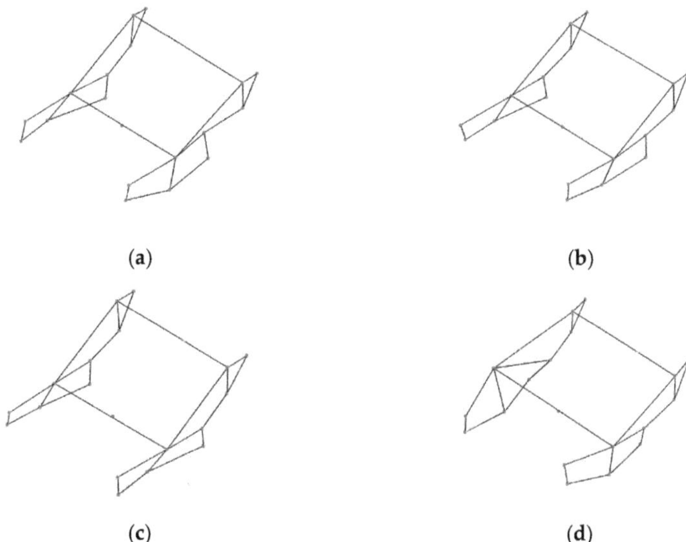

Figure 12. The first four modes of the frame in test modal analysis: (**a**) mode 1, (**b**) mode 2, (**c**) mode 3, and (**d**) mode 4.

The three directions of the frame in the three-dimensional coordinate system are shown in Figure 11. The analysis and parameter identification yield the following results for the first four mode shapes.

Mode 1: The first mode shape is primarily characterized by twisting motion in the horizontal direction. The welded structures on both sides of the frame twist around the X-axis in opposite directions. The front and middle sections of the welded structures exhibit significant deformation.

Mode 2: The second mode shape is primarily characterized by bending motion in the forward and backward directions. The welded structures on both sides of the frame exhibit displacement along the X-axis. The front and middle sections of the welded structures display significant deformation.

Mode 3: The third mode shape is primarily characterized by bending motion in the left and right directions. The welded structures on both sides of the frame exhibit displacement along the X-axis. The lateral bending of the welded structures is evident, while deformation is relatively small in other regions.

Mode 4: The fourth mode shape is primarily characterized by bending motion in the left and right directions. The welded structures on both sides of the frame bend in opposite directions around the Z-axis. The front and middle sections of the welded structures display significant deformation, and the two horizontal beams also experience a certain degree of deformation.

During modal analysis, the calculated results may have errors compared to the actual conditions due to the simplified idealized model used and the differences between the simulated and actual operating conditions. In modal testing, apart from the external signal interference caused by environmental factors, it was challenging to ensure that the excitation force applied by humans remained consistently positioned accurately at the excitation point and maintained a consistent magnitude. These factors could introduce certain errors into the test results. Therefore, it is necessary to perform modal shape correlation analysis to assess the reliability of the calculated modal analysis results in comparison with the calculated modal analysis results.

To address these challenges, it is essential to conduct modal shape correlation analysis. This analysis aims to assess the reliability of the calculated modal analysis results by comparing them with test modal analysis results [28].

As shown in Table 2, the relative errors between the calculated frequencies and test frequencies of the analyzed frame indicate that the absolute values of relative errors for each natural frequency are within 8%. This suggests that the finite element model of the frame and the results of modal analysis are accurate [29]. The relatively large errors in the first and second natural frequencies are primarily due to their small absolute values, as small absolute errors can result in relatively larger relative errors. Other contributing factors may include simplifications made in the finite element model of the frame compared to the actual model and external signal interference.

A comparison of the modal shapes from the calculated modal analysis and test modal analysis reveals a basic consistency and correlation between the two. Discrepancies in the descriptions of modal shapes for the 3rd, 10th, 11th, 12th, and 14th modes are attributed to the inability of the chosen measurement points in the test modal analysis to fully replicate the frame structure, leading to inevitable deviations in modal shapes.

Given that the lower-order modal shapes determine the dynamic characteristics of the structure [8], a focused analysis is conducted on the first four natural frequencies and corresponding modal shapes of the frame. During the operation of the residue recovery machine, the excitation forces primarily originate from the ground, the cutter shaft of the straw grinder, and the stirring shaft of the straw grinder. Ground excitation is determined by ground conditions, with typical ground excitation frequencies being below 3 Hz [30]. Actual measurements indicate that the normal operating speeds of the cutter shaft and stirring shaft are in the range of 2400 to 2500 r/min and 1200 to 1300 r/min, with excitation frequencies ranging from 40 to 41.7 Hz and 20 to 21.7 Hz, respectively.

All of the first four natural frequencies of the frame are above 3 Hz and below 40 Hz, suggesting that the influences of ground conditions and the cutter shaft are minimal. However, the fourth natural frequency of the frame is 21.314 Hz, coinciding with the excitation frequency range of the stirring shaft. This results in significant resonance during operation. To address this resonance issue, an enhanced reliability and stability of the entire machine during operation and optimization of the structural dimensions of the frame are required.

3.3. Results of Dimensional Optimization

Optimal solutions for 16 design variables were obtained through optimization calculations, as shown in Table 3. The samples from the optimization results are illustrated in Figure 13, with each curve representing a sample. Based on the optimization objectives, the optimal solution was selected from the samples.

3.4. Comparison of the Frame before and after Optimization

Using Autodesk Inventor Professional 2020, the optimal thickness values obtained from Table 3 for the 16 rectangular steel tubes were used to modify the thickness of the corresponding steel tubes on the original frame, resulting in the three-dimensional model of the optimized frame. Subsequently, in Workbench 2020 R2, a finite element model of the optimized frame was created following the same principles mentioned earlier. Static analysis and modal analysis were performed on the optimized frame to compare the calculated results with the original frame and validate the effectiveness of the optimization solution.

Table 3. Comparison of design variable values before and after optimization.

Design Variables	Before Optimization/mm	After Optimization/mm	Increment/mm
Steel Tubes 1	5	5	0
Steel Tubes 2	5	7	2
Steel Tubes 3	5	5	0
Steel Tubes 4	5	4	−1
Steel Tubes 5	5	4	−1
Steel Tubes 6	5	5	0
Steel Tubes 7	5	4	−1
Steel Tubes 8	5	7	2
Steel Tubes 9	5	7	2
Steel Tubes 10	5	6	1
Steel Tubes 11	5	6	1
Steel Tubes 12	5	5	0
Steel Tubes 13	5	5	0
Steel Tubes 14	5	5	0
Steel Tubes 15	5	5	0
Steel Tubes 16	5	5	0

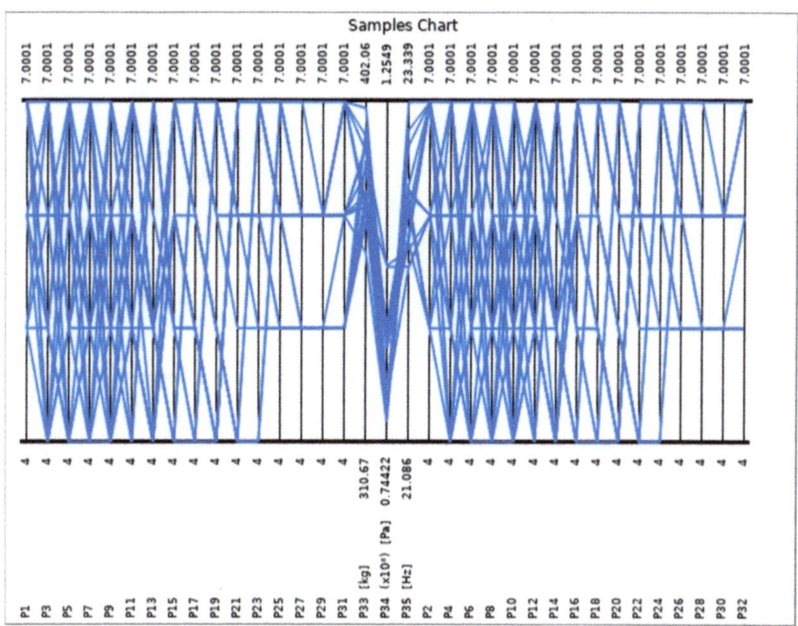

Figure 13. The samples from the optimization results.

3.4.1. Comparison of Static Analysis Results

In Mechanical, appropriate constraints and loads were applied to the optimized frame. The positions, magnitudes, directions, and loading modes of the loads were consistent with the previous description.

After applying the constraints and loads, the finite element model was solved. Post-processing the results, stress contour plots, strain contour plots, and overall deformation contour plots of the optimized frame were obtained, as shown in Figure 14.

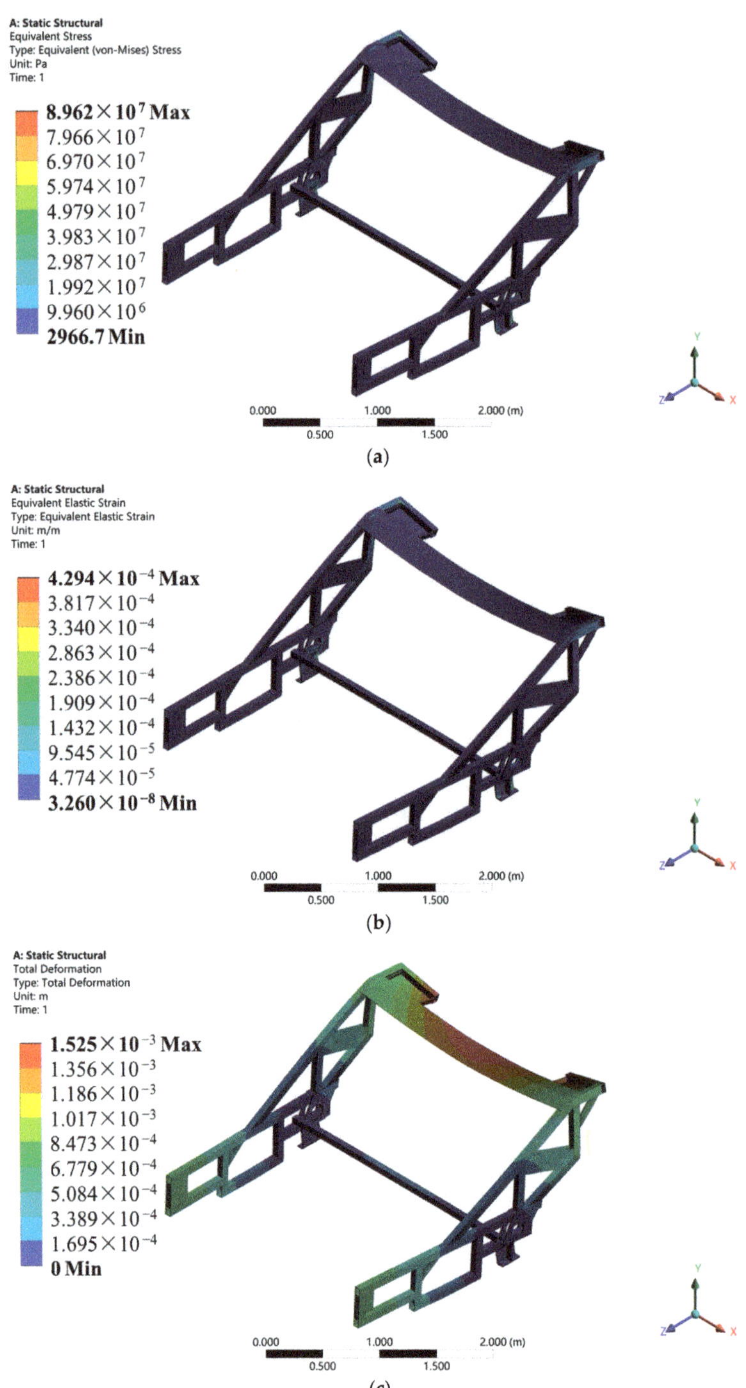

Figure 14. The results of static analysis of the optimized frame: (**a**) stress analysis contour plots; (**b**) strain analysis contour plots; and (**c**) total deformation analysis contour plots.

The static analysis results indicated that the maximum stress of the optimized frame was approximately 89.62 MPa, with an average stress of 3.83 MPa. Both the maximum and average stresses were smaller than those of the frame before optimization. The maximum overall deformation of the optimized frame was approximately 1.5 mm, which was also smaller than the deformation observed in the original frame. The mass of the optimized frame was 368.83 kg, which was an increase of 14.58 kg (4.12%) compared to the original frame. The mass increase was not significant, indicating that the optimization did not significantly affect the overall mass of the frame. The optimized frame exhibited reduced overall stress, strain, and deformation compared to the original frame. These quantities were distributed more evenly throughout the structure, satisfying both strength and stiffness requirements. This confirms the effectiveness of the structural optimization approach.

3.4.2. Comparison of Modal Analysis Results

Building upon the static analysis, the finite element model of the optimized frame was further subjected to modal analysis using Mechanical. This analysis was conducted without applying any loads or constraints. The objective was to determine the natural frequencies and corresponding mode shapes for the first four non-zero modes. The specific results are presented in Figure 15. Among them, the fourth natural frequency and mode of vibration of the optimized frame is shown in Figure 16.

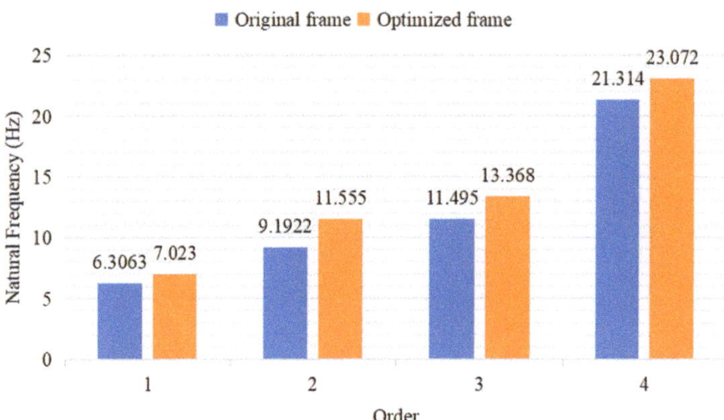

Figure 15. Comparison of first four natural frequencies before and after optimization.

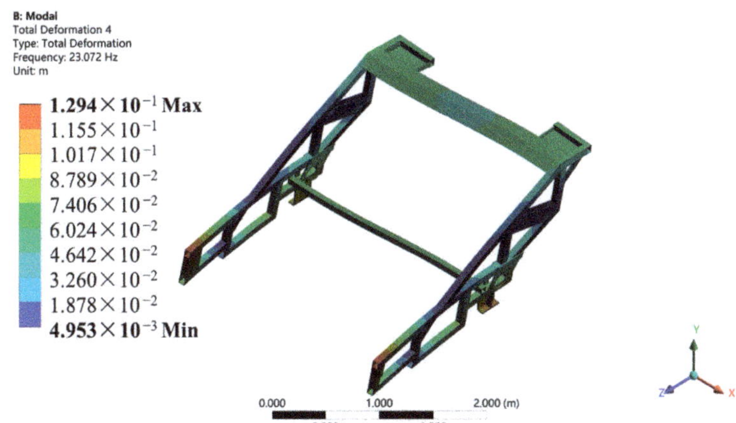

Figure 16. The fourth natural frequency and mode of vibration of the optimized frame.

Despite the increase in frame mass by 4.12%, the natural frequencies of the first four modes of the optimized frame increased by 11.36%, 25.70%, 16.29%, and 8.25%, respectively. These frequencies still remained above 3 Hz and below 40 Hz, indicating that the frame is not significantly affected by ground and harrowing machine vibration. In particular, the fourth mode frequency of the frame increased from 21.314 Hz to 23.072 Hz, effectively avoiding the vibration range of the harrowing machine's shaft. This improvement indicates a significant reduction in resonance phenomena caused by the interaction between the frame and the stirring shaft. Therefore, the improvement brought about by the optimization plan was significant.

4. Discussion

In recent years, China has conducted extensive research on vibration reduction in agricultural machinery. The main method for vibration reduction is to analyze the vibration characteristics of the frame and optimize its structure [31]. This can effectively improve the vibration problems of agricultural machinery. However, due to the limitations of optimization methods and the complexity of mechanical equipment under actual working conditions, there is a certain deviation from theoretical analysis under ideal conditions [32]. Therefore, it is difficult to completely avoid the generation of resonance problems by optimizing the frame structure.

The research findings of this paper have, to some extent, addressed the issue of significant vibrations in the 1MSF-2.0 residual film recovery machine (manufacturered by World Heavy Industries (China) Co., Ltd, Zhenjiang, China). This improvement has enhanced the reliability and stability during residual film recovery operations, ensuring the effectiveness of the recovery process. However, there are still some issues that need to be addressed.

1. The size of the residual film recovery machine is relatively large. While improving the vibration problem, the problem of large mass still exists, which will lead to increased energy consumption and have an impact on the agricultural ecological environment. Therefore, a lightweight design of the frame of the residual film recovery machine is also necessary.

2. This paper only used the method of dimensional optimization to optimize the frame design. In addition to this, material selection, shape optimization, and topology optimization are also important methods for frame optimization, each with its advantages and disadvantages [24]. It is a future research issue to comprehensively apply dimensional optimization, shape optimization, and topology optimization methods to fully exploit the space of frame optimization design and obtain better optimization solutions.

5. Conclusions

This paper conducted static and modal analyses on the frame of the residual film recovery machine and optimized the thickness of the rectangular tube of the frame, obtaining the optimal solution for 16 design variables, which improved the vibration problem of the residual film recovery machine. The optimization design of the frame in this paper ensured that there was no significant change in the frame's mass while ensuring its strength. Based on this, the fourth mode's natural frequency of the frame was adjusted to avoid the frequency range of excitation forces, effectively improving the resonance issues during the operation of the frame. This enhancement has led to an improved reliability and stability of the frame, providing support for the efficient operation of the residual film recovery machine. The research in this paper mainly includes the following aspects.

1. The three-dimensional model of the frame was established using Autodesk Inventor Professional 2020, and the finite element model of the frame was established using Workbench 2020 R2. Static analysis was performed on the finite element model of the frame. The results of static analysis showed that the maximum stress of the frame was 92.46 MPa, indicating that the strength of the frame under static load met the requirements and had a certain optimization design space.

2. The finite element model of the frame was subjected to free modal analysis using Workbench 2020 R2, and the first 14 natural frequencies and mode shapes of the frame were obtained. Among them, the first four natural frequencies with significant influence on the vibration characteristics of the frame were 6.3063 Hz, 9.1922 Hz, 11.495 Hz, and 21.314 Hz. The accuracy of the modal analysis results was verified through modal experiments, with an absolute value of relative error within 8%. The frequency range of external excitation forces was analyzed, and it was found that the excitation force of the stirring shaft could cause frame resonance, thus proposing optimization strategies.

3. Dimensional optimization of the frame was performed using Workbench 2020 R2 to control the frame mass, ensure that the frame met the requirements of strength and stiffness, and avoid the frequency range of external excitation forces around the fourth natural frequency. The final optimization scheme was obtained. After optimization, the frame met the requirements of strength and stiffness, with a mass increase of 4.12%, which was not significant. The fourth natural frequency increased by 8.25%, avoiding the frequency range of external excitation forces, and the vibration problem was improved.

4. The forces acting on the residual film recovery machine during operation are highly complex. Conducting a more comprehensive analysis of the frame's load distribution and establishing a more precise mechanical model will be crucial for future research on the 1MSF-2.0 residual film recovery machine. Additionally, given the numerous components present on the residual film recovery machine, optimizing the components outside the frame also holds significant importance.

Author Contributions: Conceptualization, X.W.; methodology, X.W.; software, T.H.; validation, T.H. and X.C.; formal analysis, T.H. and X.C.; investigation, T.H. and X.C.; resources, T.H. and X.C.; data curation, T.H. and X.C.; writing—original draft preparation, T.H.; writing—review and editing, X.W. and W.F.; visualization, X.W. and W.F.; supervision, X.W.; project administration, X.W. All authors have read and agreed to the published version of the manuscript.

Funding: This research was funded by the Key Technologies Research and Development Program (Grant NO. 2022YFD2002400), Key Scientific and Technological Projects in Key Areas of Corps (Grant NO. 2023AB014), the Agricultural Engineering Faculty of Jiangsu University (Grant NO. NZXB20200104), and Priority Academic Program Development of Jiangsu Higher Education Institutions (Grant NO. PAPD-2023).

Institutional Review Board Statement: Not applicable.

Data Availability Statement: The datasets used and/or analyzed during the current study are available from the corresponding author on reasonable request.

Conflicts of Interest: The authors declare no conflicts of interest.

References

1. Chauhan, S.; Basnet, B.; Shrestha, A.K. Innovative farming techniques for superior okra yield in Chitwan, Nepal: The benefits of plastic film mulch and pest exclusion net on soil properties, growth, quality and profitability. *Acta Ecol. Sin.* **2023**, *44*, 23–32. [CrossRef]
2. Liu, J.L.; Huang, X.Y.; Chen, H. Sustaining yield and mitigating methane emissions from rice production with plastic film mulching technique. *Agric. Water Manag.* **2021**, *245*, 106667. [CrossRef]
3. Yang, J.L.; Ren, L.Q.; Zhang, P. Can soil organic carbon sequestration and carbon management index be improved by changing the film-mulching methods in semiarid region? *J. Integr. Agric.* **2023**. [CrossRef]
4. Yang, Y.; Bai, S.H.; Ma, S.T. Residual Film Recycling Machine Research and Analysis of the Collecting Device. *Food Sci. Technol. Econ.* **2018**, *43*, 79–81. [CrossRef]
5. Ma, Z.; Zhang, Z.; Zhang, Z.; Song, Z.; Liu, Y.; Li, Y.; Xu, L. Durable Testing and Analysis of a Cleaning Sieve Based on Vibration and Strain Signals. *Agriculture* **2023**, *13*, 2232. [CrossRef]
6. Zhang, Y.P.; Hu, Z.C.; You, Z.Y.; Gu, F.W.; Wu, F.; Chen, Y.Q. Comparative study on working performance of two kinds of residue plastic film collectors. *Jiangsu Agric. Sci.* **2018**, *46*, 179–182. [CrossRef]
7. Li, Z.J.; Zheng, K.; Zheng, X.; Hao, Z.Y. Analysis of the Influence of Tooling on Constraint Modal Test of Front End Module. *Noise Vib. Control* **2019**, *39*, 150–154.
8. Li, Y.M.; Sun, P.P.; Pang, J.; Xu, L.Z. Finite element mode analysis and experiment of combine harvester chassis. *Trans. Chin. Soc. Agric. Eng.* **2013**, *29*, 38–46+301.

9. Gao, J.J.; Liu, Z.Y.; Gu, W.; Zhang, F.F.; Zhu, L.P.; Xu, J.X. Modal Analysis and Optimization of Grain Combine Harvester Undercarriage Frame. *Mach. Des. Res.* **2023**, *39*, 199–205. [CrossRef]
10. Zhang, J.M.; Yao, H.P.; Xue, J.L. Vibration characteristics analysis and suspension parameter optimization of tractor/implement system with front axle sus-pension under ploughing operation condition. *J. Terramech.* **2022**, *102*, 49–64. [CrossRef]
11. Wang, X.Z.; Cao, Y.H.; Fang, W.Q.; Sheng, H.R. Vibration Test and Analysis of Crawler Pepper Harvester under Multiple Working Conditions. *Sustainability* **2023**, *15*, 8112. [CrossRef]
12. Liu, Y.C.; Guo, J.X.; Shi, Y.; Xie, J.H.; Qian, X.G.; Han, J. Finite element analysis and optimization of the frame of residual plastic film recycling machine. *J. Chin. Agric. Mech.* **2023**, *44*, 189–194. [CrossRef]
13. Guo, J.X.; Liu, Y.C.; Wei, Y.J.; Zhou, J.; Shi, Y. Ration Characteristics Analysis and Structure Optimization of Integrated Straw Returning and Residual Film Recycling Machine. *Trans. Chin. Soc. Agric. Eng.* **2024**, *40*, 155–163. [CrossRef]
14. Liang, C.J.; Wu, X.M.; Wang, P.Z.; Li, G.C.; Tao, P.J.; Zhang, F.G. Simulation design and experimental on frame and comb teeth of the residual plastic film picker. *J. Anhui Agric. Univ.* **2019**, *46*, 203–208. [CrossRef]
15. Zhong, S.S.; Wang, Q.; Fang, H.F.; Wang, M.D. Modal Analysis of Roller in Used Mulch Shredder. *Mech. Eng.* **2018**, *4*, 43–45.
16. Zhang, P.F.; Hu, C.; Wang, X.F.; Lu, B.; Liu, C.J. Kinetic Analysis of Rotary Tillage Nail Tooth Plastic Film Recycling Machine Hook Film Unit. *J. Agric. Mech. Res.* **2018**, *40*, 14–18+25. [CrossRef]
17. Yu, Z.H.; Xu, G.M.; Wang, H.Y.; Niu, Y.P.; Cao, W.L. Force and modal analysis of film gathering device of residue film recycling machine. *China South. Agric. Mach.* **2021**, *52*, 70–71.
18. Zhang, J.; Wang, T.Y.; Yao, X.X.; Luo, W.G.; Luo, J.G. Modal analysis and physical test verification of a certain type of automobile exhaust system. *J. Guangxi Univ. Sci. Technol.* **2023**, *34*, 1–6. [CrossRef]
19. Wang, Y.L.; Dai, X.D.; Xie, Y.B. Modal Analysis of Multi-Cylinder Engine Block. *J. Xi'an Jiaotong Univ.* **2001**, *35*, 536–539.
20. Han, Z.P.; Brownjohn, J.M.W.; Chen, J. Structural modal testing using a human actuator. *Eng. Struct.* **2020**, *221*, 111113. [CrossRef]
21. Liang, J.; Zhao, D.F. Summary of the model analysis method. *Mod. Manuf. Eng.* **2006**, *8*, 139–141. [CrossRef]
22. Zhou, M.; Wei, Z.; Wang, Z.; Sun, H.; Wang, G.; Yin, J. Design and Experimental Investigation of a Transplanting Mechanism for Super Rice Pot Seedlings. *Agriculture* **2023**, *13*, 1920. [CrossRef]
23. Zhang, F.; Li, J.X.; Shi, J.S. Research on Subsynchronousr/Supersynchronous Oscillation Parameter Identification Based on Fundamental Synchrophasor: Spectrum Characteristics and Essential Issues. *Trans. China Electrotech. Soc.* **2024**. [CrossRef]
24. Cai, S.; Luo, Y.H.; Li, Q.L. State of the Art of Lightweight Technology in Agricultural Machinery and Its Development Trend. *J. Mech. Eng.* **2021**, *57*, 35–52.
25. Liao, Y.Y.; Liao, B.Y. Lightweight Design of Hydro-Generator Upper Bracket Based on Sizing Optimization. *Water Power* **2019**, *45*, 91–94+117.
26. Cao, Y.; Yu, Y.; Tang, Z.; Zhao, Y.; Gu, X.; Liu, S.; Chen, S. Multi-Tooth Cutting Method and Bionic Cutter Design for Broccoli Xylem (*Brassica oleracea* L. var. Italica Plenck). *Agriculture* **2023**, *13*, 1267. [CrossRef]
27. Wang, F.; Liu, Y.; Li, Y.; Ji, K. Research and Experiment on Variable-Diameter Threshing Drum with Movable Radial Plates for Combine Harvester. *Agriculture* **2023**, *13*, 1487. [CrossRef]
28. Zhu, M.T.; He, Z.G.; Xu, L.; Li, Z.B. Mode Analysis of Car-body and Its Correlative Research Shape. *Trans. Chin. Soc. Agric. Mach.* **2004**, *3*, 13–15+19.
29. Palumbo, A.; Polito, T.; Marulo, F. Experimental modal analysis and vibro-acoustic testing at leonardo laboratories. *Mater. Today Proc.* **2021**, *34*, 24–30. [CrossRef]
30. Zhang, K.P.; Shao, L.; Lv, J.C.; Deng, C. Modal Analysis on the Frame of Tractor. *Automob. Appl. Technol.* **2012**, *1*, 35–39.
31. Hao, S.; Tang, Z.; Guo, S.; Ding, Z.; Su, Z. Model and Method of Fault Signal Diagnosis for Blockage and Slippage of Rice Threshing Drum. *Agriculture* **2022**, *12*, 1968. [CrossRef]
32. Fang, C.; Zhao, S.Y.; Yan, G.; Qing, G.F. Multi-Objective Topology Optimization, Size Optimization and Detailed Design of Frame for an Electric Car. *Mach. Des. Manuf.* **2023**, *8*, 16–22. [CrossRef]

 *agriculture*

Article

Design and Optimization of Power Shift Tractor Starting Control Strategy Based on PSO-ELM Algorithm

Yu Qian [1], Lin Wang [2] and Zhixiong Lu [1,*]

1   College of Engineering, Nanjing Agricultural University, Nanjing 210031, China; 2020212006@stu.njau.edu.cn
2   State Key Laboratory of Intelligent Agricultural Power Equipment, Luoyang 470139, China; wanglin@ytogroup.com
*   Correspondence: luzx@njau.edu.cn

Abstract: Power shift tractors have been widely used in agricultural tractors in recent years because of their advantages of uninterrupted power during shifting, high transmission efficiency and high stability. As one of the indispensable driving states of the power shift tractor, the starting process requires a small impact and a starting speed that meets the driver's requirements. In this paper, aiming at such contradictory requirements, the starting control strategy of a power shift tractor is formulated with the goal of starting quality and the driver's intention. Firstly, the identification characteristics of the driver under three starting intentions are obtained by a real vehicle test. An extreme learning machine with fast identification speed and short training time is used to establish the basic driver's intention identification model. For the instability of the identification results of the Extreme Learning Machine (ELM), the particle swarm optimization algorithm (PSO) is used to optimize the ELM. The optimized extreme learning machine model has an accuracy of 96.891% for driver's intention identification. The wet clutch is an important part of the power shift gearbox. In this paper, the starting control strategy knowledge base of the starting clutch is established by a combination of bench tests and simulation tests. Through the fuzzy algorithm, the driver's intention is combined with the starting control strategy. Different drivers' intentions will affect the comprehensive evaluation model of the clutch (the single evaluation index of the clutch is: the maximum sliding power, the sliding power, the speed stability time, the impact degree), thus affecting the final choice of the starting clutch control strategy considering the driver's intention. On this basis, this paper studies and establishes the MPC starting controller for the power shift gearbox. Compared with the linear control strategy, the PSO-ELM-fuzzy weight starting strategy proposed in this paper can reduce the maximum sliding friction power by 45%, the sliding friction power by 69.45%, and the speed stabilization time by 0.11 s. The effectiveness of the starting control strategy considering the driver's intention proposed in this paper to improve the starting quality of the power shift tractor is verified.

Keywords: tractors; power shift transmission; starting process; driver's intention

Citation: Qian, Y.; Wang, L.; Lu, Z. Design and Optimization of Power Shift Tractor Starting Control Strategy Based on PSO-ELM Algorithm. *Agriculture* **2024**, *14*, 747. https://doi.org/10.3390/agriculture14050747

Academic Editor: Massimiliano Varani

Received: 4 April 2024
Revised: 7 May 2024
Accepted: 9 May 2024
Published: 10 May 2024

1. Introduction

As the main tool of agricultural production, agricultural tractors often need to undertake field operations such as tillage, ploughing, sowing, and harvesting [1,2]. In addition, tractors also spend a large part of their time on road transportation [3,4]. Due to the changing working environment, the work of agricultural tractors is more complicated and demanding than that of road vehicles. Agricultural tractors can be divided into step transmission and stepless transmission according to their different transmission ratios. Stepped transmissions include manual mechanical shift transmission (MT), hydraulic torque converter transmission, dual clutch transmission (DCT) and power shift transmission (PST). Electric Continuously variable transmission (CVT), hydrostatic CVT and hydro-mechanical CVT [5–8]. At present, the development of tractors is moving from traditional manual transmission and automatic transmission into newer transmission types, among which hydro-mechanical stepless transmission is a hot transmission. It combines the advantages of stepless speed regulation of the hydraulic system with the high efficiency of mechanical transmission so that the tractor can adapt to the changes in the external environment and make the engine work at its best. However, this transmission method has high requirements for manufacturing and control technologies such as shifting clutches and hydraulic control systems [9,10]. Electric continuously variable transmission is also composed of a power supply, control system and motor. The use of clean energy is friendly to the environment, but this transmission is subject to battery life [11].

The power shift transmission (PST) mainly relies on the sliding friction of the wet clutch to realize the shift without interrupting the engine power. Power shift technology solves the shortcomings of traditional mechanical gearbox parking shifts and power interruptions. Compared with the transmission mechanical shift tractor, the working efficiency of the power shift gearbox is increased by 10–20%. The power shift gearbox is divided into partial power shift and full power shift. The partial power shift is the main shift using wet clutch shift, and the auxiliary shift using electronically controlled synchronizer shift. Full power shift means all gears are wet clutch shift; this way, it requires a high-precision control system. In summary, the power shift gearbox has the advantages of high transmission efficiency, high reliability and high driving comfort and has an important position in agricultural tractors. Tanelli et al. [12] studied the shifting process and reversing process of the power shift transmission, and optimized the initial pressure of the clutch with the change in tractor speed during the shifting process as the optimization goal. Kim et al. [13] used computer simulation technology to explore the influence of the terminal pressure, forward speed, weight, shuttle gear ratio and torsional damping of the tractor on the shift performance of the power shift gearbox. As a result, the tractor weight increased the axle torque but had no effect on the power transmitted by the unit area of the clutch. Raikwar et al. [14] proposed a systematic methodology for the modeling and simulation of a power shuttle transmission system for agricultural tractors, which provides a more economical research method. Simulation is a time- and cost-effective method for the study of power shift transmission.

With the development of power shift technology, in order to meet the driver's increasing requirements for driving comfort and improve the stability of the tractor's shift or start under different working conditions, in the power shift gearbox, the electro-hydraulic control system gradually replaces the traditional mechanical shift control method. The electro-hydraulic control system refers to the pressure control and engagement and disconnection control of the wet clutch through the electro-hydraulic control valve, so as to complete the start and change control of the tractor. The electro-hydraulic control system has the advantages of fast response and high intelligence. In addition, it can be combined with an intelligent control algorithm to formulate a reasonable clutch control strategy according to different working conditions and engagement targets, so as to reduce the impact of tractor starting and section changing and improve the quality of tractor section changing and the driver's driving comfort [15].

As a key component of the power shift gearbox, the wet clutch is responsible for starting and shifting. Its engagement quality will directly determine the quality of the tractor during starting and shifting. The wet clutch uses oil pressure to push the piston, transmits torque through the friction between the friction plate and the steel plate, and performs the starting and shifting actions of the tractor. Ouyang et al. [16] established a wet clutch model that considered damping force, steady flow force and transient flow force to meet the requirements of precise control and fast response. Balau et al. [17] developed a linearized input-output model for an electro-hydraulic actuated clutch. Based on the model, a networked predictive controller was designed for the wet clutch with the aim of controlling the clutch piston displacement while decreasing the influence of the variable-time delays on the closed-loop control performances over the communication network. Zeng et al. [18] built a simulation model of a wet clutch hydraulic actuator system and proposed a model-based feedforward and PID feedback control algorithm. Based on the dynamic equation of the wet clutch piston and proportional valve body and the flow equation of the hydraulic pump, Wu et al. [19] built a clutch oil filling simulation test platform to study the influence of PWM control signals with different rising rates on the clutch oil filling process.

The tractor is controlled by the driver when driving, and the driver controls the tractor according to different working environments and needs. When the tractor is in different conditions, the driver will have different driving needs. The research of Benloucif et al. [20], Marcano et al. [21] and Li et al. [22] also shows that the driver's intention is important in the future development of human–machine cooperative driving systems, so it plays an important role in the identification of the driver's intention. There are four types of driver's intentions during tractor driving: acceleration, deceleration, steering and lane change [23]. When the tractor starts from a stationary state, the driver's intention will have: slow start demand, general start demand, and rapid start demand.

The identification of the driver's intention first needs to determine the intention identification parameters. Liu et al. [24] simulated four types of drivers' intentions during vehicle driving: lane following, left lane change, right lane change and overtaking. The average number of fixations of the driver's left rearview mirror, the average head horizontal angle, the steering wheel angle, the longitudinal acceleration and the establishment of the center line of the vehicle and the lane are used as the driver's intention identification parameters. Wang et al. [25] used accelerator pedal opening, accelerator pedal opening change rate, vehicle average acceleration and vehicle speed to identify the driver's expected acceleration. For the identification model of the driver's intention, the identification models based on time series mainly include the k-nearest neighbor model, support vector regression, and artificial neural networks. However, due to the stochastic and nonlinear nature of behavior, they thus provide unsatisfying prediction performance [26]. Yao et al. [27] used the Hidden Markov identification model, and used the driving intention characteristic parameters after cluster analysis to iteratively optimize the identification model. Wang et al. [28] used the support vector machine model to establish the driver's intention identification model, which integrates vehicle operating parameters such as vehicle speed, environmental information and the driver's visual features.

At present, the research on power shift gearboxes is mostly concentrated in the fields of transmission scheme design and characteristic analysis, and there are few studies on starting control. With the increase in tractor use scenarios, tractors often stop and start again when they are working or transporting in the field. The impact of traditional tractors when starting is relatively high, which will have a great impact on the physical and mental health of agricultural machinery drivers who need to drive tractors for a long time. At the same time, research on the improvement of tractor starting quality in agricultural machinery is missing. The starting process of the power shift tractor is a process of 'human–machine–ground' interaction. The starting control considering the driver's intention will improve the starting quality of the tractor, which is of great significance to the performance improvement of the power shift gearbox. In this paper, the research goal is to improve the starting quality of the power shift tractor. Considering the driver's starting intention (starting demand), the optimal control strategy of the main starting part 'wet clutch' is studied.

The main contents of this paper are as follows:

(1) The identification parameters of the driver's starting intention and the classification of the driver's starting intention are determined by a real vehicle test. (2) The PSO algorithm is used to optimize ELM, and the identification model of the PST tractor driver's starting intention is established. (3) The bench test of clutch oil filling characteristics is carried out, and the knowledge base of the PST starting clutch control strategy is established. (4) In order to establish the comprehensive evaluation index of PST starting quality, fuzzy recognition is used to establish the mapping between the driver's starting intention and the weight of each engagement quality evaluation index in the starting clutch. (5) The PST starting simulation test platform is established, and the starting quality under the

starting control strategy proposed in this paper and the ordinary starting control strategy is compared based on the MPC controller. The purpose of this paper is to provide a theoretical basis for the formulation of a PST starting control strategy. This paper aims to improve the starting performance of a power shift tractor and the working comfort of agricultural machinery drivers.

2. Materials and Methods

2.1. Three Kinds of Driver Starting Intention Test

When the power shift tractor starts, the driver's actions mainly include stepping on the accelerator pedal and pushing the gear lever (starting gear: the power shift tractor generally starts at low speed). The driver's starting intention is generally divided into three types [29–32]. (1) Slow start. At this time, the driver steps on the accelerator pedal at a slower speed, hoping to start smoothly. (2) General start. At this time, the driver stampedes on the accelerator pedal speed in general, conventional starting speed and the starting time requirements in general; (3) Rapid start. The driver quickly steps on the accelerator pedal, and the change rate of the pedal opening is large. This starting mode has a higher requirement for the starting time. The depth of the driver's stampede on the accelerator pedal and the stampede speed are direct manifestations of the urgency of the start. At the same time, starting is a process of 'human–vehicle' interaction. The driver's different operations on the throttle will affect the response of the power shift tractor. The impact of the tractor is the most significant starting intention feedback index (where the impact is the first derivative of the longitudinal acceleration of the tractor). In summary, the accelerator pedal opening, the accelerator pedal opening change rate and the longitudinal acceleration are selected as the identification parameters of the driver's starting intention.

Let the experienced driver start the tractor at the test site. The tractor model used in the test is DF Ward 854 (China Jiangsu Changzhou Dongfeng Agricultural Machinery Group Co., Ltd., Changzhou, China) (Table 1). In order to measure the accelerator pedal opening signal and acceleration signal, the pedal opening sensor and acceleration sensor are installed on the tractor, Figure 1 shows the test process. A total of 60 groups of starting intention experiments were carried out (Figure 2), including 20 groups of slow starting intention, general starting intention and fast starting intention. The identification parameters under different starting intentions are obtained: throttle opening, throttle opening change rate and longitudinal acceleration.

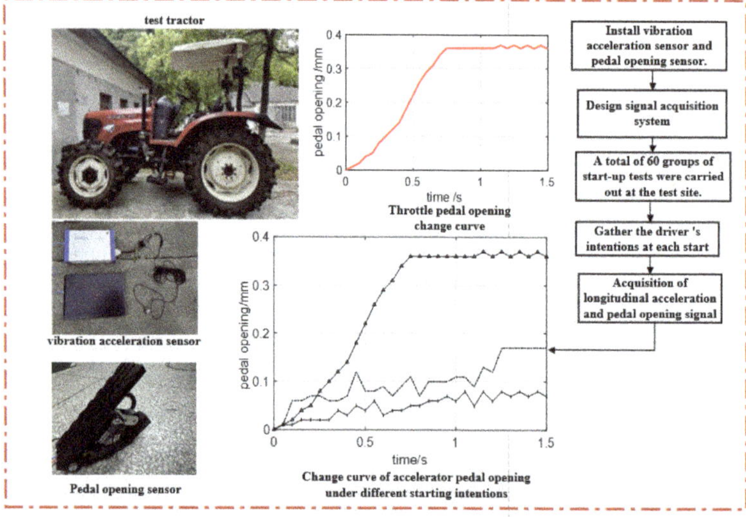

Figure 1. Driver's starting intention acquisition test.

Table 1. Comprehensive parameters of DF Wode 854 tractor.

Parameter	Value
Power shift tractor machine type	Wheel
Weight (kg)	3570
Take-off output (kW)	$\geq$56.3
Rated tractive effect (kN)	$\geq$20
Rear cross member gauge (mm)	1620–2020
The working quality of whole machine (kg)	3800
Engine type	LRC4108T
Rated speed (r/min)	2300

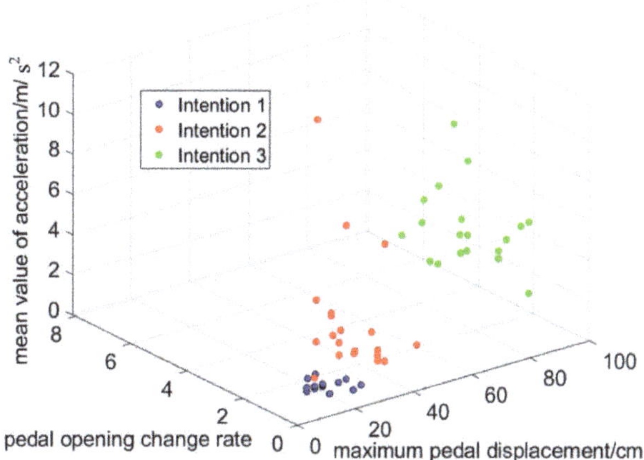

Figure 2. Sixty sets of starting intention test data.

2.2. PSO Optimizes the ELM Identification Model of Driver's Starting Intention

2.2.1. ELM Identification Model

Extreme Learning Machine (ELM) is a single hidden-layer feedforward neural network composed of an input layer, a hidden layer and an output layer [33–35]. The essence of ELM is based on the linear solution process of random feature mapping, which can greatly reduce the complexity of the algorithm and improve the learning rate. Compared with the traditional neural network algorithm, which iteratively calculates and adjusts parameters to optimize the structure, the ELM algorithm is more suitable for the starting scene of the power shift tractor because of its short training time and small computational complexity.

The neural network model of ELM is shown in Figure 3. Input weight, output weight and threshold are used to establish the relationship between the two layers. Users can set the activation function and the number of hidden layer neurons. For Q group training data (X_i, Y_i), $X_i = [x_{i1}, x_{i2}, \ldots x_{in}] \in R^n$, $Y_i = [y_{i1}, y_{i2}, \ldots y_{im}] \in R^m$, L is the number of nodes in the hidden layer, $f(x)$ is the activation function. A single-layer feedforward network is modeled as follows.

$$\sum_{i=1}^{L} \beta_i f(w_i \cdot x_i + b_i), 1 \leq j \leq N \tag{1}$$

w_i is the weight between the input layer node and the ith hidden layer node. β_i is the weight between the ith hidden layer node and output layer node. b_i is the threshold of the ith hidden layer node. Q_j is the output of neural network. w_i, β_i and Q_j are vector values, $w_i = [w_{i1}, w_{i2}, \ldots w_{in}]^T$, $\beta_i = [\beta_{i1}, \beta_{i2}, \ldots \beta_{im}]^T$, $Q_j = [Q_{j1}, Q_{j2}, \ldots Q_{jn}]^T$. By learning the training data, the output value of the ELM model can approach the real value with faster speed and smaller error.

$$\sum_{i=1}^{L} \beta_i f(w_i \cdot x_i + b_i) = y_j, 1 \leq j \leq N \tag{2}$$

The output matrix of ELM is set to H, and the model can be rewritten as

$$H\beta = Y \tag{3}$$

H is the output value obtained from the Neural network. The learning goal of ELM is to minimize the output error. At this time, it can be solved by the least square method.

$$\left\| H\hat{\beta} - Y \right\| = \frac{min}{\beta} \left\| H\beta - Y \right\| \tag{4}$$

The optimal value of β obtained by matrix operation is

$$\hat{\beta} = H^{\dagger}T \tag{5}$$

In the formula, $H^{\dagger}$ is the generalized inverse matrix of matrix H.

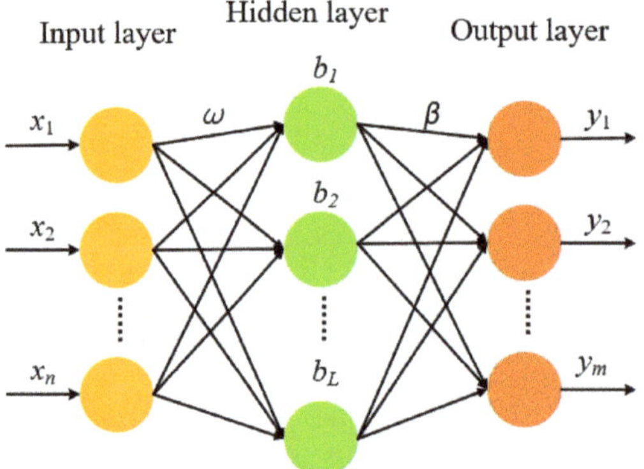

Figure 3. Extreme learning machine neural network model.

Based on the ELM algorithm, the driver's starting intention identification model is constructed. The number of hidden layer nodes is selected to be 100, and the sigmoid function is selected as the activation function. The expression is:

$$sig(x) = \frac{1}{1 + e^{-x}} \tag{6}$$

2.2.2. PSO-ELM Identification Model

In the particle swarm optimization algorithm model, the particles continuously optimize the target through the sharing and updating of group information. The particle velocity, position calculation formula and update formula [36–38] are:

$$V_t = \omega_s \bullet V_{t-1} + c_1 r_1 (p_{best} - x(t)) + c_2 r_2 (g_{best} - x(t))$$
$$\omega_s = \omega_{st} - (\omega_{st} - \omega_{en}) * \frac{K}{T} \tag{7}$$

Formula (7) is the particle update formula, V_t is the velocity of the particle at the next moment, ω_s is the inertia weight, V_{t-1} is the velocity of the particle at this time, c_1 and c_2 are individual learning factor and social learning factor, respectively. r_1 and r_2 are two unequal numbers in $[0, 1]$, respectively. p_{best} is the historical optimal position of the particle, and g_{best} is the group optimal position. $x(t)$ is the position of the particle at this time, ω_{st} is the initial inertia weight of the particle, ω_{en} is the final inertia weight after the iteration is completed, K is the current number of iterations, and T is the total number of iterations. The variable inertia weight is introduced. In the early stage of iteration, the inertia weight of individual particles is large, which makes the particles have strong exploration abilities. As the iteration progresses, the inertia weight of the individual particles gradually decreases, so that the particles quickly fly to the global optimal solution.

The input weights of the ELM algorithm are randomly generated by the system, and there is a certain instability. In this paper, the PSO algorithm is used to optimize the input weights of the ELM identification model. The particle swarm optimization algorithm has the advantages of strong versatility and fast convergence speed. For the study of drivers' starting intention recognition in this paper, the application scenarios that need faster recognition speed are more matched. The algorithm optimization flow chart of the particle swarm optimization algorithm to optimize the ELM model is shown in Figure 4:

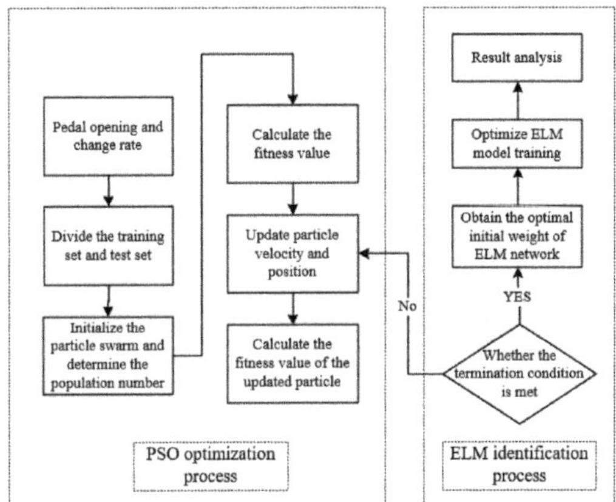

Figure 4. The flow chart of ELM optimized by PSO algorithm.

2.3. The Formulation of Clutch Control Strategy under Different Driving Intentions

2.3.1. Weight Determination of Starting Quality Evaluation Index Based on Fuzzy Recognition

The evaluation of clutch engagement quality includes maximum sliding friction power, sliding friction work, speed stability time and impact degree.

(1) Maximum sliding power y_1: The maximum sliding power generated during the clutch sliding process.

$$y_1 = \frac{da}{dt} = \frac{d^2v}{dt^2} = \frac{1}{\delta m} \frac{\eta i_g i_0}{r} \frac{dT_{cl}}{dt} \tag{8}$$

where δ is the rotational mass conversion coefficient of the power shift tractor; η is the transmission efficiency; i_g is the transmission ratio of the gearbox, i_0 is the total transmission ratio of the main reducer and the wheel reducer; T_{cl} is the torque transmitted by the clutch.

(2) Friction power y_2: During the clutch engagement process, there is a speed difference between the master and slave parts, resulting in energy consumption. The generation of sliding friction work will make the friction material on the surface of the friction plate wear and reduce the torque transmission performance of the clutch.

$$y_2 = \int_{t0}^{t1} T_c(\omega_e - \omega_c)dt \tag{9}$$

In the formula, t_0 is the moment when the clutch produces torque, and t_1 is the moment when the clutch is fully combined to eliminate the speed difference.

(3) Speed stability time y_3: When the power shift tractor reaches the target speed, it is determined that the power shift tractor starts to complete, and the time period when the clutch starts to fill oil to no speed difference is defined as the speed stability time.

(4) Impact degree y_4: the change rate of longitudinal acceleration of a power shift tractor. The greater the impact value, the stronger the sudden frustration of the power shift tractor.

$$y_4 = \max(T_c(\omega_e - \omega_c)) \tag{10}$$

The throttle opening and the throttle opening change rate are the driver's actions, representing the driver's starting intention; the longitudinal acceleration of the vehicle represents the reaction of the tractor to the driver's action. It is reasonable to combine the driver's subjective intention and the tractor's response to start the control. The fuzzy identification system is used to determine the weight of the four clutch engagement quality indicators in the comprehensive evaluation index (Figure 5).

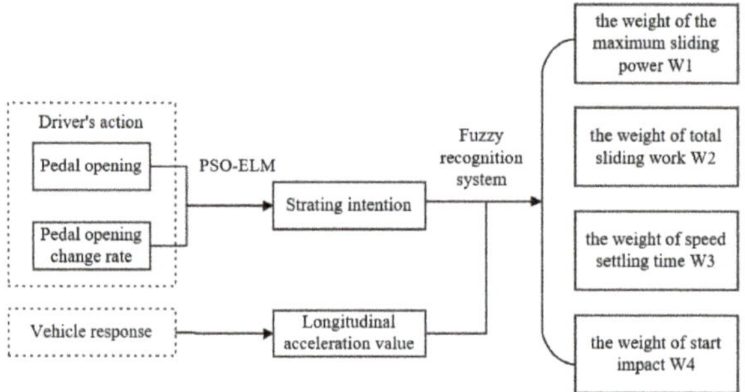

Figure 5. W1–W4 determination method flow chart.

The weight fuzzy identification process [39] is as follows: The fuzzy linguistic variables of the driver's intention are selected as {negative large, negative small, zero, positive small, positive large}, the corresponding fuzzy subsets are {NB, NS, ZE, PS, PB}, and the domain is [–6, 6]; The fuzzy linguistic variables of the driver's intention are selected as {negative large, negative small, zero, positive small, positive large}. The fuzzy linguistic variable of longitudinal acceleration is selected as {small, medium and large}, and the domain is [–6, 6]. The four weight fuzzy linguistic variables are selected as {small, medium and large}, and the domain is [0, 1]; The membership function selects the triangle membership function. The formulation of fuzzy control rules. The weight values of the maximum sliding friction power, sliding friction work, vehicle speed stabilization time and impact degree are shown in Figure 6.

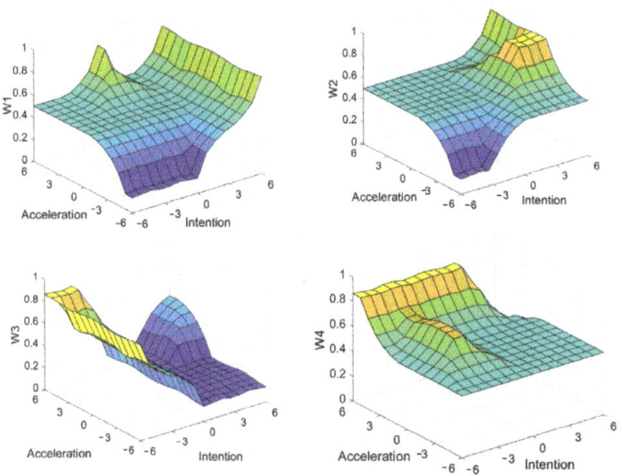

Figure 6. W1–W4 value.

2.3.2. Establishment of Starting Clutch Control Strategy Knowledge Base

The wet clutch control strategy knowledge base is established based on the wet clutch test bench (Figure 7), which was independently developed by the research group [40], and the SimulationX wet clutch engagement quality simulation test platform.

Figure 7. Wet clutch test bench. 1—hydraulic control station. 2—eddy current dynamometer. 3—drive shaft. 4—wet clutch box. 5—speed torque sensor. 6—motor.

The establishment steps of the wet clutch control strategy knowledge base are as follows:
① Through the wet clutch test bench, 20 sets of clutch oil pressure curves are obtained.
② The engagement quality of wet clutch under 20 control strategies is explored on the clutch engagement quality simulation platform, and the knowledge base of clutch control strategy is formed.

a. Bench test of starting clutch

The variable frequency speed-regulating motor (YXVF315L2-4 type, produced by Anhui Wannan Motor Co., Ltd., Wannan, China) is used as the power source of the test bench. The rated torque is 1286 Nm, and the speed range is 0–1450 r/min. The output power of the motor is adjusted by the Delixi frequency converter so as to change the speed of the input end of the wet clutch. A speed and torque sensor (ZJ-2000A, Jiangsu Lanling Electromechanical Technology Co., Ltd., Nantong, China) is installed on the driving shaft of the wet clutch box. The speed range is 0–3000 r/min, and the torque range is 0.01–2000 Nm. An eddy current dynamometer (CW2000B, Jiangsu Lanling Electromechanical Technology Co., Ltd., Nantong, China) was installed at the output end of the wet clutch box to simulate the load at the output shaft end of the clutch and integrate a speed and torque sensor. The hydraulic control station of the clutch mainly includes the main oil relief valve, the proportional pressure reducing valve, the oil pressure sensor, the oil pump and the oil pump motor, which are used to control the combination and disconnection of the wet clutch.

b. Simulation test platform for starting clutch based on SimulationX.

Because of the noise interference of the test bench, the impact degree is increased, and the calculated clutch engagement quality is distorted. Therefore, in this paper, we choose to carry out the engagement quality test on the wet clutch simulation test platform based on SimulationX (Figure 8), (ESI ITI, GmbH, Dresden, Germany) [41–43]. Before that, we first verified the accuracy of the simulation test platform. For these test conditions, the relative error between the simulation results and the test results of the maximum output speed of the gearbox is 5.3%, which verifies the correctness of the simulation model. The working condition of the test is that a power shift tractor (8.5 t) runs on the ground without hanging agricultural machinery. And the engine speed is 1070 r/min (engine minimum stable speed), and the load is 85.17 Nm.

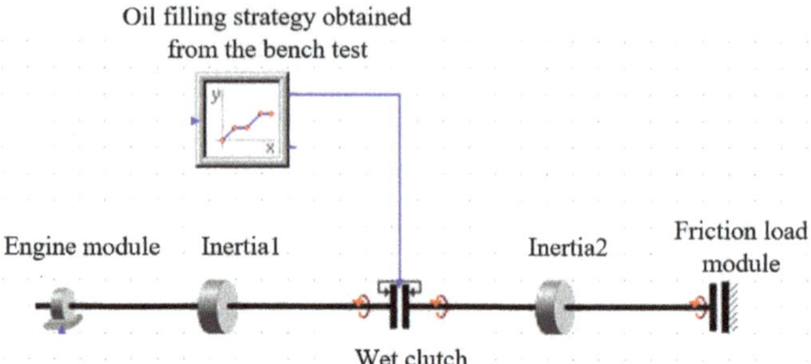

Figure 8. Wet clutch engagement quality simulation test platform.

2.3.3. Establishment of Comprehensive Evaluation Index for Starting under Different Driver Intentions

In order to eliminate the influence of the order of magnitude of different indicators on the final results, the data should be first normalized and substituted into the overall evaluation function of the starting quality to calculate (Figure 9).

$$F = [y_1, y_2, y_3, y_4], \ \min F = W_1 y_1 + W_2 y_2 + W_3 y_3 + W_4 y_4 \tag{11}$$

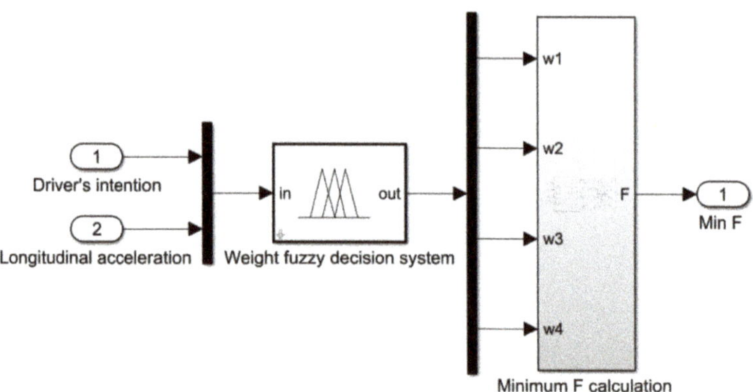

Figure 9. Total evaluation function calculation process.

The comprehensive evaluation index values corresponding to each strategy are calculated, and the clutch control strategy corresponding to the minimum value is the optimal control strategy under the current driver's intention.

2.4. Start-Up Control Simulation Platform Establishment

2.4.1. Design of MPC Controller for Starting Clutch

The mechanism of model predictive control is to predict the future output of the system according to the prediction model at each sampling time of the system and to correct the prediction model according to the current actual measured system output. The open-loop optimization problem in a certain time domain is solved online, and the first element of the obtained control sequence is applied to the controlled object. At the next sampling time, the above process is repeated to re-solve [44–46].

In this paper, a PST starting clutch oil pressure controller is designed based on the model predictive controller. The control goal is to achieve precise control of the clutch oil pressure [47]. The incremental form of the controller model is derived. The state space model is discretized to obtain a discrete incremental state space model:

$$\begin{cases} \triangle x(k+1) = A \triangle x(k) + B_u \triangle u(k) + B_d \triangle d(k) \\ y_c(k) = C_c \triangle x(k) + y_c(k-1) \end{cases} \tag{12}$$

In the formula:

$$\begin{cases} \triangle x(k) = x(k) - x(k-1) \\ \triangle u(k) = u(k) - u(k-1) \\ \triangle d(k) = d(k) - d(k-1) \end{cases} \tag{13}$$

$x(k)$ is a state variable, $u(k)$ is a control input, $d(k)$ is a measurable interference input, $y_c(k)$ is the controlled output. The optimization problem can be described as:

$$J = \left\| T_y(Y_p(k+1|k) - R_e(k+1)) \right\|^2 + \left\| T_u \triangle U(k) \right\|^2 \tag{14}$$

In the formula, $Y_p(k+1|k)$ is the predictive output, $\triangle U(k)$ is the control increment sequence, $R_e(k+1)$ is the given input, m and p is the control time domain and prediction time domain.

$$Y_p(k+1|k) = \begin{bmatrix} y(k+1|k) \\ y(k+2|k) \\ \vdots \\ y(k+p|k) \end{bmatrix}, \triangle U(k) = \begin{bmatrix} \triangle u(k) \\ \triangle u(k+1) \\ \vdots \\ \triangle u(k+m-1) \end{bmatrix}, R_e(k+1) = \begin{bmatrix} r(k+1) \\ r(k+2) \\ \vdots \\ r(k+p) \end{bmatrix} \tag{15}$$

And,

$$y(k+i|k) = [\Delta\omega(k+i|k)] \tag{16}$$

$$\triangle u(k+i|k) = \begin{bmatrix} \Delta T_e(k+i|k) \\ \Delta T_c(k+i|k) \end{bmatrix} \tag{17}$$

$$r(k+i) = \Delta\omega_{\text{ref}}(k+i|k) \tag{18}$$

The weighting matrix are:

$$T_y = \begin{bmatrix} \gamma_{y,1} & 0 & \cdots & 0 \\ 0 & \gamma_{y,2} & \cdots & 0 \\ \vdots & \vdots & \vdots & \vdots \\ 0 & 0 & \cdots & \gamma_{y,p} \end{bmatrix}, \; T_u = \begin{bmatrix} \gamma_{u,1} & 0 & \cdots & 0 \\ 0 & \gamma_{u,2} & \cdots & 0 \\ \vdots & \vdots & \vdots & \vdots \\ 0 & 0 & \cdots & \gamma_{u,m} \end{bmatrix} \tag{19}$$

And

$$\gamma_{u,i} = \begin{bmatrix} \gamma_{T_e,i} & 0 \\ 0 & \gamma_{T_c,i} \end{bmatrix} \tag{20}$$

T_y is the weight factor of the output sequence. The larger the value is, the smaller the error of clutch oil pressure tracking is. T_u is the weight factor of the control signal. The larger the value, the smaller the cost and the smaller the consumption when the controller moves. T_y and T_u are mutually restricted, so it is necessary to coordinate these two values according to the control requirements in the actual system.

The constraint conditions are:

$$u_{\min}(k+i) \le u(k+i) \le u_{\max}(k+i), \; i = 0, 1, \ldots, m-1 \tag{21}$$

$$\triangle u_{\min}(k+i) \le \triangle u(k+i) \le \triangle u_{\max}(k+i), \; i = 0, 1, \ldots, m-1 \tag{22}$$

$$y_{c\,\min}(k+i) \le y_c(k+i) \le y_{c\,\max}(k+i), \; i = 0, 1, \ldots, p \tag{23}$$

The optimal control sequence is obtained by solving, and the optimization results are applied to the system.

$$\triangle U(k) = [\triangle u(k), \triangle u(k+1), \dots, \triangle u(k+m-1)] \tag{24}$$

According to the analysis of the starting process and the starting control strategy of the power shift tractor, the speed differences between the clutch $\Delta\omega$ and the clutch and the torque transmitted by the clutch T_{cl} are selected as the state variables, and the first derivative of the torque transmitted by the clutch is taken as the control variable. The state space equation of the PST starting process is:

$$\begin{cases} \dot{x} = Ax + Bu + D \\ y = Cx \end{cases} \tag{25}$$

Among them, $A = \begin{bmatrix} 0 & -(\frac{1}{I_e} + \frac{1}{I_0}) \\ 0 & 0 \end{bmatrix}, B = \begin{bmatrix} 0 \\ 1 \end{bmatrix}, C = \begin{bmatrix} 1 & 0 \end{bmatrix}, D = \begin{bmatrix} \frac{i_1}{I_e}T_e + \frac{1}{i_v i_0 I_c}T_f \\ 0 \end{bmatrix},$

$x = [\Delta\omega, T_{cl}], u = \dot{T}_{cl}, \Delta\omega = \omega_e - \omega_c.\omega_e$ is the speed of the active end of the clutch, ω_c is the speed of the driven end of the clutch. I_e is the equivalent rotational inertia of the active end of the clutch, I_0 is the equivalent rotational inertia of the clutch driven end, i_1 is the transmission ratio of PST starting section, T_e is the output torque of the engine, I_c is the equivalent rotational inertia of the whole machine, $i_v i_0$ is the transmission ratio of the central drive and the final drive, T_f is the resistance moment.

According to the principle of model predictive control, the Formula (25) is discretized and rewritten into an augmented format. The discretization step size = 0.05 s, and the predictive model of starting clutch control is obtained as follows:

$$\begin{cases} x(k+1) = A_1 x(k) + B_1 u(k) + D_1 \\ y = C_1 x(k) \end{cases} \tag{26}$$

Among them,

$$A1 = \begin{bmatrix} 1 & -(\frac{\Delta t}{I_e} + \frac{\Delta t}{I_0}) \\ 0 & 1 \end{bmatrix}, B1 = \begin{bmatrix} 0 \\ \Delta t \end{bmatrix}, C1 = \begin{bmatrix} 1 & 0 \end{bmatrix}, D1 = \begin{bmatrix} \frac{i_1 \Delta t}{I_e}T_e + \frac{\Delta t}{i_v i_0 I_c}T_f \\ 0 \end{bmatrix}$$

When the PST tractor starts, it is necessary to ensure that the oil-filling process of the wet clutch is consistent with the target process as much as possible, and the change range of the control quantity cannot be too large. The objective function of the controller is set as follows:

$$\begin{cases} \min_{\Delta u(k)} J = J_1 + J_2 \\ J_1 = \|y(k+1) - R(k+1)\|^2 \\ J_2 = \|\Delta u(k)\|^2 \end{cases} \tag{27}$$

Among them, $y(k+1)$ is the prediction oil pressure (MPa); $R(k+1)$ is oil pressure target trajectory.

When the PST tractor starts, the change of the oil filling pressure is too fast to produce the starting impact. The constraint condition of the controller is set to:

$$\Delta u_{\min}(k) \leq \Delta u(k) \leq \Delta u_{\max}(k), \forall k \geq 0 \tag{28}$$

According to the prediction model, objective function and constraint conditions, the MPC tracking controller of the starting clutch is designed as follows (Figure 10):

The oil pressure tracking control flow is as follows: The system's current state error is input into the system error model, and the system's predicted output can be obtained. After the solution of the objective function is completed in each control cycle, the control error increment sequence is obtained. The first amount in the control sequence is used as the actual control error increment of the system at the current time. The control error at the current time can be obtained by combining the control error at the previous time, and then the reference control amount is corrected to obtain the current actual control amount, $u(k)$.

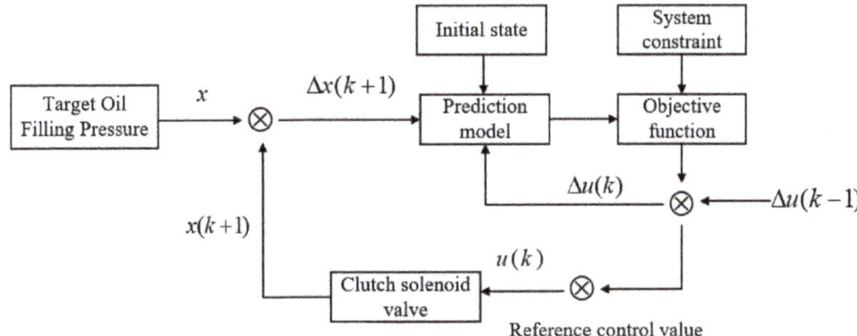

Figure 10. Oil pressure tracking control based on MPC.

2.4.2. Construction of Starting Simulation Platform for Power Shift Gearbox

The proportional pressure electromagnetic of the starting clutch has strong nonlinearity. In order to simplify the system modeling, this paper studies the following two simplifications: (1) ignore the influence of hydraulic oil leakage during valve movement; (2) the hydraulic pressure of the pressure feedback chamber is approximately equal to the output pressure of the proportional voltage valve; (3) ignore the volume of the proportional hydraulic chamber and the inflow of hydraulic oil flow.

Using AMEsim and Simulink to build a joint simulation platform to simulate and analyze the PST start-up process (Figure 11).

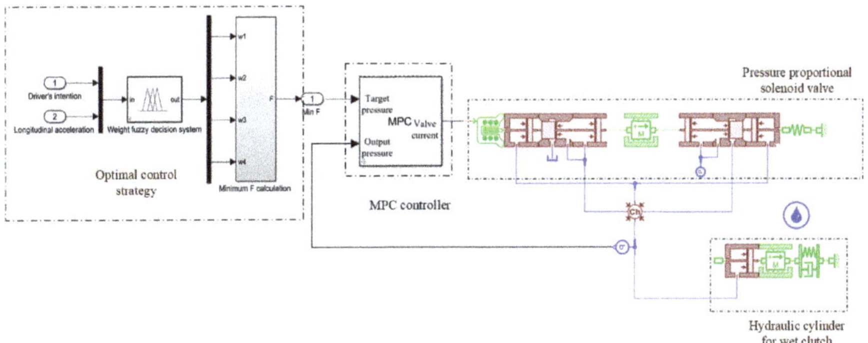

Figure 11. Construction of electronic control simulation platform for starting clutch.

3. Results

3.1. Accuracy Verification and Comparison of PSO-ELM Identification Model for Driver's Starting Intention

From the 60 sets of starting tests, 48 sets of data, including the pedal-opening signal and acceleration signal, were randomly selected as training data, and 12 sets of tests were used as test data. Because of the dimension and order of magnitude of the data, it is difficult to guarantee the accuracy of the trained neural network for the recognition results of the new sample data. Therefore, it is necessary to normalize the data of each node first. The validation data are imported into the trained PSO-ELM identification model.

The experimental resultsshow (Figure 12) that the error between the output value and the true value is less than 0.15, and the circle is an integer value. It can be seen from

Figure 5 that the R2 of the prediction set of the PSO-ELM model is 0.96891, and the R_2 of the ELM model prediction set is 0.91593, indicating that the PSO-ELM model has a better degree of nonlinear fitting. The simulation mean square error mse of the PSO-ELM model is 0.01.

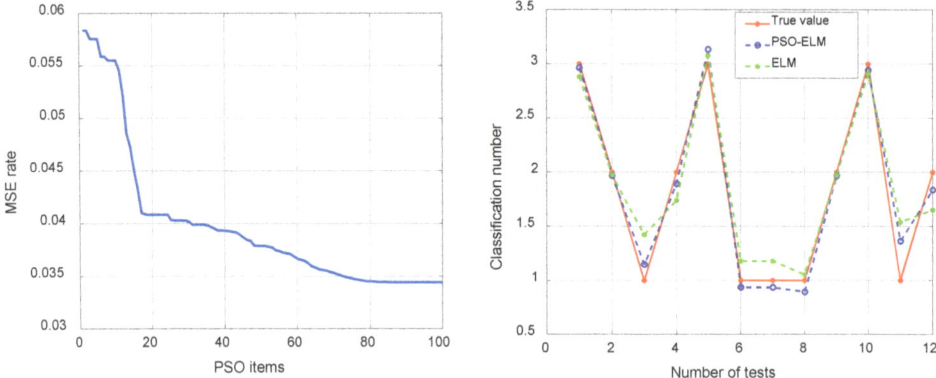

Figure 12. The identification results of ELM and PSO-ELM compared with the real values.

Among the 12 sets of verification data (Table 2), the ELM model correctly identified 6 sets of intentions, and the recognition accuracy was 50%. The PSO-ELM model correctly identified 11 sets of intentions and only one set of errors, and the recognition accuracy was 91.67%, which was 41.67% higher than that of ELM. In driving intention recognition, the PSO-ELM identification model performs better.

Table 2. Comparison of model prediction accuracy.

Test Number	1	2	3	4	5	6	7	8	9	10	11	12
True value	3	2	1	2	3	1	1	1	2	3	1	2
ELM value	3	2	1.42	1.73	3	1.17	1.17	1	2	3	1.54	1.65
PSO-ELM value	3	2	1	2	3	1	1	1	2	3	1	1.83

3.2. Tractor Starting Control Strategy Considering Driver's Intention

3.2.1. Starting Clutch Engagement Oil Pressure Range to Determine the Test Results

It can be seen from the Table 3 that when the clutch oil pressure is less than 1.1 MPa, the clutch is not fully engaged, and the maximum output speed of the gearbox does not reach the theoretical output speed (Figure 13). At this time, the bench will emit obvious abnormal noise, and the speed drop during the speed rise is a sign that the clutch is not fully engaged.

Table 3. The maximum output speed of gearbox under 6 kinds of oil pressure.

Oil Pressure of Wet Clutch/MPa	Maximum Output Speed of the Gearbox/r/min	Theoretical Output Speed/r/min
0.5	128.52	297
0.6	161.09	297
0.7	191.7	297
0.8	225.61	297
0.9	268.82	297
1	293	297

When the oil pressure is greater than 4.9 MPa, the clutch engagement will produce a significant impact during the test, and a large noise will be generated. Figure 14 shows the oil filling process of the clutch when the oil pressure is 5 MPa. It can be seen from the diagram that the oil pressure rises very fast, and the clutch oil filling process is too fast, which will affect the clutch engagement speed and increase the impact of the clutch engagement.

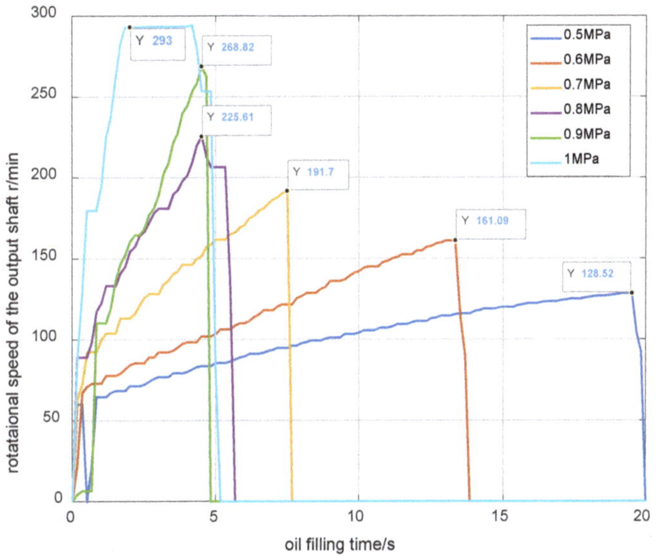

Figure 13. The output speed change curve of gearbox under 6 kinds of oil pressure.

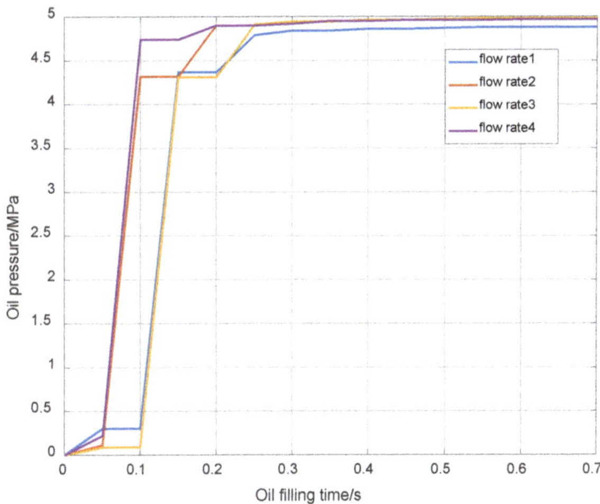

Figure 14. Five megapascals oil pressure change curve of clutch.

Therefore, in this paper, the range of oil filling pressure is set to 1.1–4.9 MPa when establishing the knowledge base of starting clutch control strategy. In the process of a single test, the opening of the flow valve is first adjusted, and then the oil filling pressure of the clutch is set in the host computer to record the change in the oil filling pressure of the clutch with time. By changing the oil filling pressure (1.1, 2.1, 3.1, 4.1, 4.9 MPa) and the opening of the flow valve (30%, 50%, 70%, 90%), 20 sets of clutch oil filling strategies are obtained (Figure 15 is the clutch oil filling control strategy of 1.1 MPa and 2.1 MPa).

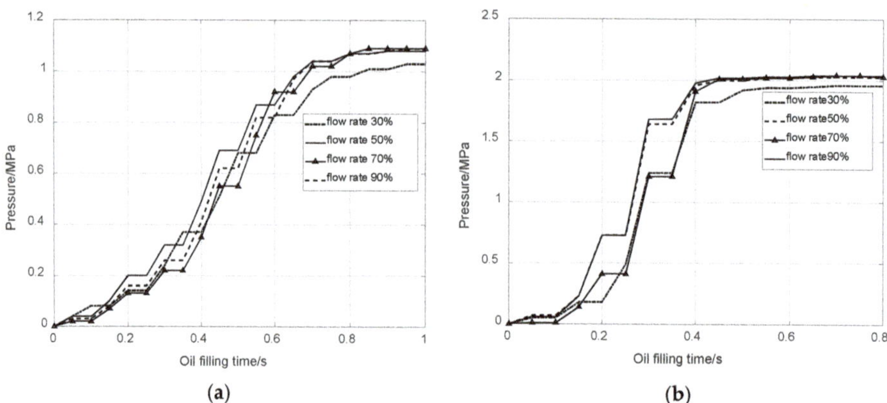

Figure 15. Wet clutch oil filling control strategy. (**a**) 1.1 MPa; (**b**) 2.1 MPa.

3.2.2. Accuracy Verification Results of Wet Clutch Simulation Test Platform

Simulation test platform for starting clutch based on SimulationX: Because of the noise interference of the test bench, the impact degree is increased, and the calculated clutch engagement quality is distorted. Therefore, in this paper, we choose to carry out the engagement quality test on the wet clutch simulation test platform based on SimulationX (ESI ITI, GmbH, Germany) [1–3]. Before that, we first verified the accuracy of the simulation test platform. For these test conditions, the relative error between the simulation results and the test results of the maximum output speed of the gearbox is 5.3%, which verifies the correctness of the simulation model (Figure 16).

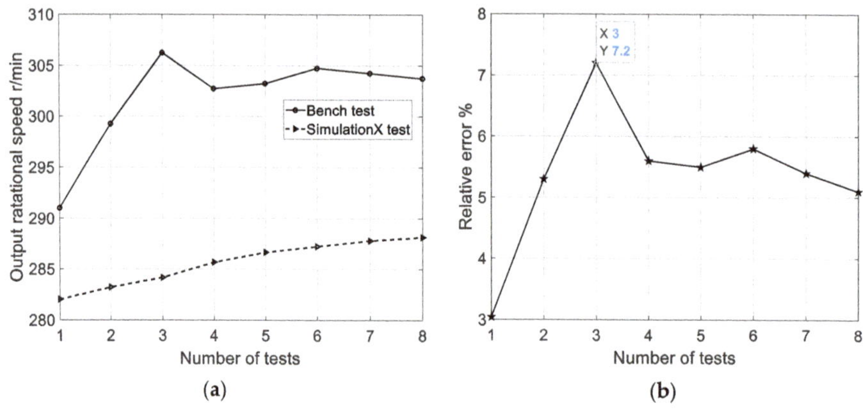

Figure 16. The simulation platform is compared with the test bench to verify the results. (**a**) Simulation output speed and bench test output speed; (**b**) Relative error.

The working condition of the test is that a certain type of power shift tractor (8.5 t) runs on the ground without hanging agricultural machinery, and the engine speed is 1070 r/min (engine minimum stable speed) and load is 85.17 Nm.

3.2.3. The Knowledge Base of Tractor Starting Clutch Control Strategy

Through the clutch simulation test platform, the knowledge base of the starting clutch control strategy is obtained, as shown in Table 4:

Table 4. Starting clutch control strategy knowledge base.

Test Number	Oil Pressure/MPa	Flow Valve Opening%	Maximum Sliding Power J	Friction Power J	Speed Stability Time s	Impact m/s^3
1	1.1	0.3	575,489.94	56,400.61	0.20	584,899.50
2	1.1	0.5	755,803.53	57,981.71	0.25	937,238.46
3	1.1	0.7	981,585.26	55,403.02	0.38	1,258,678.03
4	1.1	0.9	698,689.29	60,328.37	0.31	1,090,579.35
5	2.1	0.3	833,068.79	49,755.37	0.18	1,064,541.15
6	2.1	0.5	888,340.15	36,262.43	0.12	1,446,325.92
7	2.1	0.7	1,022,171.96	38,896.36	0.17	1,624,146.11
8	2.1	0.9	964,428.49	36,996.41	0.13	1,273,260.25
9	3.1	0.3	1,665,466.82	36,756.97	0.14	1,986,833.58
10	3.1	0.5	379,151.35	35,698.86	0.11	598,887.57
11	3.1	0.7	1,222,831.56	33,686.07	0.12	1,685,159.26
12	3.1	0.9	546,644.01	19,182.57	0.06	543,346.56
13	4.1	0.3	835,046.18	21,643.97	0.06	1,080,757.57
14	4.1	0.5	379,151.35	35,698.86	0.11	598,887.57
15	4.1	0.7	417,575.59	24,198.42	0.06	681,833.18
16	4.1	0.9	325,316.41	6429.21	0.07	504,150.58
17	4.9	0.3	829,032.18	68,997.07	0.24	1,066,653.49
18	4.9	0.5	806,755.85	60,128.87	0.28	1,084,836.86
19	4.9	0.7	379,151.35	4798.62	0.02	535,396.52
20	4.9	0.9	474,832.52	45,256.81	0.11	751,212.03

3.3. Verification of Starting Clutch Control Effect Based on MPC Controller

Taking the step signal within 100 s of sampling time as the target value, the control effect of the starting clutch MPC controller is verified.

4. Discussion

(1) According to the comparison of the identification results of the two identification methods in Table 2, it can be found that the prediction simulation error of the PSO-ELM model is 38.05% lower than that of the ELM model. Among them, with the help of the PSO algorithm in the multi-dimensional solution space, a large number of random particles will search for the position of the current optimal particle and then update their speed and position to achieve the goal of quickly finding the optimal solution to the problem. The PSO-ELM identification model can avoid the network falling into the local optimum and find the global optimal model parameter solution, thereby improving the accuracy of the model. Compared with the unoptimized ELM, the PSO-ELM identification model has a better identification effect on the identification of the driver's starting intention.

(2) In Figure 17, it can be seen that the MPC controller can better follow the target value. There is a certain tracking error in the initial stage of control, and the tracking error decreases rapidly in the later stage, and the oil pressure output value is stabilized at the target value. On the simulation experiment platform, the starting quality of the starting control strategy determined by the PSO-ELM-fuzzy weight method and the conventional starting control strategy is compared.

By using the PSO-ELM-fuzzy weight starting control strategy (Table 5), compared with the linear control strategy, the maximum sliding friction power y1 is reduced by 45%, the sliding friction power is reduced by 69.45%, the speed stabilization time is shortened by 0.11 s, and the impact degree is increased by 0.003%. Compared with the linear control strategy, the starting control strategy proposed in this paper takes the driver's intention as one of the important criteria for the formulation of the starting control strategy. For different driver intentions, the optimal starting control strategy under the current intention is proposed. In summary, the PSO-ELM-fuzzy weight starting control strategy proposed in this paper has better starting quality.

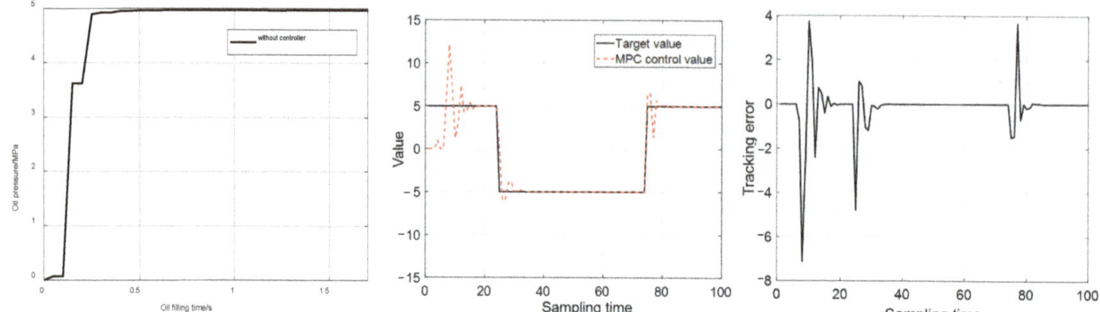

Figure 17. MPC controller square wave signal tracking control.

Table 5. Comparison between PSO-ELM-fuzzy weight starting strategy and linear control strategy.

Content	Maximum Sliding Power J	Friction Power J	Speed Stability Time s	Impact m/s^3
PSO-ELM-fuzzy weight starting strategy	417,575.59	24,198.42	0.06	681,833.18
Linear control strategy	762,047.30	79,213.59	0.17	679,402.27

5. Conclusions

In this paper, aiming at the starting quality of a power shift tractor, a tractor starting control strategy considering the driver's intention is proposed. Firstly, the identification modeling data samples of the driver's starting intention are obtained by vehicle testing. After the driver's intention identification model is established, we match the optimal control strategy for different starting intentions. Therefore, we establish a knowledge base for starting control strategies through the joint bench test and simulation test platform. After the establishment of the control strategy knowledge base and the driver's intention identification model, we chose the fuzzy weight method to determine the connection between the driver's intention and the control strategy. At the same time, the MPC starting controller of the power shift gearbox is established to verify the control effect of the starting control strategy proposed in this paper. The following conclusions are drawn from this study:

(1) The ELM driver intention identification model improved by the PSO algorithm has a prediction accuracy of 91.67%. Compared with the ELM identification model, the prediction accuracy is improved by 41.67%, and the ELM model optimized by PSO can avoid the network falling into local optimum.

(2) The wet clutch has a minimum oil filling threshold and a maximum oil filling threshold. The minimum oil filling threshold is determined by the structural size and working condition of the wet clutch, and the clutch has sliding friction when the value is less than this value. The maximum oil filling threshold can be obtained from the test. If the value is greater than this value, the clutch engagement speed is too fast, resulting in a large impact and poor engagement quality. Through experiments, the oil filling pressure range of the wet clutch studied in this paper is 1.1 MPa–4.9 MPa.

(3) Compared with the linear control strategy, the maximum sliding friction power is reduced by 45%, and the sliding friction power is reduced by 69.45%. The speed stabilization time is shortened by 0.11 s, and the impact degree is in-creased by 0.003%. In summary, the PSO-ELM-fuzzy weight starting strategy proposed in this paper has better starting quality.

This paper studies the starting process of power shift tractors from the perspective of 'people–vehicle–ground'. It is expected to improve the starting quality of power shift tractors, improve the working environment of tractor drivers, and protect their physical and mental health. At the same time, it is expected to provide research help for tractors to realize intelligent automatic driving.

Author Contributions: Methodology, Y.Q. and L.W.; software, Y.Q.; validation, Y.Q.; investigation, Y.Q.; resources, Z.L. and L.W.; writing—original draft preparation, Y.Q.; writing—review and editing, Y.Q. and Z.L.; supervision, Z.L.; and project administration, Z.L. and Y.Q. All authors have read and agreed to the published version of the manuscript.

Funding: This research was funded by the Open Project of the State Key Laboratory of Intelligent Agricultural Power Equipment (SKLIAPE2023019), National Key Research and Development Plan (2022YFD2001202) and the Postgraduate Research & Practice Program of Jiangsu Province (KYCX21_0572).

Institutional Review Board Statement: Not applicable.

Data Availability Statement: The data presented in this study are available on demand from the corresponding author or first author at (luzx@njau.edu.cn or 2020212006@stu.njau.edu.cn).

Acknowledgments: We thank the anonymous reviewers for providing critical comments and suggestions that improved the manuscript.

Conflicts of Interest: The authors declare no conflicts of interests.

References

1. Mairghany, M.; Yahya, A.; Adam, N.M.; Mat Su, A.; Aimrun, W.; Elsoragaby, S. Rotary tillage effects on some selected physical properties of fine textured soil in wetland rice cultivation in Malaysia. *Soil Tillage Res.* **2019**, *194*, 104318. [CrossRef]
2. AI-Sager, S.M.; Almady, S.S.; Marey, S.A.; AI-Hamed, S.; Aboukarima, A.M. Prediction of specific fuel consumption of a tractor during the tillage process using an artificial neural network method. *Agronomy* **2024**, *14*, 492. [CrossRef]
3. Vu, N.-L.; Messier, P.; Nguyễn, B.-H.; Vo-Duy, T.; Trovão, J.; Desrochers, A.; Rodrigues, A. Energy-optimization design and management strategy for hybrid electric non-road mobile machinery: A case study of snowblower. *Energy* **2023**, *284*, 129249. [CrossRef]
4. Mederle, M.; Urban, A.; Fischer, H.; Hufnagel, U.; Bernhardt, H. Optimization potential of a standard tractor in road transportation. *Landtechnik* **2015**, *70*, 194–202.
5. Zhou, X.Y.; Qin, D.T.; Hu, J. Multi-objective optimization design and performance evaluation for plug-in hybrid electric vehicle powertrains. *Appl. Energy* **2017**, *208*, 1608–1625. [CrossRef]
6. Mattetti, M.; Maraldi, M.; Sedoni, E.; Molari, G. Optimal criteria for durability test of stepped transmissions of agricultural tractors. *Biosyst. Eng.* **2019**, *100*, 145–155. [CrossRef]
7. Molari, G.; Sedoni, E. Experimental evaluation of power losses in a power-shift agricultural tractor transmission. *Biosyst. Eng.* **2008**, *100*, 177–183. [CrossRef]
8. Savaresi, S.M.; Taroni, F.L.; Predivi, F.; Bittaniti, S. Control system design on a power-split CVT for high-power agricultural tractors. *IEEE-ASME Trans. Mechatron.* **2004**, *9*, 569–579. [CrossRef]
9. Xia, Y.; Sun, D.Y.; Qin, D.T.; Zhou, X.Y. Optimization of the power-cycle hydro-mechanical parameters in a continuously variable transmission designed for agricultural tractors. *Biosyst. Eng.* **2020**, *193*, 12–24. [CrossRef]
10. Macor, A.; Rossetti, A. Optimization of hydro-mechanical power split transmissions. *Mech. Mach. Theory* **2011**, *46*, 1901–1919. [CrossRef]
11. Cheng, Z. High nonlinearity of BEV's stepped automatic transmission design objectives and its optimal solution by a novel ISA-RSA. *Energy* **2023**, *282*, 128834. [CrossRef]
12. Tanelli, M.; Panzani, G.; Savaresi, S.M.; Pirla, C. Transmission control for power-shift agricultural tractors: Design and end-of-line automatic tuning. *Mechatronics* **2011**, *21*, 285–297. [CrossRef]
13. Kim, D.C.; Kim, K.U.; Park, Y.J.; Huh, J.Y. Analysis of shifting performance of power shuttle transmission. *J. Terramechanics* **2007**, *44*, 111–122. [CrossRef]
14. Raikwar, S.; Tewari, V.K.; Mukhopadhyay, S.; Verma, C.R.B.; Rao, M.S. Simulation of components of a power shuttle transmission system for an agricultural tractor. *Comput. Electron. Agric.* **2015**, *114*, 114–124. [CrossRef]
15. Sun, C.Y.; Si, T.F.; Zhu, M.H.; Han, M.Y.; Hou, Z.H.; Zhang, Z.J. Research on the optimization of power shift quality based on the oil pressure segment control strategy of wet clutch. *J. Chin. Agric. Mech.* **2022**, *43*, 106–112.
16. Ouyang, T.C.; Li, S.Y.; Huang, F.; Zhou, F.; Chen, N. Mathematical modeling and performance prediction of a clutch actuator for heavy-duty automatic transmission vehicles. *Mech. Mach. Theory* **2019**, *136*, 190–205. [CrossRef]
17. Balau, A.E.; Caruntu, C.F.; Lazar, C. Simulation and control of an electro-hydraulic actuated clutch. *Mech. Syst. Signal Process.* **2011**, *25*, 1911–1922. [CrossRef]

18. Zeng, X.H.; Chen, H.X.; Dong, B.B.; Song, D.F. Modeling and Simulation of the Hydraulic Actuator System of Wet Clutch. *Mach. Tool Hydraul.* **2021**, *49*, 120–125.

19. Wu, J.P.; Wang, L.Y.; Li, L.; Zhou, Q.J. The study on the influence of control signal on charge characteristics of wet clutch. *Chin. Hydraul. Pneum.* **2016**, *02*, 62–66.

20. Benloucif, A.; Nguyen, A.T.; Sentouh, C.; Popieul, J.C. Cooperative trajectory planning for haptic shared control between driver and automation in highway driving. *IEEE Trans. Ind. Electron.* **2019**, *66*, 9846–9857. [CrossRef]

21. Marcano, M.; Díaz, S.; Pérez, J.; Irigoyen, E. A review of shared control for automated vehicles: Theory and applications. *IEEE Trans. Hum. Mach. Syst.* **2020**, *50*, 475–491. [CrossRef]

22. Li, M.J.; Cao, H.T.; Song, X.L.; Huang, Y.J.; Wang, J.Q.; Huang, Z. Shared control driver assistance system based on driving intention and situation assessment. *IEEE Trans. Ind. Inform.* **2018**, *14*, 4982–4994. [CrossRef]

23. Huang, T.; Fu, R.; Sun, Q.Y.; Deng, Z.J.; Liu, Z.F.; Jin, L.S.; Khajepour, A. Driver lane change intention prediction based on topological graph constructed by driver behaviors and traffic context for human-machine co-driving system. *Transp. Res. Part C Emerg. Technol.* **2024**, *160*, 104497. [CrossRef]

24. Liu, Z.Q.; Wu, X.G.; Ni, G.; Zhang, T. Driving intention recognition based on HMM and SVM cascade algorithm. *Automot. Eng.* **2018**, *40*, 858–864.

25. Wang, Q.L.; Tang, X.Z.; Wang, P.Y.; Tian, L.Y.; Sun, L. Driving intention identification method for hybrid vehicles based on Neural Network. *Trans. Chin. Soc. Agric. Mach.* **2012**, *43*, 32–36.

26. Hong, W.C.H.; Dong, Y.C.; Zheng, F.F.; Wei, S.Y. Hybrid evolutionary algorithms in a SVR traffic flow forecasting model. *Appl. Math. Comput.* **2011**, *217*, 6733–6747. [CrossRef]

27. Yao, Y.; Zhao, X.H.; Wu, Y.P.; Zhang, Y.L.; Rong, J. Clustering driving behavior using dynamic time warping and hidden Markov model. *J. Intell. Transp. Syst.* **2019**, *25*, 249–262. [CrossRef]

28. Wang, X.; Guo, Y.; Bai, C. Driver's intention identification with the involvement of emotional factors in two-lane roads. *IEEE Trans. Intell. Transp. Syst.* **2020**, *22*, 6866–6874. [CrossRef]

29. Xie, F.; Lou, J.T.; Zhao, K. A research on vehicle trajectory prediction method based on behavior recognition and curvature constraints. *Automot. Eng.* **2019**, *41*, 1036–1042.

30. Lethaus, F.; Baumann, M.; Koester, F. A comparison of selected simple supervised learning algorithm to predict driver intent based on gaze data. *Neurocomputing* **2013**, *121*, 108–130. [CrossRef]

31. Zhang, L.J.; Tang, X.; Meng, J.D. Driver's head and face visual feature extraction for driving intention recognition. *Automob. Technol.* **2019**, *521*, 18–24.

32. Deng, Q.; Wang, J.; Hillebrand, K. Prediction performance of lane changing behaviors: A study of combining environmental and eye-tracking data in a driving simulator. *IEEE Trans. Interlligent Syst.* **2019**, *21*, 3561–3570. [CrossRef]

33. He, F.; Zhang, X.X.; Chen, W.R. Fault diagnosis method of PEMFC system based on P-L dual feature extraction. *Acta Energiae Solaris Sin.* **2024**, *45*, 492–499.

34. Huang, M.T.; Xing, F.F.; Chen, X.B.; Lu, M. Short-term PV power prediction based on K-means clustering and extreme learning machine combination algorithm. *Water Resour. Power* **2024**, *42*, 216–220.

35. Liu, Y.D.; Ma, J.H.; Wang, X.G. H-beam steel structure prediction expert system based on ELM. *Forg. Stamp. Technol.* **2024**, *49*, 241–248.

36. Sevim, S.Y.; Mladen, T. Estimation of daily potato crop evapotranspiration using three different machine learning algorithms and four scenarios of available meteorological data. *Agric. Water Manag.* **2022**, *228*, 105875.

37. Yam, K.; Ji, Z.; Lu, H. Fast and accurate classification of time series data using extended ELM: Application in fault diagnosis of air handling units. *IEEE Trans. Syst. Man Cybern. Syst.* **2017**, *49*, 1349–1356.

38. Liu, X.; Fan, X.; Guo, Y. Multi-objective optimization of GFRP injection molding process parameters, using GA-ELM, MOFA, and GRA-TOPSIS. *Trans. Can. Soc. Mech. Eng.* **2021**, *46*, 37–49. [CrossRef]

39. Meng, F.; Chen, H.Y.; Zhang, T.; Zhu, X.Y. Clutch fill control of an automatic transmission for heavy-duty vehicle applications. *Mech. Syst. Signal Process.* **2015**, *64–65*, 16–28. [CrossRef]

40. Lu, Z.X.; Wang, Y.T.; Wang, L.; Zhao, Y.R.; Wang, X.W.; Zhou, J.B. Prediction of HMCVT wet clutch friction pair temperature based on IGWPSO-SVM. *Trans. Chin. Soc. Agric. Mach.* **2023**, *54*, 407–415.

41. Kyuhyun, S.; Hwayoung, L.; Jiwon, Y.; Chanho, C.; Sung-Ho, H. Effectiveness evaluation of hydro-pneumatic and semi-active cab suspension for the improvement of ride comfort of agricultural tractors. *J. Terramechanics* **2017**, *69*, 23–32.

42. Savin, V.; Christian, H.; Olav, E.; Ronny, S. Cosimulation of a direct-acting riser-tensioner system—Validation with field measurements and sample simulations. *Ocean. Eng.* **2023**, *276*, 114241.

43. Chen, X.; Chen, F.Q.; Zhou, J.; Li, L.H.; Zhang, Y.L. Cushioning structure optimization of excavator arm cylinder. *Autom. Constr.* **2015**, *53*, 120–130. [CrossRef]

44. Leung, J.; Permenter, F.; Kolmanovsky, I.V. A stability governor for constrained linear–quadratic MPC without terminal constraints. *Automatica* **2024**, *164*, 111650. [CrossRef]

45. Zhang, J.J.; Feng, G.H.; Yan, X.H.; He, Y.D.; Liu, M.N.; Xu, L.Y. Cooperative control method considering efficiency and tracking performance for unmanned hybrid tractor based on rotary tillage prediction. *Energy* **2024**, *288*, 129874. [CrossRef]

46. Robat, A.B.; Arezoo, K.; Alipour, K.; Tarvirdizadeh, B. Dynamics modeling and path following controller of tractor-trailer-wheeled robots considering wheels slip. *ISA Trans.* **2024**, *03*, 004.
47. Xu, Y.T.; Chen, H.; Ji, D.D.; Xu, F. MPC controller design and FPGA implementation for start-up of vehicles. *Control Engineeing China* **2015**, *22*, 785–792.

MDPI AG

Grosspeteranlage 5

4052 Basel

Switzerland

Tel.: +41 61 683 77 34

Agriculture Editorial Office

E-mail: agriculture@mdpi.com

www.mdpi.com/journal/agriculture